A HISTORY

OF

ELECTRIC TELEGRAPHY,

TO THE YEAR 1837.

F. & N. Spon: London & New York.

Frontispiece.

See page 123

OF

ELECTRIC TELEGRAPHY,

TO THE YEAR 1837.

*CHIEFLY COMPILED FROM ORIGINAL SOURCES, AND
HITHERTO UNPUBLISHED DOCUMENTS*

BY

J. J. FAHIE,

MEMBER OF THE SOCIETY OF TELEGRAPH-ENGINEERS AND ELECTRICIANS, LONDON;
AND OF THE INTERNATIONAL SOCIETY OF ELECTRICIANS, PARIS.

"Their line is gone out through all the earth,
And their words to the end of the world."
Psalms xix. 4.

LONDON:

E. & F. N. SPON, 16, CHARING CROSS.

NEW YORK: 35, MURRAY STREET.

1884.

𝔇𝔢𝔡𝔦𝔠𝔞𝔱𝔢𝔡

TO

LATIMER CLARK, ESQUIRE,

M.I.C.E., F R G.S., F.M S., PAST PRES. S.T.E. AND E.,

IN ACKNOWLEDGMENT OF MANY KINDNESSES,

BY HIS OBLIGED FRIEND,

THE AUTHOR.

LONDON, *February* 1884.

PREFACE.

PLUTARCH, in the opening sentences of his Life of Demosthenes, says : — "Whosoever shall design to write a history, consisting of materials which must be gathered from observation, and the reading of authors not easy to be had nor writ in his own native language, but many of them foreign and dispersed in other hands : for him it is in the first place and above all things most necessary to reside in some city of good note and fame, addicted to the liberal arts, and populous, where he may have plenty of all sorts of books, and, upon inquiry, may hear and inform himself of such particulars as, having escaped the pens of writers, are yet faithfully preserved in the memories of men ; lest otherwise he publish a work deficient in many things, and those such as are necessary to its perfection."

Had we seen this passage a few years ago, the following pages had, probably, never been written, and there would be no need for this preface. The work was begun and brought to a very forward state, not in some city of good note and fame, where plenty of books were to be had, but in what has been rightly called "the confines of the earth—the hot regions of Persia," and under circumstances which, we think, will bear relating.

In our youthful days we contracted two habits, which have been ever since the bane or the solace (we hardly know which to call them) of our existence, *viz.*, a taste for writing, and a taste for scraps. The *Cacoëthes Scribendi* first attacked us, and we can recall letters in the local papers on various topics of local interest, all of which were written early in our teens. When about sixteen years of age we commenced a history of the old castles and churches which abound (in ruins) in and about our native place, the said history being intended to serve also as a guide for tourists who were constantly visiting the neighbourhood. With great industry we got together, in time, some two hundred pages (foolscap) of writing; but the work was never completed. For years we hawked the MS. about, latterly never looking at it, having come to regard it as a standing reproach for time and money misspent; and at last, in a fit of remorse, we gave the papers to the flames in 1875.

Soon after joining the telegraph service, in 1865, our archæological bent took another turn, and we now began to collect books and scraps on electricity, magnetism, and their applications—particularly to telegraphy, and with the same industrious ardour as before. In December 1867, we entered the Persian Gulf Telegraph Department under the Government of

India, where, having a good deal of spare time on our hands, we indulged our habits to the full. In 1871, having amassed a large number of notes, scraps, &c., on submarine telegraphy, we began a work on the history and working of the Persian Gulf cables, of which we had then had over three years' practical experience.

Gradually this developed itself into an ambitious treatise, which we styled " Submarine Telegraphs, their Construction, Submersion, and Maintenance, including their Testing and Practical Working." Of this some three hundred pages (foolscap) are now lying " submerged " in the depths of our trunk, to be, perhaps, " recovered " at some future day—if, haply, they do not share the fate of our History of Ruins !

Unfortunately for us, at least from a book-selling point of view, our old taste for archæology, after lying dormant for years, reasserted itself, and, about six years ago, we found ourselves in the design of writing a history of telegraphy from the time of Adam down to our own ! For this we had a pile of notes and paper cuttings—the accumulation of a dozen years, but few books (books are heavy and awkward baggage for one of our necessarily semi-nomadic life). However, with our materials we built up a tolerably fleshy skeleton (if we may so speak), which, on our arrival in England at the close of 1882, after nearly fifteen years' absence, we showed to some friends.

They advised us to fill up the gaps and bring out

our book immediately. The first was easy of accomplishment, with the use of the splendid technical libraries of Mr Latimer Clark and of the Society of Telegraph-Engineers and Electricians, and with an occasional reference to the British Museum ; but to find a publisher, *that* was not so easy. Publishers, now as always, fight shy of Dryasdust, and the two or three whom we tried asked us to bring them something new, for, owing to the machinations of us, Electrical Engineers, the world was going at lightning speed, and had no time to look back.

Ultimately we paid a visit to the Editor of *The Electrician*, told him of our discomfiture, showed him our MSS., and repeated an offer that we had made him years before, from Persia, but which he then declined, *viz.*, to publish our articles from week to week in his paper. The Editor did not take long to decide ; he would only, however, accept the electrical portion, the non-electric part which deals with fire-, flag-, and semaphore- signalling, acoustic, pneumatic, and hydraulic telegraphs, &c., &c., being, he said, unsuited for his journal. On the principle that half a loaf is better than no bread, we concluded arrangements there and then, and parted with our new-found friend with feelings which time has but intensified.

The present volume is a collection, with very few alterations, of the articles which have regularly appeared in *The Electrician* for the last twelve months. Of these alterations the only ones worth mentioning will be found in our chapters on Mr. Edward Davy ; we have made our account of his electro-chemical recording telegraph a little fuller, and have added some new matter lately acquired (1) from recent letters of Mr. Davy himself, (2) from an examination of the private papers of the late Sir William Fothergill Cooke—a privilege for which we are indebted to our kind friend, Mr. Latimer Clark, and (3) from Mr. W. H. Thornthwaite, of London, an old pupil of Edward Davy, whose very interesting reminiscences, we feel sure, will be scanned with pleasure by all our readers.

Now as to the plan of the work. We have divided the history of electricity into three parts, (1) static, or frictional, electricity, (2) dynamic, or galvanic, electricity, and (3) electro-magnetism and magneto-electricity. We have brought our account of each part down to the year 1837, confining ourselves to a notice of such facts and principles only as are employed in the various telegraphic proposals that follow. These, in their turn, are divided into three classes, electrical, galvanic (chemical), and electro-magnetic ; and each class, treated chronologically, follows naturally the corresponding part of the history of electricity. The whole is preceded by a full account of what we have called a *foreshadowing of the electric telegraph,* and is followed by an appendix, containing (A) a clear and correct statement of Professor Joseph Henry's little-known connection with electric telegraphy, which is too important to be omitted, but for which we could not conveniently find room in the body of the work, and (B) a few pages supplementary of our chapters on Edward Davy.

In limiting ourselves to the year 1837, we have done so advisedly, for, to attempt even the barest outline of what has been accomplished since then would occupy volumes. Our object has been, as it were, to make a *special* survey of a river from its rise away in some tiny spring to its mouth in the mighty ocean, marking down, as we came along, those of the tributary streams and such other circumstances as specially interested us. Arrived at the mouth, the traveller who wishes for further exploration has only to chose his pilot; for, fortunately, there is no lack of these. We have Highton, Lardner, Sabine, and Culley in England; Shaffner, Prescott, and Reid in America; Moigno, Blavier, and du Moncel in France; Schellen, and Zetzsche in Germany; Saavedra in Spain, and many others in various parts of the world whose names need not be specially mentioned.

As we have in the body of the work given full references for every important statement, it will not be necessary to acknowledge here the sources of our information; indeed it would be simply impossible to do so within the limits of a preface which we feel is already too long. Like Molière, we have taken our materials wherever we could find them, and it is no exaggeration to say that in pursuit of our subject we have laid many hundreds of volumes under tribute ; some have given us clues, some have been mines of wealth, others have yielded nothing at all, while, what was worse, a goodly number were of the *ignis fatuus* kind—false accounts, false dates, false references, false everything—which worried us considerably, and over which we lost much precious time.

We gladly, however, take this opportunity of thanking Messrs. Ispolatoff (Russia), D'Amico (Italy), Aylmer (France), Sömmerring (Germany), and Collette (Holland), for their assistance, of which, as they will see, we have made good use in the text. To our friend, Mr. Latimer Clark, our debt is too heavy for liquidation and must remain. He has not only given us the free use of his magnificent library, but has aided and encouraged us with his advice and sympathy, and, in the most generous manner, has placed at our disposal all his private notes. These, we need hardly say, have been of great use to us, and would have been of greater still had we seen them at an earlier stage of our researches.

As we have to return almost immediately to "the confines of the earth," the preparation of the index has been kindly undertaken by our friend, Mr. A. J. Frost, Librarian of the Society of Telegraph-Engineers and Electricians, whose name will be a sufficient guarantee for the accuracy and completeness of the work. In tendering him our cordial thanks for this assistance, we have much pleasure in recording our appreciation of the zeal, ability, and unvarying courtesy with which he performs the duties of his office. His bibliographical knowledge is great and *special,* and has at all times been freely placed at our disposal.

Our book, we hope, will give the *coup de grâce* to many popular errors. Thus, we show that Watson, Franklin, Cavendish, and Volta did not suggest electric telegraphs (pp. 60, 66, and 82); that Galvani was not the first to observe the fundamental phenomenon of what we now call *galvanism* (pp. 175–9); that his experiments in this field were not suggested by a preparation of frog-broth (pp. 180–3); that not Daniell but Dobereiner and Becquerel first employed two-fluid cells with membranous or porous partitions (p. 215); that not Sömmerring but Salvá first proposed a galvanic (chemical) telegraph (p. 220); that not Schilling but Salvá first suggested a submarine cable (p. 105); that Romagnosi did not discover electro-magnetism (p. 257); that not Ritter but Gautherot first described the secondary battery (p. 267); that not Cumming nor Nobili but Ampère first invented the astatic needle (p. 280); that not Seebeck but Dessaignes first discovered thermo-electricity (p. 297); that not Thomson but Gauss and Weber first constructed the mirror galvanometer (p. 319); that the use of the earth circuit in telegraphy was clearly and intelligently suggested by an Englishman long before Steinheil made his *accidental* discovery of it (p. 345); and that not Cooke and Wheatstone, nor Morse, but

Henry in America and Edward Davy in England first applied the principle of the relay—a principle of the utmost importance in telegraphy (pp. 359, 511, and 515).

There may be some amongst our readers who will not thank us for upsetting their belief on these and many other points of lesser importance, and who may even call us bad names, as did Professor Leslie on a former occasion, and *à propos* of somebody's quoting Swammerdam's and Sulzer's experiments (pp. 175 and 178) as suggestive of galvanism. Leslie says :— " Such facts are curious and deserve attention, but every honourable mind must pity or scorn that invidious spirit with which some unhappy jackals hunt after imperfect and neglected anticipations with a view of detracting from the merit of full discovery" (*Ency. Brit.*, 8th edition, vol. i. p. 739). For our part we can honestly say that in drawing up our history we have not been influenced by any such views ; our sole object has been to tell the truth, the whole truth, to

> " nothing extenuate,
> Nor set down aught in malice."

It is possible, however, that with the best intentions we may, either by omission or commission, be guilty of some unfairness ; and if our readers will only show us wherein we have transgressed, we will be ready to make the *amende* if they will kindly afford us an opportunity—in a second edition.

We began our preface with an apology, we will end it with an appeal. We borrowed the one from Plutarch, Newton shall supply the other. At the close of the preface to his immortal *Principia* he says :—" I earnestly entreat that all may be read with candour, and that my labours may be examined not so much with a view to censure as to supply their defects."

THE AUTHOR.

LONDON, *February* 1884.

CONTENTS.

————•◦•————

CHAPTER I.

b

CHAPTER VII.

CHAPTER VIII.

CHAPTER IX.

CHAPTER X.

CHAPTER XI.

CHAPTER XII.

CHAPTER XIII.

A

HISTORY

OF

ELECTRIC TELEGRAPHY

TO THE YEAR 1837.

———·———

CHAPTER I.

FORESHADOWING OF THE ELECTRIC TELEGRAPH.

"Whatever draws me on,
Or sympathy, or some connatural force,
Powerful at greatest distance to unite,
With secret amity, things of like kind,
By secretest conveyance."

Milton, *Paradise Lost*, x. 246. 1667.

AMONGST the many flights of imagination, by which
genius has often anticipated the achievements of her
more deliberate and cautious sister, earth-walking
reason, none, perhaps, is more striking than the story
of the sympathetic needles, which was so prevalent in
the sixteenth, seventeenth, and eighteenth centuries,
and which so beautifully foreshadowed the invention
of the electric telegraph.* This romantic tale had

* "In the dream of the Elector Frederick of Saxony, in 1517, the
curious reader may like to discern another dim glimmering, a more
shadowy foreshadowing, of the electric telegraph, whose hosts of iron

B

reference to a sort of magnetic telegraph, based on the sympathy which was supposed to exist between needles that had been touched by the same magnet, or loadstone, whereby an intercourse could be maintained between distant friends, since every movement imparted to one needle would immediately induce, by sympathy, similar movements in the other. As a history of telegraphy would be manifestly incomplete without a reference to this fabulous contrivance, we propose to deal with it at some length in the present chapter.

For the first suggestions of the sympathetic needle telegraph we must go back a very long way, probably to the date of the discovery of the magnet's attraction for iron. At any rate, we believe that we have found traces of it in the working of the oracles of pagan Greece and Rome. Thus, we read in Maimbourg's *Histoire de l'Arianisme* (Paris, 1686)* :—

---

and copper 'pens' reach to-day the farthest ends of the earth. In this strange dream Martin Luther appeared writing upon the door of the Palace Chapel at Wittemburg. The pen with which he wrote seemed so long that its feather end reached to Rome, and ran full tilt against the Pope's tiara, which his holiness was at the moment wearing. On seeing the danger, the cardinals and princes of the State ran up to support the tottering crown, and, one after another, tried to break the pen, but tried in vain. It crackled, as if made of iron, and could not be broken. While all were wondering at its strength a loud cry arose, and from the monk's long pen issued a host of others "—*Electricity and the Electric Telegraph*, by Dr. George Wilson, London, 1852, p. 59 ; or D'Aubigné's *History of the Reformation*, chap. iv. book iii.

* English translation of 1728, by the Rev. W. Webster, chap. vi.

"Whilst Valens [the Roman Emperor] was at Antioch in his third consulship, in the year 370, several pagans of distinction, with the philosophers who were in so great reputation under Julian, not being able to bear that the empire should continue in the hands of the Christians, consulted privately the demons, by the means of conjurations, in order to know the destiny of the emperor, and who should be his successor, persuading themselves that the oracle would name a person who should restore the worship of the gods. For this purpose they made a three-footed stool of laurel in imitation of the tripos at Delphos, upon which having laid a basin of divers metals they placed the twenty-four letters of the alphabet round it; then one of these philosophers, who was a magician, being wrapped up in a large mantle, and his head covered, holding in one hand vervain, and in the other a ring, which hung at the end of a small thread, pronounced some execrable conjurations in order to invoke the devils; at which the three-footed stool turning round, and the ring moving of itself, and turning from one side to the other over the letters, it caused them to fall upon the table, and place themselves near each other, whilst the persons who were present set down the like letters in their table-books, till their answer was delivered in heroic verse, which foretold them that their criminal inquiry would cost them their lives, and that the Furies were waiting for the emperor at Mimas, where he was to die of a horrid kind of death [he was subsequently burnt alive by the Goths]; after which the enchanted ring turning about again over the letters, in order to express the name of him who should succeed the emperor, formed first of all these three characters, TH E O ; then having added a D to form THEOD the ring stopped, and was not seen to move any more; at which one of the assistants cried out in a transport of joy, ' We must not doubt any longer of it ; Theodorus is the person whom the gods appoint for our emperor.' "

If, as it must be admitted, the *modus operandi* is not here very clear, we can still carry back our subject to the same early date, in citing an experiment on magnetic attractions which was certainly popular in the days of St. Augustine, 354–430.

In his *De Civitate Dei*, which was written about 413, he tells us that, being one day on a visit to a bishop named Severus, he saw him take a magnetic stone and hold it under a silver plate, on which he had thrown a piece of iron, which followed exactly all the movements of the hand in which the loadstone was held.   He adds that, at the time of his writing, he had under his eyes a vessel filled with water, placed on a table six inches thick, and containing a needle floating on cork, which he could move from side to side according to the movements of a magnetic stone held under the table.*

Leonardus (Camillus), in his *Speculum Lapidum,*

* Basileæ, 1522, pp. 718-19.

&c., 1502, *verbo* MAGNES, refers to this experiment as one familiar to mariners, and Blasius de Vigenere, in his annotations of Livy, says that a letter might be read through a stone wall three feet thick, by guiding, by means of a loadstone or magnet, the needle of a compass over the letters of the alphabet written in the circumference.*

From such experiments as these the sympathetic telegraph was but a step, involving only the supposition that the same effects might be possible at a greater distance, but when, or by whom, this step was first taken it is now difficult to say. It has been traced back to Baptista Porta, the celebrated Neapolitan philosopher, and in all probability originated with him ; for in the same book in which he announces the conceit he describes the above experiment of St. Augustine, and other "wonders of the magnet"; adding that the impostors of his time abused by these means the credulity of the people, by arranging around a basin of water, on which a magnet floated, certain words to serve as answers to the questions which superstitious persons might put to them on the future.†

* *Les Cinq Premiers Livres de Tite Live*, Paris, 1576, vol. i. col. 1316.

† While it is generally admitted that magnetism has conferred incalculable benefits on mankind (witness only the mariner's compass), we have never yet seen it stated that it has at the same time contributed more to our bamboozlement than any other, we might almost say all, of the physical sciences. With the charlatans in all ages and nations, its mysterious powers have ever been fruitful sources of imposture, sometimes harmless, sometimes not. Thus, from the iron crook of the

He then concludes the 21st chapter with the following words, which, so far as yet discovered, contain the first clear enunciation of the sympathetic needle telegraph : —"Lastly, owing to the convenience afforded by the magnet, persons can converse together through long distances." * In the edition of 1589 he is even more explicit, and says in the preface to the seventh book : " I do not fear that with a long absent friend, even though he be confined by prison walls, we can communicate what we wish by means of two compass needles circumscribed with an alphabet."

The next person who mentions this curious notion was Daniel Schwenter, who wrote under the assumed name of Johannes Hercules de Sunde.   In his *Stega-nologia et Steganographia,* published at Nürnberg in 1600, he says, p. 127 :— " Inasmuch as this is a wonderful secret I have hitherto hesitated about divulging it, and for this reason disguised my remarks in the first edition of my book so as only to be under-

---

Greek shepherd Magnes, and the magnetic mountains of the geographer Ptolemy, to the magnetic trains of early railway enthusiasts ; from the magnetically protected coffin of Confucius to the magnetically suspended one of Mahomed ; from the magnetic powders and potions of the ancients, and the metal discs, rods, and unguents of the old magnetisers, to the magnetic belts of the new—the modern panacea for all the ills that flesh is heir to ; from the magnetic telegraphs of the sixteenth century to the Gary and Hosmer perpetual motors of the nineteenth, *et hoc genus omne ;* all these impostures are, or were, based entirely on the (supposed) force of magnetic attraction, to which must be added an unconscionable amount of ignorance or credulity.

* *Magiæ Naturalis,* p. 88,' Naples, 1558.

stood by learned chemists and physicians. I will
now, however, communicate it for the benefit of the
lovers of science generally." He then goes on to
describe, in true cabalistic fashion, the preparation of

FIG. 1.

De Sunde's dial as given in Schott's *Schola Steganographica.*

the two compasses, the needles of which were to be
made diamond-shaped from the same piece of steel
and magnetised by the same magnet, or rather,
magnets, for there were four: 1, Almagrito; 2,
Theamedes; 3, Almaslargont; 4, Calamitro; which

imparted south, north, east, and west-turning properties respectively to the needles. The compass-
cards were divided off into compartments, each containing four letters of the alphabet, and each letter
was indicated by the needle pointing, from one to
four times, to the division in which it stood. Thus,
the letter C would be indicated by three movements
of the needle to the first division of the card. The
needles were actuated by bar magnets, or chadids,
and attention was called by the ringing of a tiny
bell, which was so placed in the way of the needle
that at each deflection of the latter it was struck,
and so continued to ring until removed by the
correspondent.

The next and most widely known relation of
the story occurs in the *Prolusiones Academicæ,** of
Famianus Strada, a learned Italian Jesuit, first
published at Rome in 1617, and often reprinted
since. Although the idea did not originate with
Strada (for he seems to attribute it to Cardinal
Bembo, who died about 1547), he was certainly, as
Sir Thomas Browne quaintly says, "The *æolus*
that blew it about," for his *Prolusiones* had long
been a favourite classic, while the passage referring
to the loadstone has, if we may say so, been continually going the rounds of the newspapers. It
is quoted more or less fully in many authors of
the seventeenth and eighteenth centuries, famous

* Lib. ii., prol. 6.

amongst whom are Hakewill,* Addison,† Akenside,‡ and "Misographos." §

The references to it in the present century are simply too numerous to mention. The following is the latest English version, which, with the original Latin, appeared in the *Telegraphic Journal*, for November 15, 1875 :—

"There is a wonderful kind of magnetic stone to which if you bring in contact several bodies of iron or dial-pins, from thence they will not only derive a force and motion by which they will always try to turn themselves to the bear which shines near the pole, but, also, by a strange method and fashion between each other, as many dial-pins as have touched that stone, you will see them all agree in the same position and motion, so that if, by chance, one of these be observed at Rome, another, although it may be removed a long way off, turns itself in the same direction by a secret law of its nature. Therefore try the experiment, if you desire a friend who is at a distance to know anything to whom no letter could get, take a flat smooth disc, describe round the outside edges of the disc stops, and the first letters of the alphabet, in the order in which boys learn them, and place in the centre, lying horizontally, a dial-pin that has touched the magnet,

---

* *An Apologie or Declaration of the Power and Providence of God in the Government of the World*, 1630.

† *Spectator*, No. 241, 1711, and *Guardian*, No. 119, 1713.

‡ *The Pleasures of Imagination*, 1744.

§ *The Student; or, the Oxford and Cambridge Miscellany*, 1750.

so that, turned easily from thence, it can touch each separate letter that you desire.

"After the pattern of this one, construct another disc, described with a similar margin, and furnished with a pointer of iron—of iron that has received a motion from the same magnet. Let your friend about to depart carry this disc with him, and let it be agreed beforehand at what time, or on what days, he shall observe whether the dial-pin trembles, or what it marks with the indicator. These things being thus arranged, if you desire to address your friend secretly, whom a part of the earth separates far from you, bring your hand to the disc, take hold of the movable iron, here you observe the letters arranged round the whole margin, with stops of which there is need for words, hither direct the iron, and touch with the point the separate letters, now this one, and now the other, whilst, by turning the iron round again and again throughout these, you may distinctly express all the sentiments of your mind.

"Strange, but true! the friend who is far distant sees the movable iron tremble without the touch of any one, and to traverse, now in one, now in another direction ; he stands attentive, and observes the leading of the iron, and follows, by collecting the letters from each direction, with which, being formed into words, he perceives what may be intended, and learns from the iron as his interpreter. Moreover, when he sees the dial-pin stop, he, in his turn, if he thinks of any things to answer, in the same manner by the letters being touched separately writes back to his friend.

"Oh, I wish this mode of writing may become in use, a letter would travel safer and quicker, fearing no plots of robbers and retarding rivers. The prince, with his own hands, might despatch business for himself. We, the race of scribes, escaped from an inky sea, would dedicate the pen to the Shores of Magnet."

The Starry Galileo had his say on the same subject, and, as we may expect, said it well : "You remind me," says he, "of one who offered to sell me a secret art, by which, through the attraction of a certain magnetic needle, it would be possible to converse across a space of two or three thousand miles. And I said to him that I would willingly become the purchaser, provided only that I might first make a trial of the art, and that it would be sufficient for the purpose if I were to place myself in one corner of the room and he in the other. He replied that, in so short a distance the action would be scarcely discernible ; whereupon I dismissed the fellow, saying that it was not convenient for me just then to travel into Egypt, or Muscovy, for the purpose of trying the experiment, but that if he chose to go there himself, I would remain in Venice and attend to the rest."*

* *Dialogus de Systemate Mundi*, 1632, p. 88. It is curious that Kepler appears to have believed in the efficacy of the sympathetic telegraph. See Fournier's *Le Vieux-Neuf*, Paris, 1857, vol. i. p. 200.

Cardinal Richelieu's system of espionage was so perfect that he was regarded (and feared) by his contemporaries as a dabbler in "diabolical magic." He was supposed to have possessed either a magic mirror, in which he could see all that went on in the world, or the equally magic magnetic telegraph. *À propos* of this, we find the following passage in the *Letters writ by a Turkish Spy*, a work which has been attributed by the elder Disraeli to John Paul Marana :—" This Cardinal said, on another time, that he kept a great many courtiers, yet he could well enough spare them ; that he knew what passed in remote places as soon as what was done near him. He once affirmed he knew in less than two hours that the King of England had signed the warrant for the execution of ———. If this particular be true, this minister must be more than a man. Those who are his most devoted creatures affirm he has in a private place in his closet a certain mathematical figure, in the circumference of which are written all the letters of the alphabet, armed with a dart, which marks the letters, which are also marked by their correspondents ; and it appears that this dart ripens by the sympathy of a stone, which those who give and receive his advice keep always at hand, which hath been separated from another which the Cardinal has always by him ; and it is affirmed that with such an instrument he gives and receives immediately advices."*

The learned physician, Sir Thomas Browne, has

* Thirteenth letter, dated Paris 1639, vol. i.

some cautiously worded sentences on the mythical telegraph, which are worth quoting. "There is," he says, " another conceit of better notice, and whispered thorow the world with some attention ; credulous and vulgar auditors readily believing it, and more judicious and distinctive heads not altogether rejecting it. The conceit is excellent, and, if the effect would follow somewhat divine ; whereby we might communicate like spirits, and confer on earth with Menippus in the moon. And this is pretended from the sympathy of two needles, touched with the same loadstone, and placed in the center of two abecedary circles, or rings, with letters described round about them, one friend keeping one, and another the other, and agreeing upon an hour wherein they will communicate. For then, saith tradition, at what distance of place soever, when one needle shall be removed unto any letter, the other by a wonderful sympathy, will move unto the same. But herein I confess my experience can find no truth, for having expressly framed two circles of wood, and, according to the number of the Latine letters, divided each into twenty-three parts, placing therein two stiles, or needles, composed of the same steel, touched with the same loadstone and at the same point. Of these two, whenever I removed the one, although but at the distance of but half a span, the other would stand like Hercules pillars, and, if the earth stand still, have surely no motion at all." *

* *Pseudodoxia Epidemica*, book ii. chap. 3.

The *Scepsis Scientifica* of Joseph Glanvill, published in 1665, and which, by the way, secured his admission to the Royal Society, contains, perhaps, the most remarkable allusion to the then prevalent telegraphic fancy.   Glanvill, albeit very superstitious, was an ardent and keen-sighted philosopher, and held the most hopeful views as to the discoveries that would be made in after-times.   In the following passages he clearly foretells, amongst other wonders, the discovery and extension of telegraphs :—

"Should those heroes go on as they have happily begun, they'll fill the world with wonders.   And I doubt not but posterity will find many things that are now but rumours verified into practical realities.   It may be, some ages hence, a voyage to the southern unknown tracts, yea, possibly the moon, will not be more strange than one to America.   To them that come after us it may be as ordinary to buy a pair of wings to fly into the remotest regions as now a pair of boots to ride a journey.   *And to confer at the distance of the Indies by sympathetic conveyances may be as usual to future times as to us in a literary correspondence.*"— C. xix.

"That men should confer at very distant removes by an extemporary intercourse is a reputed impossibility, yet there are some hints in natural operations that give us probability that 'tis feasible, and may be compast without unwarrantable assistance from dæmoniack correspondence.   That a couple of needles equally toucht by the same magnet being set in two dyals exactly proportion'd to each other, and circumscribed by the letters of the alphabet, may affect this magnale hath considerable authorities to avouch it. The manner of it is thus represented. Let the friends that would communicate take each a dyal ; and having appointed a time for their sympathetic conference, let one move his impregnate needle to any letter in the alphabet, and its affected fellow will precisely respect the same. So that would I know what my friend would acquaint me with, 'tis but observing the letters that are pointed at by my needle, and in their order transcribing them from their sympathised index as its motion directs : and I may be assured that my friend described the same with his, and that the words on my paper are of his inditing.

" Now, though there will be some ill contrivance in a circumstance of this invention, in that the thus impregnate needles will not move to, but avert from each other (as ingenious Dr. Browne in his *Pseudodoxia Epidemica* hath observed), yet this cannot prejudice the main design of this way of secret conveyance, since 'tis but reading counter to the magnetic informer, and noting the letter which is most distant in the abecedarian circle from that which the needle turns to, and the case is not alter'd. Now, though this desirable effect possibly may not yet answer the expectation of inquisitive experiment, *yet 'tis no despicable item, that by some other such way of magnetick efficiency it may hereafter with success be attempted,* when magical history shall be enlarged by riper inspections, and 'tis not unlikely but that present discoveries might be improved to the performance."—C. xxi.

At the end of this chapter we give a list of references, as complete as we could make it, which will be useful to those of our readers who may wish to pursue the subject. It will also be instructive from another point of view, for it illustrates, in a very complete way, what Professor Tyndall has so well called the "menial spirit" of the old philosophers.[*] Notwithstanding that some of the more enlightened authors endeavoured laboriously to disprove the story, it was, for the most part, blindly and unquestioningly repeated, by one writer after another—credulous and vulgar auditors, as Sir Thomas Browne says, readily believing it, and more judicious and distinctive heads not altogether rejecting it, amongst whom we are tempted to reckon the learned knight himself.

Of those who stoutly and, at an early period, combatted the story, Fathers Cabeus and Kircher deserve

---

[*] " The seekers after natural knowledge had forsaken that fountain of living waters, the direct appeal to nature by observation and experiment, and had given themselves up to the remanipulation of the notions of their predecessors. It was a time when thought had become abject, and when the acceptance of mere authority led, as it always does in science, to intellectual death. Natural events, instead of being traced to physical, were referred to moral causes ; while an exercise of the phantasy, almost as degrading as the spiritualism of the present day, took the place of scientific speculation."—Tyndall's *Address to the British Association at Belfast,* 1874.

to be mentioned—the one for the excellence, and the other for the vehemence of his observations. Those of the former are particularly remarkable, as containing a hazy definition of the " lines of force" theory —a theory which Faraday has turned to such good account in his *Experimental Researches*. Cabeus, as well as we can understand him, says, in his tenth chapter :—" The action by which compass needles are mutually disturbed is not brought about by sympathy, as some persons imagine, who consider sympathy to be a certain agreement, or conformity, between natures or bodies which may be established without any communication. Magnetic attractions and repulsions are physical actions which take place through the instrumentality of a certain quality, or condition, of the intervening space, and which [quality] extends from the influencing body to the influenced body. I cannot admit any other mode of action in magnetic phenomena ; nor have I ever seen in the whole circle of the sciences any instance of sympathy or antipathy [at a distance]. * * *

"That which is diffused as a medium [or, that quality, or condition, of the intervening space] is thin and subtle, and can only be seen in its effects ; nor does it affect *all* bodies, only such as are either conformable with the influencing body, in which case the result is a perfecting change [or sympathy = attraction], or non-conformable, in which case the result is a corrupting change [or antipathy = repulsion]. This

C

quality is, I repeat, thin and subtle, and does not sensibly affect *all* intermediate [*i. e.*, neighbouring] bodies, although it may be disseminated through them. It only shows a sensibly good or bad effect according to the natures of the bodies opposed to one another.

"Bodies, therefore, are not moved by sympathy or antipathy, unless it be, as I have said, through the medium of certain essences [forces] which are uniformly diffused. When these reach a body that is suitable, they produce certain changes in it, but do not affect, sensibly, the intervening space, or neighbouring non-kindred bodies. Thus, the sense of smell is not perceived in the hand, nor the sense of hearing in the elbow, because, although these parts are equally immersed in the essences [or forces], they are not suitable, or kindred, in their natures to the odoriferous, or acoustic, vibrations." *

Kircher scouts the notion in no measured terms ; after soundly rating the propagators of the fable on their invention of the terms *chadid, almagrito, theamedes, almaslargont,* and *calamitro*—vile jargon, which, he says, was coined in the devil's kitchen—he thus delivers himself :—" I do not recollect to have ever

---

* *Philosophia Magnetica,* &c., chap. x.   *A brief letter from a young Oxonian to one of his late fellow pupils upon the subject of Magnetism,* London, 1697, contains, at page 10, a "draught" which illustrates very well the arrangement of magnetic lines of force, and which differs but little from the graphic representations of the present day. The curious little pamphlet is one of many gems in Mr. Latimer Clark's library.

met anything more stupid and silly than this idiotic conception, in the enunciation of which I find as many lies and impositions as there are words, and a crass ignorance of magnetic phenomena withal. In their craving after something wonderful and unknown they have manufactured a secret by means of barbarous and high-sounding words and by imitating the forms of recondite science, with the result that even they themselves cannot understand their own words." *

Many of the authors, who describe the sympathetic needle (dial) telegraph, speak also of another form, which seems to have been especially believed in by the Rosicrusians and Magnetisers of the last two centuries. It was supposed that a sympathetic alphabet could be marked on the flesh, by means of which people could correspond with each other, and communicate all their ideas with the rapidity of volition, no matter how far asunder. From the arms, or hands, of two persons intending to employ this method of correspondence a piece of flesh was cut, and mutually transplanted while still warm and bleeding. The piece grew to the new arm, but still retained so close a sympathy with its native limb, that the latter was always sensible of any injury done to it. Upon these transplanted pieces of flesh were tattooed the letters of the alphabet, and whenever a communication was to be made it was only necessary to prick with a magnetic needle the letters upon the arm composing

* *Magnes, sive de Arte Magnetica,* book ii. part iv. chap. 5.

the message ; for whatever letter the one pricked, the same was instantly pained on the arm of the other.*

List of authors of the sixteenth, seventeenth, and eighteenth, centuries, who either describe the sympathetic needle and sympathetic flesh telegraphs, or make a passing allusion to one or both of them ; chiefly compiled from Mr. Latimer Clark's list of books shown at the Paris Electrical Exhibition of 1881, and from the catalogues of the British Museum. As far as possible, only first editions quoted in full :—

1558   PORTA (GIAN B.).   *Magiæ Naturalis, &c. Libri IIII.*
        8vo. (See page 90.   Other editions : Antwerp, 1561,
        8vo.; Lugduni, 1561, 16mo. ; Venetia, 1560, 8vo. ;
        and 1665, 12mo. ; Coloniæ, 1562, 12mo.)   Neapoli, 1558.

1570   PARACELSUS (*i. e.,* Bombast Von Hohenheim).   *De*
        Secretis naturæ mysteriis, &c.   8vo. (Speaks only
        of sympathetic flesh telegraph.   Numerous editions
        in British Museum.)   Basileæ, 1570.

1586   VIGENERE (BLAISE DE).   *Traicté des Chiffres, ou*
        Secretes Manieres d'Escrire. (Quoted in *L'Élec-*
        tricien of Jan. 15, 1884, p. 95.)   Paris, 1586.

1589   PORTA (GIAN B).   *Magiæ Naturalis, &c. Libri XX.*
        Folio. (See preface to Book VII. for first clear
        mention of sympathetic needle telegraph.   Other
        editions : Francofurti, 1607, 8vo.; Napoli, 1611,

---

* Upon this delusion is founded Edmund About's curious novel, *Le Nez d'un Notaire*, in which he relates the odd results of sympathy between the notary's nose and the arm of the man from whom the flesh was taken.   But it is not in novels only, that we read of instances of the marvellous power of sympathy in these enlightened days ; witness the story of *The Sympathetic Snail Telegraph* of Messrs. Biat and Benoit, which went the rounds of the newspapers forty years ago, and which the curious—we were going to say sympathetic—reader will find fully described in *Chambers's Edinburgh Journal,* for February 15, 1851.

4to.; Hanoviæ, 1619, 8vo.; Lugduni, 1644 and 1651,
12mo.; London, 1658, 4to.; and Amstelodami, 1664,
12mo.) Neapoli, 1589.

1599 PANCIROLLUS (G.). *Rerum Memorabilium*, &c. 8vo.
(See Book II. [Nova Reperta], chap. xi., Notes.
This author refers to Scaliger [*Exotericarum exer-
citationem*, &c., exercit. 131], and Bodin [*Methodus
ad facilem Historiarum*, &c., chap. vii.], but they
only speak of magnetic sympathy at great distances,
without any reference to telegraphy. Other editions:
two 8vo., Ambergæ, 1607 and 1612; four Francofurti, 1622, 1629–31, 1646, and 1660; Lyon, 1617;
and London, 1715) Ambergæ, 1599.

1600 DE SUNDE (J. H.) (*i.e.*, Daniel Schwenter). *Stegano-
logia et Steganographia*. 8vo. (See p. 127. Janus
Hercules de Sunde is an assumed name. Hiller
in the preface to his *Mysterium Artis Steganographicæ*
1682, says that it is a synonym for Daniel Schwenter
Noribergense; and again on p. 287, quoting Schwenter,
he adds in parenthesis, "is est Hercules de Sunde."
Other edition: Nurnberg, 1650, 12mo.) Nurnberg, 1600.

1609 DE BOODT (ANSELMUS B.). *Gemmarum et Lapidum
Historia*, &c. 4to. (See Book II. Other editions:
Lugduni, 1636, 8vo.; Lyon, 1644, 8vo.; and again
Lugduni, 1647, 8vo.) Hanoviæ, 1609.

1610 ARGOLUS (ANDREAS). *Epistola ad Davidem Fabricium
Frisium*. (He made what he calls a "Stenographic
Compass," and held many agreeable conversations
by its means with one of his friends.)

In Ephemeridæ Patavii, **1610.**

1610 ARLENSIS (PETRUS), of Scudalupis. *Sympathia
Septem Metallorum*, &c. 8vo. (See chap. 2. This
writer, a noted astrologer and alchemist, was the
friend and fellow-citizen of Porta, to whom he seems
to attribute the first conception of the sympathetic
needle telegraph. His *Sympathia* was first published
at Rome, but immediately suppressed in order that
its grand secrets might not become known. It next
appeared at Madrid in folio. The Paris ed. of **1610**
was reissued at Hamburg in 1717.) Parisiis, 1610.

1617 STRADA (FAMIANUS). *Prolusiones Academicæ*, &c.
8vo. (See Lib. II., Prol. VI. Other editions:

Lugduni, 1617, and 1627, sm. 8vo.; Audomari,
1619, 12mo.; Mediolani, 1626, 16mo.; Oxoniæ,
1631, 8vo.; and again Oxoniæ, 1745, 8vo.)   Romæ, 1617.

1624 VAN ETTEN (H.), (*i. e.*, Leurechon Jean). *La Récréation
Mathématique*, &c. 8vo. (See p. 94. This author
is the first to give a drawing of the dial. H. Van
Etten was a *nom de plume*. See *Notes and Queries*,
1st series, vol. xi. p. 516. Other editions: Paris,
1626; Lyon, 1627; and three London, 1633, 1653,
and 1674. To the two latter is added a work of
Oughtred, the editor, whose name is so conspicuous
on the title-page, that rapid cataloguers make him
the author. Ozanam founded his *Recreations* on
Van Etten; Montucla made a new book of Ozanam
by large additions; and Hutton did the same by
Montucla, so that Hutton's well-known work is at
the end of a chain, of which Van Etten's is at the
beginning. *Notes and Queries*, 1st series, vol. xi. p.
504.)   Pont-à-Mousson, 1624

1629 CABEUS (NICOLAS). *Philosophia Magnetica*, &c.
Folio. (See p. 302.)   Coloniæ, 1629.

1630 HAKEWILL (GEORGE). *An Apologie or Declaration
of the Power and Providence of God*, &c. Folio.
(See p. 285. This is second edition; a first appeared
in [?] 1627, and a third in 1635.   London and Oxford, 1630.

1630 MYDORGE (CLAUDE). *Examen du livre des Récréations Mathématiques*, &c. 12mo. (See Problem 74,
pp. 140–44. This is a critically revised edition of
Van Etten. Another edition, Paris, 1638.)   Paris, 1630.

1631 KIRCHER (ATHANASIUS). *Ars Magnesia*, &c. 4to.
(See pp. 35 and 36.)   Herbipoli, 1631.

1632 GALILEO (G.). *Dialogus de Systemate Mundi*, &c.
4to. (See p. 88. Editions innumerable in British
Museum catalogue.)   Fiorenza, 1632.

1636 SCHWENTER (DANIEL). *Deliciæ Physico-Mathematicæ*.
(See p. 346. This work is based on Van Etten's,
supra. Two other 4to. editions appeared at Nurnberg, 1651–3 and 1677.)   Nurnberg, 1636.

1638 FLUDD (ROBERT). *Philosophia Moysaica*, &c.   Folio.
(See Sec. II., Lib. II., Memb. II., Cap. V., and Sec.
II., Lib. III., *passim*. An edition in English
appeared in London, 1659.)   Goudæ, 1638.

1641 KIRCHER (ATHANASIUS). *Magnes, sive de Arte Magnetica.* Sm. 4to. (See p. 382. Other editions: Coloniæ, 1643, 4to.; and Romæ, 1654, folio.)
Romæ, 1641.

1641 WILKINS (JOHN). *Mercury, or the secret and swift messenger, showing how a man with privacy and speed may communicate his thoughts to a friend at any distance.* 12mo. (See p. 147. Another edition in 1694.)
London, 1641.

1643 SERVIUS (PETRUS). *Dissertatio de Unguento Armario, Sive De Naturæ Artisque Miraculis.* (See para. 65, p. 68 This work is printed in Rattray's *Theatrum*, &c., *infra*.)
Romæ, 1643.

1646 BROWNE (SIR THOMAS). *Pseudodoxia Epidemica, or Enquiries into very many received tenents, and commonly presumed truths.* 4to. (See p. 76. Numerous editions in the British Museum.)
London, 1646.

1657 TURNER (ROBT). *Ars Notoria. The Notary Art of Solomon, showing the cabalistical key of magical operations,* &c. 18mo. (See p. 136.)
London, 1657.

1657-9 SCHOTT (GASPAR). *Magia Universalis Naturæ et Artis,* &c. 4 vols. 4to. (See vol. iv. p. 49. Copied from De Sunde and Kircher. Other edition: Bambergæ, 1677, 4to.)
Herbipoli, 1657-9.

1661 HENRION (DENIS) and MYDORGE (CLAUDE). *Les Récréations Mathématiques, avec l'examen de ses problèmes,* &c. Premièrement reveu par D. Henrion, depuis par M. Mydorge, Cinquième et dernière ed. 12mo. (See Problem 74, pp. 158-61. This is only a revised edition of Mydorge's *Van Etten,* of 1630.)
Paris, 1661.

1661 GLANVILL (J.). *The Vanity of Dogmatising, and an Apology for Philosophy.* 8vo. (See p. 202.) London, 1661.

1662 WESTEN (WYNANT VAN). *Het eerste Deel van de Mathematische Vermaeck,* &c. 8vo. Three parts. (See p. 125, Part I. This is an enlarged Dutch edition of Van Etten's, *supra*.)
Arnhem, 1662.

1662 RATTRAY (SYLVESTER). *Theatrum Sympatheticum Auctum, exhibens Varios Authores de Pulvere Sympathetico,* &c. 4to. (See p. 546, see Petrus Servius, *supra*.)
Norimbergæ, 1662

1663 HELVETIUS (J. F.). *Theatridium Herculis Triumph-*
 antis, &c. 8vo. (See pp. 11 and 15.) Haye, 1663.

1665 GLANVILL (JOSEPH). *Scepsis Scientifica; or, Confest*
 Ignorance the Way to Science, &c. 4to. (See p. 150.)
 London, 1665.

1665 SCHOTT (GASPAR). *Schola Steganographica,* &c. 4to.
 (See pp. 258-64. Description from De Sunde's,
 supra, with an elaborate drawing of the dial. Copper-
 plate title-page bears date 1665, printed title-page
 dated 1680.) Norimbergæ, 1665.

1676 HEIDEL (W. E.). *Johannis Trithemii, &c., Stegano-*
 graphia que Hucusqu: a nemine intellecta, &c. 4to.
 (See p. 358.) Moguntiæ, 1676.

1679 MAXWELL (WILLIAM). *De Medicina Magnetica, &c.*
 Lib. III. 12mo. (See chaps. 11, 12, and 13.)
 Francofurti, 1679.

1684 DE LANIS (FRANCISCUS). *Magisterium Naturæ et*
 Artis, Opus Physico-Mathematicum. 3 vols. (See
 vol. iii. p. 412.) Brixiæ, 1684–96.

1684 MARANA (G. P.) (or The Turkish Spy). *L'Espion du*
 Grand Seigneur, &c. 12mo. (See vol. i., 13th letter,
 dated Paris, 1639. Six other editions in British
 Museum.) ? Paris, 1684, &c.

1689 BLAGRAVE (JOSEPH). *Astrological Practice of Physick,*
 &c. 12mo. (See p. 112.) London, 1689.

1689 DE RENNEFORT (SOUCHU). *L'Aiman Mystique.* 12mo.
 Paris, 1689.

1696 DE VALLEMONT (PIERRE LE LORRAIN). *La Physique*
 Occulte, ou traité de la Baguette Divinatoire, &c.
 12mo. (See p. 32 of Appendix. Other editions :
 Paris and Amsterdam, 1693, 12mo.; and Amsterdam,
 1696, 12mo.) Paris, 1696.

1701-2 LE BRUN (PIERRE). *Histoire Critique des Pratiques*
 Superstitieuses. 2 vols. 12mo. (See vol. i. p. 294.
 Other editions : Amsterdam, 1733–36 ; and Paris,
 1750–1.) Rouen, 1701–2.

1711-13 ADDISON (JOSEPH). *The Spectator,* No. 241, for
 1711. (See p. 206. See also *The Guardian,* No. 119,
 for 1713.) London, 1711–13.

1718 DU PETIT ALBERT. *Secrets Merveilleux de la Magie*
 Naturelle et Cabalistique. (See p. 228. Other editions : Lyon, 1743 and 1762 ; and Paris, 1815.) Lyon, 1718.

1723 SANTANELLI (F.). *Philosophiæ Reconditæ, sive Magicæ Magneticæ,* &c. 4to. (See chap. xiv.) Coloniæ, 1723.

1730 BAILEY (NATHAN). *Dictionarium Britannicum,* &c. Folio. See word "Loadstone." Another London edition of 1736.) London, 1730.

1744 AKENSIDE (MARK). *The Pleasures of Imagination.* (See Book III., verses 325–37.) London, 1744.

1750–1 "MISOGRAPHOS." *The Student; or, the Oxford and Cambridge Monthly Miscellany.* 2 vols. (See vol. 1. p. 354. A translation of Strada's verses.) Oxford, 1750–1.

1762 DIDEROT. *Memoirs. Correspondance et ouvrages inédits de Diderot.* (See p. 278. Diderot, in his letter to Madame Volland of 28th July, 1762, alludes to Comus [Ledru] and his supposed telegraph.) Paris, 1841.

1769 GUYOT. *Nouvelles Récréations Physiques et Mathématiques.* 4 vols. 8vo. (See vol. i. p. 17. At p. 134 there is a full description, with illustrations, of what was probably Comus's apparatus. Two other Paris editions of 1786 and 1799) Paris, 1769.

1788 BARTHÉLEMY (JEAN JACQUES). *Voyage du Jeune Anacharsis en Grèce,* &c. 4to. (Quoted in *Journal of the Society of Arts,* May 20, 1859, p. 472 · twelve other editions (of which three are English translations) in the British Museum. See also *Correspondance Inédite du Madame du Deffand,* vol. ii. p. 99.) Paris, 1788.

1795 EDGEWORTH (RICHARD LOVELL). *Essay on the Art of Conveying Secret and Swift Intelligence.* Published in the *Transactions of the Royal Irish Academy.* (See vol. vi. p. 125.) Dublin, 1797.

1797 GAMBLE (J.). *An Essay on the Different Modes of Communication by Signals,* &c. 4to. (See p. 57.) London, 1797.

CHAPTER II.

STATIC, OR FRICTIONAL, ELECTRICITY—HISTORY IN RELATION TO TELEGRAPHY.

> " Thales call,
> He, whose enquiring mind paused musingly
> On the mysterious power, to action roused
> By amber rubbed.   This power (to him) a spirit,
> Woke from its slumbers by all-wondrous art."
>
> Oersted's *The Soul in Nature*,
> p. 157 of Bohn's edition.

THE science of electricity is a comparatively modern creation, dating only from the commencement of the seventeenth century.   It owes nothing, or almost nothing, to antiquity, and, in this respect, forms a remarkable contrast to most of the other branches of human knowledge—notably those of astronomy and mechanics, heat and light.   The vast discoveries, says Lardner, which have accumulated respecting this extraordinary agent, by which its connection with, and influence upon, the whole material universe—its relations to the phenomena of organised bodies—the part it plays in the functions of animal and vegetable vitality—its subservience to the uses of man as a mechanical power—its intimate connection with the chemical constitution of material substances—in fine, its application in almost every division of the sciences, and every department of the arts, have been severally demonstrated, are exclusively and peculiarly due to the spirit of modern research, and, in a great degree, to the labours of the present age.*

Yet it is not that, in this case, nature had concealed her secrets with more than her usual coyness, for we find, scattered through the writings of the ancients, many observations on a class of phenomena, which, if rightly examined, must have led to the establishment of electricity as a department of physics.

That amber acquires, by friction, the power of attracting light bodies, such as bits of straw, wood, and dry leaves, is a fact which is probably as old as the discovery of the substance itself. Thales, one of the seven wise men of Greece, described the property six hundred years before Christ, and not as if it were with him a new phenomenon, but rather as a familiar illustration of his philosophical tenets.† Aristotle, Pliny, and other Greek and Roman writers, also record the fact, and even sometimes mention luminous appearances attending the friction.‡ Theophrastus, B.C. 321, on the authority of Diocles, speaks of the *lapis lyncurius*, supposed to be our modern tourma-

* *Manual of Electricity, Magnetism, and Meteorology*, vol. i. p. 2.

† He ascribed to amber some living principle, some soul, which could be roused to action by friction, and, in the spirit of the age, it was declared sacred. For the same reason, the loadstone was venerated, it being supposed to possess an immaterial spirit under the influence of which it attracted iron.—Aristotle, *De Anima*, i. 2.

‡ Pliny, book xxxvii. chap. iii.

line, as possessing the same property as amber, adding that it attracts not only straws and leaves, but copper also, and even iron, if it be in small particles.*

The emission of sparks from the human body, when submitted to friction, had also been noticed, as in the case of Servius Tullius, the sixth King of Rome, whose locks were frequently observed to give off sparks under the operations of the toilette. Eustathius, Bishop of Thessalonica, A.D. 1160, cites another instance in his *Commentarii ad Homeri Iliadem*, that of a certain ancient philosopher, who, occasionally, when changing his dress, emitted sparks, and, sometimes, even entire flames, accompanied by crackling noises. He also mentions the case of Walimer, a Gothic chief, who flourished A.D. 415, who used to give off sparks from his body.†

The Greeks and Romans were not the only people

---

* *De Lapidibus*, p. 124, Hill's edition.

† In *Iliad*, E, p. 515, Roman ed. We do not notice the frequent allusions in the pages of Cæsar, Livy, Plutarch, and others, to flames at the points of the soldiers' javelins, at the tops of the masts of ships, and, sometimes, even on the heads of the sailors themselves; for all these phenomena, though now known to be of the same nature as those described in the text, were then regarded simply as manifestations of the gods. See a very interesting example of this in Plutarch's *Life of Timoleon*, vol. iii. p. 16, Dacier's edition. For much interesting information on this subject, see Dr. William Falconer's " Observations on the Knowledge of the Ancients respecting Electricity," in vol. iii. *Memoirs of the Literary and Philosophical Society of Manchester*, 1790; also Tomlinson's *The Thunderstorm*, p. 96. In the early ages of the Church, the Popes were often reckoned as magicians, Gregory VII. being held in especial awe, because when he pulled off his gloves fiery sparks issued from them.

of antiquity to whom these phenomena were familiar. Thus, in the Persian language amber is called *Káh-rubá*, or attractor of straw, as the magnet is called *Ahang-rubá*, or attractor of iron. In the old Persian romance, *The Loves of Majnoon and Leila*, the lover says of his adored one, " She was as amber, and I but as straw ; she touched me, and I shall ever cling to her." In the writings of Kuopho, a Chinese physicist of the fourth century, we read, " The attraction of a magnet for iron is like that of amber for the smallest grain of mustard seed. It is like a breath of wind, which mysteriously penetrates through both, and communicates itself with the rapidity of an arrow."

Humboldt,* after referring to this interesting fact, tells us how he himself had observed, with astonishment, on the woody banks of the Orinoco, in the sports of the natives, that the excitement of electricity by friction was known to these savage races. Children, he says, may be seen to rub the dry, flat, and shining seeds, or husks, of a trailing plant until they are able to attract threads of cotton and pieces of bamboo cane.

Such phenomena, says Lardner, in the work from which we lately quoted,† attracted little attention, and provoked no scientific research. Vacant wonder was the most exalted sentiment they raised ; and they accordingly remained, while centuries rolled away,

---

* *Cosmos*, London, 1849 ed., vol. i. p. 176. † Vol. i. p. 4.

barren and isolated facts upon the surface of human knowledge. The vein whence these precious fragments were detached, and which, as we have shown, *cropped out* sufficiently often to challenge the notice of the miner, continued unexplored; and its splendid treasures were reserved to reward the toil and crown the enterprise of modern times.

Without going the length of asserting that electrical phenomena were entirely neglected during the long night of the middle ages, it seems certain that, with the exception of the discovery of the electrical property of jet, little advance was made up to the close of the sixteenth century. Then it was that Dr. Gilbert, of Colchester, for the first time collected the scattered fragments, and, with many valuable observations of his own, shaped them into the nucleus of a new science, to which he gave the name Electricity, from the Greek word ἤλεκτρον, signifying amber. In his great work, *De Magnete,** published in the year 1600, he described the only three substances known up to his time as susceptible of electrical excitation, and added a variety of others, such as spars, jems, fossils, glasses, and resins, which enjoyed, equally with them, the power of attracting not only light

---

* This book, although mainly devoted to magnetism, has many pages on electricity; and, besides its intrinsic value, is interesting as containing the first publications on our subject. William Gilbert was a member of the College of Physicians, London, and became Physician in Ordinary to Queen Elizabeth, who, conceiving a high opinion of his learning, allowed him an annual pension to enable him to prosecute his studies. He died in 1603.

bodies, like feathers and straws, but all solid and fluid matter, as metals, stones, water, and oil.

He also observed some of the circumstances which affect the production of electricity, such as the hygrometric state of the atmosphere. Thus, he noticed that when the wind blew from the north and east, and was dry, the body could be excited by a brisk and light friction continued for a few minutes, but that when the wind was from the south and moist, it was difficult, and sometimes impossible, to excite it at all. In order to test the condition of the various substances experimented upon, Gilbert made use of a light needle of any metal, balanced, and turning freely on a pivot, like the magnetic needle, to the extremities of which he presented the bodies after excitation.

Some of Gilbert's deductions were curiously fallacious. In pointing out, for instance, the distinction between magnetic and electric attraction, he affirmed that magnets and iron mutually attracted each other, but that when an electric was excited it alone attracted, the substances attracted remaining inactive. He noticed also, as a special distinction between magnetism and electricity, that the former repelled as well as attracted, whilst the latter only attracted.*

The few references to electricity in the works of Sir Francis Bacon, Nicolas Cabeus, Kenelm Digby, Gassendi, Descartes, Thomas Browne, and others, may be passed over in silence, as they are chiefly

* *De Magnete*, lib. ii. cap. 2-4.

theoretical, and did not contribute in any way to the advancement of the science.*

The celebrated Robert Boyle, to whom some of the other physical sciences owe such great obligations, directed much of his attention to the subject of electricity, and has left us an account of his experiments, in a small work, entitled *Experiments and Notes about the Mechanical Origine or Production of Electricity*, London, 1675. By means of a suspended needle, he discovered that amber retained its attractive virtue after the friction which excited it had ceased; and though smoothness of surface had been regarded as advantageous for excitation, yet he found a diamond, which, in its rough state, exceeded all the polished ones, and all the electrics that he had tried, it having been able to move the needle three minutes after he had ceased to rub it. He found also that heat and "tersion" (*i. e.*, the cleaning or wiping of any body) increased the electrical effect; and that if the attracted body were fixed, and the attracting one movable, their approach would take place all the same, thus disproving one of Gilbert's deductions. To Dr. Gilbert's list of electrics, he added several new ones, as glass of antimony, white sapphire, white amethyst, carnelian, &c.

Like all his predecessors, Boyle (in whom, by the way, the theorising faculty was particularly strong)

* Jacob Böhmen, the Teutonic Theosopher, who lived 1575–1624, and who wrote largely on astrology, philosophy, chemistry, and divinity, has some pages on electricity. See *Notes and Queries*, July 28, 1855, p. 63.

speculated, in his turn, on the cause of electrical phenomena ; but it seems that he, as well as they, could find no better explanation than that offered by the Ionic sage, twenty-three centuries before. The supposition was that the excited body threw out a glutinous or unctuous effluvium, which laid hold of small bodies in its path, and, on returning to its source, carried them along with it.* The *Philosophical Transactions* of this period contain some learned disquisitions in support of this (now strange) hypothesis, and even experiments are described which were considered as conclusive of its correctness.†

Otto Guericke, burgomaster of Magdeburg, and inventor of the air-pump, was contemporary with Boyle, and to him we owe some most important advances. In 1671, he constructed the first electrical machine, by means of which he was able to produce electricity in far greater quantities than had hitherto been possible from the friction of glass or sulphur rods. With this machine, which consisted of a globe of

---

* Boyle is sometimes said to have been the first, in modern times, to observe the electric light—an assertion which seems to be based upon his observation, in 1663, of the light which some diamonds gave out, in the dark, after being rubbed. But it is doubtful if this was not an optical rather than an electrical effect, an instance of what may be called latent light, and therefore belonging to the class of phenomena, of which the celebrated Bologna stone, discovered in 1602 by the quondam shoemaker Casiorolus, was the first recorded example, as Balmain's luminous paint is the last. For much interesting information on this subject, see Sir D. Brewster's *Letters on Natural Magic.*

† *Phil. Trans.,* for 1699, vol. xxi. p. 5.

D

sulphur,* mounted on a revolving axis, and excited by the friction of a cloth held in the hand, he discovered

FIG. 2.

The First Electrical Machine, copied from p. 148 of Otto Guericke's
Experimenta Nova, &c.

the " hissing noise and gleaming light " which accompany strong electrification.

* Sulphur, it may be remarked, was a favourite electric with early experimenters, as it was imagined that electricity was emitted with the sulphurous effluvium produced by the friction.   In the construction of his machine, Guericke, for example, cast the sulphur in a glass globe, which he afterwards broke, so as to expose the sulphur to the action of the rubber, little imagining that the glass globe itself would have answered his purpose just as well.

To him also belongs the discovery of the property of electrical repulsion. He ascertained that a feather, when attracted to an excited electric, was instantly repelled, and was incapable of a second attraction, until it had been touched by the finger or some other body. He also observed that a feather, when thus repelled, always kept the same side towards the excited electric—a fact the correspondence of which with the position of the moon towards the earth, induced him and other philosophers to assume that the revolution of the moon round the earth might be explained on electrical principles. Again, in the observation that a substance becomes electric by being merely brought near to another electrified body, Guericke discovered the fact, though not the principle, of induction.*

Newton, about the same time, published another effect of induction, *viz.:* one side of a glass plate being electrified, the other side will also be electrified, and will attract any light bodies within its influence. Laying upon a table a disc of glass two inches broad, in a brass hoop or ring, so that it might be one-eighth of an inch from the table, and then rubbing it briskly, little pieces of paper, laid upon the table under the glass, moved nimbly to and fro, and twirled about in the air, continuing these motions for a considerable time after he had ceased rubbing. Upon sliding his

---

* *Experimenta Nova Magdeburgica*, Amstelodami, 1672, lib. iv. cap. 15.

finger over the glass, though he did not agitate it, nor, by consequence, the air beneath, he observed that the papers, as they hung under the glass, would receive some new motion, inclining this way or that, according to the direction of his finger.

The Royal Society had ordered this experiment to be repeated at their meeting of December 16, 1675, and, in order to ensure its success, had obtained a full account of it from its distinguished author. The experiment, however, failed, and the secretary requested the loan of Sir Isaac's apparatus, inquiring, at the same time, whether or not he had guarded against the papers being disturbed by the air which might have somewhere stolen in? In replying, on the 21st of December, Newton advised them to rub the glass " with stuff whose threads may rake its surface, and if that will not do, rub it with the finger ends to and fro, and knock them as often upon the glass." Following these directions, the Society succeeded, on January 31, 1676, when they used a scrubbing brush of short hog's bristles, and the heft of a knife made with whalebone ! *

In the 8th and 27th queries at the end of his treatise on Optics, Newton has introduced the subject of electricity in such a manner as to convey some notion of the theoretical views which he had been led to form.   He says (8th query) :—" A globe of glass about eight or ten inches in diameter being put into a

* See Brewster's *Life of Sir Isaac Newton*, pp. 307–8 : or Birch's *History of the Royal Society*, vol. iii. pp. 260–70.

frame where it may be swiftly turned round, its axis will, in turning, shine where it rubs against the palm of one's hand applied to it; and if at the same time a piece of white paper be held at the distance of half an inch from the glass, the electric vapour, which is excited by the friction of the glass against the hand, will, by dashing against the paper, be put into such an agitation as to emit light, and make the paper appear livid like a glow-worm. In rushing out of the glass, it will even sometimes push against the finger so as to be felt." And again, in the 27th query, he says :—" Let him also tell me how an electric body can, by friction, emit an exhalation so rare and subtile, and yet so potent, as by its emission to cause no sensible diminution of the weight of the electric body, and to be expanded through a sphere whose diameter is above two feet, and yet to be able to agitate and carry up leaf copper, or leaf gold, at the distance of above a foot from the electric body." *

Between 1705 and 1711, Hauksbee made many

* These appear to be the only published observations of the great Sir Isaac on electrical matters; but it would seem that, in moments of leisure from weightier business, he bestowed an occasional glance on the infant science. This will be apparent from the following extract from an autograph letter, which Mr Latimer Clark has lately unearthed, and which will be found in full in *The Electrician* Journal, for April 16, 1881 :—" I have been much amused by ye singular φενομενα resulting from bringing of a needle into contact with a piece of amber or resin fricated on silke clothe. Ye flame putteth me in mind of sheet lightning on a small (how very small) scale." Although this letter is dated "London, December 15, 1716," it would seem from the wording that Newton was unaware of similar comparisons instituted several years before, by Hauksbee and Wall.

valuable and interesting observations, of which we must content ourselves with a brief *résumé*, referring our readers for fuller accounts to the original papers in the *Philosophical Transactions*, or to Priestley's excellent *History and Present State of Electricity*, pp. 15–23, 5th ed.   In 1705, he showed that light could be produced by passing common air through mercury, contained in a well-exhausted glass receiver.   The air, rushing through the mercury, blew it against the sides of the glass, and made it appear like a body of fire, consisting of an abundance of glowing globules.   In repeating this experiment with about three pounds of mercury, and making it break into a shower, by dashing it against the crown of another glass vessel, flashes resembling lightning, of a very pale colour, and distinguishable from the rest of the produced light, were thrown off from the crown of the glass in all directions.*   Hauksbee likewise showed that considerable light may be produced by agitating mercury in a partially exhausted tube ; and that even in the open air numerous flashes of light are discoverable by shaking quicksilver in any glass vessel.

* Electric light *in vacuo* was first observed by Picard in 1675   While carrying a barometer from the Observatory to Porte St. Michel in Paris, he observed light in the vacuous portion.   Sebastien and Cassini observed it afterwards in other barometers.   John Bernouilli, in 1700, devised a "mercurial phosphorus" by shaking mercury in a tube which had been exhausted by an air-pump.   This was handed to the King of Prussia—Frederick I.—who awarded it a medal, of forty ducats' value. The great mathematician wrote a poem in honour of the occasion.— Tyndall's *Notes on Electricity*.

In a subsequent series of experiments on the light produced by the attrition of bodies *in vacuo*, he showed that glass, when thus excited, emitted light in as strange a form as lightning, particularly when he used a rubber that had been previously drenched in spirits of wine. In all these experiments Hauksbee had no notion of the electrical origin of the light, and in saying that it resembled lightning he was only using a simile, without any suspicion of a closer connection.

Like Sir Isaac Newton, Hauksbee employed a glass globe machine, as he thought that this material was capable of more powerful effects. When exhausted of air, and turned briskly, the application of his hand would produce a strong light on the inside ; and, by re-admitting the air, light appeared on the outside also. By bringing an exhausted globe near to an excited one, he found that a light was produced in the former, which soon disappeared ; but which immediately re-appeared, with great beauty, on a further excitation.

The following experiment must at that time, and indeed for long after, have been considered one of great singularity. Having coated one half of the inside of a glass globe with sealing-wax, which in some places was an eighth of an inch thick, and therefore quite opaque, he exhausted it and put it in motion. On applying his hand, for the purpose of excitation, its outline soon became distinctly visible on the concave surface of the wax, thus making it seem to be transparent, although before excitation it would only just allow the flame of a lighted candle to be seen through it in the dark. The same result was obtained when pitch, or common brimstone, was substituted for the sealing-wax.

Besides light and crackling noises, Hauksbee also noticed that an electrified body was able to produce a sense of pain (the electric shock) in the hand, or face, that touched it—an observation which is also claimed for his friend, Dr. Wall.

This latter philosopher is, however, best known as being the first to suspect the identity of lightning and electricity. The happy thought was suggested to him, as he tells us in a paper read before the Royal Society in 1708, by the sparks and crackling sounds produced by the friction of a large stick of amber against a woollen cloth. "Upon drawing," he says, "the piece of amber swiftly through the woollen cloth, and squeezing it pretty hard with my hand, a prodigious number of little cracklings was heard, every one of which produced a little flash of light ; but when the amber was drawn gently and slightly through the cloth, it produced a light, but no crackling. By holding a finger at a little distance from the amber, a crackling is produced, with a great flash of light succeeding it ; and what is very surprising, on its eruption it strikes the finger very sensibly, where-soever applied, with a push or puff like wind. The crackling is full as loud as that of charcoal on fire ;

nay, five or six cracklings, or more, according to the quickness of placing the finger, have been produced from one single friction, light always succeeding each of them. Now I doubt not but on using a longer and larger piece of amber, both the cracklings and light would be much greater. This light and crackling seem in some degree to represent thunder and lightning." *

So far, experimenters had worked without any system, and without in the least comprehending the principles on which the effects they produced depended. Highly important as were all their observations, the true foundations of electricity as a science cannot, therefore, be said to have been laid until Stephen Gray, a pensioner of the Charter-house, London, gave to the world that justly celebrated series of experiments which, begun in 1720, only ended with his last breath in 1736.† As from this point the domain widens, we will confine ourselves in the rest of this chapter to noticing only such discoveries of Gray and succeeding philosophers as bear intimately on our subject.

In February 1729, Gray discovered the principle of electric conduction and insulation, and in doing

---

* Hutton's *Phil. Trans. Abridged*, vol. v. p. 409.

† This remarkable man was (so to speak) dying when his last experiments were made, and, unable to write himself, he dictated an account of them to Dr. Mortimer, the secretary of the Royal Society, the day before his death.—See *Phil. Trans.*, vol. xxxix. p. 400, 1735–36, or Hutton's *Abridgment*, vol. viii. p. 110.

so might almost be said to have invented electric telegraphy, of which it is the very *alpha* and *omega.* This important discovery was made in the following manner :—Wishing to excite in metals, as had already been done in glass, resin, &c., the power of attraction and repulsion, he tried various methods, such as rubbing, heating, and hammering ; but all to no end. At last an idea occurred to him that, as a glass tube, when rubbed in the dark, communicated its light freely to bodies, so it might communicate a power of attraction, which, at this time, was considered the only absolute proof of the presence of electricity. In order to test this, he took a glass tube, 3 feet 5 inches long and 1 inch diameter, and filled up the ends with pieces of cork to keep out the dust when the tube was not in use. His first experiment was to ascertain if there was any difference in its power of attraction when the tube was stopped at both ends by the corks, and when left entirely open ; but he could perceive no sensible difference. Then holding a feather over against the end of the tube, he found it would fly to the cork, being attracted by it as readily as by the tube itself. He concluded from this that the electric virtue, conferred on the tube by friction, passed spontaneously to the cork.

It then occurred to him [*] to inquire whether this

---

[*] We follow in this and the next three paragraphs Lardner's *Manual of Electricity, Magnetism,* &c., vol. i. pp. 8–9. See also Priestley's *History of Electricity*, pp. 24–39.

transmission of electricity would be made to other substances besides cork. With this view he obtained a deal
rod about four inches in length, to one end of which
he attached an ivory ball, and inserted the other in
the cork, by which the glass tube was stopped. On
exciting the tube, he found that the ivory ball attracted
and repelled the feather even more vigorously than the
cork. He then tried longer rods of deal, and pieces
of brass and iron wire, with like results. Finally
he attached to one end of the tube a piece of common packthread, and, suspending from its lower end
the ivory ball and various other bodies, found that all
of them were capable of acquiring the electric state
when the tube was excited. Experiments of this
kind were made from the balconies of his house
and other elevated stations.

With a true philosophic spirit, he now determined
to inquire what circumstances attending the *manner*
of experimenting produced any real effect upon the
results ; and, first, whether the *position* or *direction* of
the rods, wires, or cords, by which the electricity was
transmitted from the excited tube, affected the phenomena. For this purpose he extended a piece of
packthread in a horizontal direction, supporting it at
different points by other pieces of similar cord, which
were attached to nails driven into a wooden beam, and
which were, therefore, in a vertical position. To one
end of the horizontal cord he attached the ivory ball,
and to the other he tied the end of the glass tube. On

exciting the tube he found that no electricity was transmitted to the ball, a circumstance which he rightly ascribed to its escape by the vertical cords, the nails supporting them, and the wooden beam.

Soon after this (June 30, 1729), Gray was engaged in repeating his experiments at the house of Mr. Wheeler, who was afterwards associated with him in these investigations, when that gentleman suggested that threads of silk should be used to support the horizontal line of cord, instead of pieces of packthread. It does not appear that this suggestion of Wheeler proceeded from any knowledge, or suspicion, of the electric properties of silk ; and still less does it appear that Gray was acquainted with them ; for, in assent-ing to the proposition of his friend, he observed, that "silk might do better than packthread on account of its smallness, as less of the virtue would probably pass off by it than by the thickness of the hempen line which had been previously used."

They accordingly (July 2, 1729) extended a pack-thread through a distance of about eighty feet in a horizontal direction, supporting it by threads of silk. To one end they attached the ivory ball, and to the other the glass tube. When the latter was excited, the ball immediately became electric, as was mani-fested by its attracting metallic leaf held near it. Next day, they extended their experiments to lines of packthread still longer, when the silk threads used for its support were found to be too weak, and were broken. Being under the (erroneous) impression that the escape of the electricity was prevented by the fineness of the silk, they now substituted for it thin brass wire, which they expected, being still finer than the silk, would more effectually intercept the electricity; and which, from its nature, would have all the necessary strength. The experiment, however, completely failed. No electricity was conveyed to the ivory ball, the whole having escaped by the brass wire, notwithstanding its fineness. They now saw that the silk threads intercepted the electricity, because they were *silk*, and not because they were *fine*.

Having thus accidentally discovered the property of insulation, they proceeded to investigate its generalisation, and found that it was enjoyed by resin, hair, glass, and some other substances.

In fact, it soon became apparent that in this respect all matter may be said to belong to one of two classes, the one like the packthread and brass wire, favouring the dissipation, or carrying away, of the electric power, and the other like the silk and glass opposing it.*

* Soon after this, in August 1729, Gray discovered that when the electrified tube was brought near to any part of a *non-electric* or conducting body, without touching it, the part most remote from the tube became electrified. He thus fell upon the fact, which afterwards led to the principle of induction. The science, however, was not yet ripe for that great discovery, and Gray, like Otto Guericke before him, and Wilson and Canton after him, continued to apply the principles of induction without the most remote suspicion of the rich mine whose treasures lay beneath his feet, and which it was one of the glories of Franklin to bring to light.

Armed with this knowledge, Gray and Wheeler, in July 1729, had the great satisfaction of being able to transmit the electric power through as much as 765 feet of packthread, supported by loops of silk ; and in August 1730, through 886 feet of wire. It is curious to observe that in these experiments, as, indeed, in all others on electrical conduction, we have all the essentials—crude, of course—of a perfect telegraph, the insulated line, the source of electricity in the rubbed glass, the indicating instrument in the down feather, and the earth, or return circuit, the function of which, however, was not then suspected.

While Gray and Wheeler were pursuing their investigations in England, Dufay, of the Academy of Sciences, and Intendant of the Royal Botanic Gardens, was actively engaged in Paris, in a similar manner. The researches of this philosopher, so celebrated as the originator * of the double-fluid theory of electricity, embraced the period between 1733 and 1737. He added largely to the class of bodies called *electrics*, by showing that all substances, except metals, and bodies in the soft or liquid state, might be made electric, by first heating them, and then rubbing them on any kind of cloth ; and as regards even these

* He can hardly be called its author—at all events in its present form. For Symmer's claims to this honour, the reader is referred to Priestley's *History of Electricity*, p. 227. The writer of the article Electricity in the *Encyclopædia Britannica*, 7th edition, says, but we know not on what authority, that this important discovery was simultaneously and independently made by Dufay in France and by White in England.

exceptions, he showed that they, and, generally, *all* bodies, solid and liquid, could be electrified, if only the precaution were taken of first placing them on glass stands.

In repeating Gray's experiments with the pack-thread, he perceived that they succeeded better after wetting the line, and, with the aid of this fact, he was able to transmit the electric power along a cord of nearly 1300 feet, which he supported at intervals on glass tubes.

His discovery of the dual character of electricity was, like most of the other capital discoveries hitherto made, entirely due to chance. A piece of gold leaf having been repelled by an excited glass rod, Dufay pursued it with an excited rod of sealing-wax, expecting that the effect would be the same. His astonishment, therefore, was great on seeing the gold leaf fly to the wax, and, on repeating the experiment, the same result invariably followed; the gold leaf, when repelled by glass, was attracted by resin, and, when repelled by resin, was attracted by glass. Hence Dufay concluded that there were two distinct kinds of electricity, and, as one was produced from glass, and the other from resin, he distinguished them by the names *vitreous* and *resinous*.

In repeating Otto Guericke's experiments, Dufay discovered another general law, which enabled him to explain a number of observations that hitherto were obscure and puzzling. This law is, that an electrified body attracts those that are not so, and repels them as soon as they become electric by contact with itself. Thus, gold leaf is first attracted by the excited tube, and acquires an electricity by the contact, in consequence of which it is immediately repelled. Nor is it again attracted while it retains this electric quality; but, if now it chance to light on some other body, it straightway loses its electricity, and is then re-attracted by the tube, which, after having given it a new charge, repels it a second time, and so on, as long as the tube itself retains any electricity.*

The study of electricity was next taken up, in 1737, by Desaguliers, who, though born in France in 1683, early removed to England, and died in London in 1744. Two years before his death he published a *Dissertation Concerning Electricity*,† which is remarkable as being the first book on the subject in the English language. Desaguliers' investigations were mainly concerned with the relative conducting powers of various bodies, but he otherwise did good and useful work, by methodising the information that had already accumulated, and by improving in some

* Priestley's *History of Electricity*, pp. 40–50.

† As a reason for his engaging in this pursuit so late in life, Desaguliers makes the curious assertion that he was debarred from doing so earlier by the peculiar temper of Stephen Gray, who would have abandoned the field entirely if he saw that anything was done in apparent opposition or rivalry to himself.—Brewster's *Edinburgh Encyclopædia*, *verbo* Electricity, p. 415.

It is difficult to reconcile this passage with the following, which we extract from Desaguliers' *Dissertation*, p. 47:—"Indeed, a few electrical experiments, made by Mr. Gray and myself many years ago, important respects the nomenclature. Thus, the labours of Gray, Wheeler, Dufay, and himself, had shown that all matter was divisible into two great classes, these he now proposed to distinguish by the names *Electrics*, or bodies in which electricity could be excited by friction, and *Non-electrics*, or those in which it could not be excited, but which could receive it from an electric. He also first employed the words *Conductor* and *Non-conductor* in the same sense as they are used at the present day.

In the *Philosophical Transactions*, for 1739, vol. xli. p. 209, will be found his experiments on the transmission of electricity, which were made at H.R.H. the Prince of Wales's house at Cliefden, on April 15, 1738. " Having heard that electricity had been carried along a hempen string five or six hundred feet, but having only seen it done when the string was carried backwards and forwards in a room, by silk supporters, Dr. D. wished to try it with a packthread stretched out at full length ; for which purpose, having joined a piece

---

are mentioned in the first volume of my *Course of Experimental Philosophy*, pp. 17–21."

The following lines by the poet Cawthorn depict the neglect and indigence into which Desaguliers fell in his old age :—

> "Can Britain   *   *   *   *   *
>   *   *   permit the weeping muse to tell
> How poor neglected Desaguliers fell ?
> How he, who taught two gracious kings to view
> All Boyle ennobled, and all Bacon knew,
> Died in a cell, without a friend to save,
> Without a guinea, and without a grave ? "
>
> *The Vanity of Human Enjoyments*, v. 147–54.

E

of catgut to one end of a string, he fastened it to a door ; and having also tied another catgut to the other end of the string, he fastened it at the other end of the house. At the places where the packthread was joined to the catgut he left eighteen inches of the thread hanging down, and fastened a *lignum vitæ* handle of a burning-glass to one, while he applied a rubbed tube to the other. He made the electricity run to the *lignum vitæ*, but with some difficulty, which he attributed to the sizing, being an animal substance, that still adhered to the thread as it was new ; therefore, he caused the thread to be wet with a sponge from one end to the other, to wash off the size ; then was the electricity from the tube communicated very soon and very strongly ; for the thread of trial was drawn by the *lignum vitæ* at the distance of a foot.

"Afterwards, having joined more packthread together, he made a string of 420 feet long, which he supported at intervals by pieces of catgut. The string was previously dipped in a pail of water, but great care was taken that the catgut should not be wet. Then he applied the rubbed tube at one end, while an assistant held the thread of trial near the handle at the other, whereupon it was strongly attracted, though the wind was very high, and blowed in the contrary direction to that in which the electricity ran.

" He first tried the experiment with the packthread dry, but then it would not succeed at that distance." *

* Hutton's *Phil. Trans. Abridged,* vol. viii. p. 357.

Up to this time, and until some years later, experiments on the transmission of electricity to a distance excited no attention outside a very narrow circle of scientific men, and even amongst these, they served only to illustrate the two great electrical properties of bodies—conduction and insulation—without evoking the slightest suspicion of their practical value. The whole subject of electricity now, however, began to attract general attention, especially amongst the Germans, and the first consequence was considerable improvement in the power and efficiency of electrical apparatus. About 1741, Professors Hausen, of Leipsic, and Boze, of Wittemburg, revived the use of the glass globe machine, first introduced many years before, by Newton and Hauksbee, but which, after their time, had been supplanted, to the great detriment of the science, by the glass tube and silk rubber of Gray. Boze also added, for the first time, the prime conductor, which consisted of an oblong cylinder of tin or iron. This was at first held in position by a man, who was insulated, by standing on cakes of resin, but it was subsequently suspended by silken cords, and, in order to facilitate the passage of the electricity, a number of linen strings were added, which served the purpose, though very imperfectly, of the metal points now employed. Professor Winkler, of Leipsic, next substituted a fixed woollen cushion in place of the hand for exciting the globe, and lastly, in 1742, Gordon, a Scotch Benedictine monk, and Professor of Natural

Philosophy at Erfurt, substituted a glass cylinder for the globe, and otherwise so increased the power of the machine, that he was able to kill small birds at the end of an iron wire 200 ells (250 yards) long.[*]

These various improvements were followed, in October 1745, by the discovery of the Leyden Jar. This invention is one of the vexed questions of the science, being claimed, and perhaps with equal justice, for Von Kleist, dean of the Cathedral at Kamin, in Pomerania ; for Musschenbröck, the celebrated professor of Leyden ; and for Cuneus, a rich burgess of that town. Von Kleist appears to have been first, in point of priority of publication ; but his account of the discovery was so obscurely worded, that it was impossible for some time to verify it. The following is an extract from his letter on the subject, which was addressed to Dr. Lieberkuhn, of Berlin, on the 4th November, 1745, and by him communicated to the Berlin Academy :—

"When a nail, or a piece of thick brass wire, is put into a small apothecary's phial and electrified, remarkable effects follow ; but the phial must be very dry or warm. I commonly rub it over beforehand with a finger on which I put some pounded chalk. If a little mercury, or a few drops of spirit of wine, be put into it, the experiment succeeds the better. As soon as this phial and nail are removed from the electrifying glass, or the prime conductor to which it has been

* Priestley's *History of Electricity*, pp. 64-67.

exposed is taken away, it throws out a pencil of flame so strong, that with this burning instrument in my hand I have taken above sixty steps in walking about my room. When it is electrified strongly, I can take it into another room, and there fire spirits of wine with it.

"If, whilst it is electrifying, I put my finger, or a piece of gold which I hold in my hand, to the nail, I receive a shock which stuns my arms and shoulders. A tin tube, or a man, placed upon electrics, is electrified much more strongly by this means than in the common way. When I present this phial and nail to a tin tube which I have, fifteen feet long, nothing but experience can make a person believe how strongly it is electrified. Two thin glasses have been broken by the shock. It appears to me very extraordinary that when this phial and nail are in contact with either conducting or non-conducting matter, the strong shock does not follow. I have cemented it to wood, glass, sealing-wax, metal, &c., which I have electrified without any great effect. The human body, therefore, must contribute something to it. This opinion is confirmed by observing that, unless I hold the phial in my hand, I cannot fire spirits of wine with it."

In January 1746, Cuneus made the same discovery, and apparently in the same accidental way. It having been observed by Musschenbrock and his colleagues, Cuneus and Allamand, that electrified bodies speedily lost their virtue, which was supposed to be abstracted by the air itself, and by vapours and effluvia suspended in it, they imagined that if they could surround them with any insulating substance, so as to exclude the contact of the atmosphere, they could communicate a more intense electrical power, and could preserve that power for a longer time.* Water appeared one of the most convenient recipients for the electrical influence, and glass the most effectual and easy insulating envelope. It appeared, therefore, very obvious, that water enclosed in a glass bottle must retain the electricity given to it, and that by such means a greater charge or accumulation of electric force might be obtained than by any expedient before resorted to.

In the first experiments made in conformity with these views, no remarkable results were obtained. But it happened on one occasion that Cuneus held the glass bottle in his right hand, while the water contained in it communicated by a wire with the prime conductor of a powerful machine. When he considered that it had received a sufficient charge, he applied his left hand to the wire to disengage it

* In a paper read before the Royal Society in 1735, Stephen Gray has these curiously prophetic words :—" Though these effects of the fire and explosion of electricity communicated to a metallic rod are at present only minute, it is probable that in time there may be found out *a way to collect a greater quantity of the electric fire*, and consequently to increase the force of that power, which by several of these experiments, if we are permitted to compare small things with great, seems to be of the same nature with that of thunder and lightning."—Priestley, p. 54.

from the conductor. He was instantly struck with a convulsive shock, which filled him with the utmost consternation, and made him let fall the flask. Musschenbröck and others quickly repeated the experiment, and with like results.*

In describing these, in a letter to Réaumur, Musschenbröck said he felt himself struck in the arms, shoulders, and breast, so that he lost his breath, and was two days before he recovered from the effects of the blow, and the terror. He added that he would not repeat the experiment for the whole kingdom of France. Boze, on the other hand, seems to have coveted electrical martyrdom, for he is said to have expressed a wish to die by the shock (the name by which this phenomenon was known), that the account of his death might furnish an article for the Memoirs of the French Academy. Allamand, the associate of Musschenbröck, took the shock from a common beer-glass, and lost the use of his breath for some minutes, and then felt so intense a pain along his right arm, that he feared permanent injury from it. Professor Winkler, on undergoing the experiment for the first time, suffered great convulsions, his blood was agitated, and fearing an ardent fever, he had recourse to cooling medicines. His wife, also, with a courage only equalled by her curiosity, twice subjected herself to the shock, and was so enfeebled thereby that she could hardly walk, and on trying it

* Priestley's *History of Electricity*, pp. 75-8.

again, a week later, it gave her bleeding at the nose.*

An account of these extraordinary effects soon got abroad, and spread ·over Europe with the rapidity almost of the spark itself. The experiments were repeated everywhere, and excited the wonder of all classes towards what was regarded as "a prodigy of nature and philosophy." Indeed, so popular did they become, that great numbers of *impromptu* electricians wandered over every part of Europe, and enriched themselves by gratifying the universal curiosity at so much per shock.

But as soon as these first feelings of wonder had abated, philosophers set themselves seriously to study the powers of the new machine ; and the circumstances which influenced the force of the shock first engaged their attention.

Musschenbrock observed that if the glass were wet on the outer surface the success of the experiment was impaired. Dr. (afterwards Sir William) Watson, apothecary and physician, of London, next proved that, while the force of the shock was increased by diminishing the thickness of the glass, it was independent of the power of the machine by which the glass was charged.

* Priestley, pp. 78–9. It is no doubt to the "uncontrolled use of the imagination in science," that we must, in a great measure, attribute these first effects of an experiment with which electricians are now so familiar, and which every school boy and girl undergo nowadays from motives of curiosity or amusement.

By further repeating and varying the experiment, Watson found that the force of the charge depended on the extent of the external surface of the glass in contact with the hand of the operator. It next occurred to Dr. Bevis that the hand might be efficient merely as a conductor of electricity, and in that case that the object might be more effectually and conveniently attained by coating the exterior of the phial with sheet lead or tin-foil. This expedient was completely successful, and the phial, so far as related to its external surface, assumed its present form.

Another important step in the improvement of the Leyden jar was also due to the suggestion of Dr. Bevis. It appeared that the force of the charge increased with the magnitude of the jar, but not in proportion to the quantity of water it contained. It was conjectured that it might depend on the extent of the surface of glass in contact with water; and that as water was considered to play the part merely of a conductor in the experiment, metal, which was a better conductor, would be at least equally effectual. Three phials were therefore procured and filled to the usual height with shot instead of water. A metallic communication was made between the shot contained in each of them, and the result was a charge of greatly augmented force. This was, in fact, the first electric battery.

Dr. Bevis now saw that the seat of the electric

influence was the surface of contact of the metal[*]
and the glass, and rightly inferred that the form of a
bottle or jar was not, in any way, connected with the
principle of the experiment. He, therefore, took a
common pane of glass, and having coated the opposite
faces with tin-foil, to within an inch of the edge,
obtained as strong a charge from it as from a phial
having the same extent of coated surface.   Dr.
Watson being informed of this, coated large jars,
made of thin glass, on the inside and outside with
silver leaf, extending nearly to the top of the jars, the
effects of which fully corroborated the anticipations
of Dr. Bevis, and established the law that the force
of the charge was proportional to the extent of coated
surface, and to the thinness of the glass.[†]

Experiments on the transmission and velocity of
electricity, to which the new discovery lent a fresh
and fascinating interest, were now resumed. Daniel
Gralath, early in 1746, was the first to transmit the

---

[*] This question was very beautifully settled a year or two later by the
celebrated Benjamin Franklin. He charged a jar, and then insulating
it, removed the cork and the wire by which the electricity was conveyed from the machine to the inside of the jar.   On examining
these he found them free from electricity. He next carefully decanted
the water from the charged jar into another insulated vessel.   On
examining this it was also found to be free from electricity. Other
water in its natural state was now introduced into the charged jar to
replace that which had been decanted, and on placing one hand on the
outside coating, and the other in the water, he received the shock **as**
forcibly as if no change had been made in the jar since it was first
charged.—Priestley, p. 144.

[†] Priestley, pp. 82-7.

shock to a distance, which he did by discharging a battery, composed of several jars, through a chain of twenty persons, with outstretched arms. In May 1746, Joseph Franz, at Vienna, discharged a jar through 1500 feet of iron, and, in the following July, Winkler charged, as well as discharged, a battery of three jars through an insulated wire, thirty ells long, and laid along the bank of the river Pleisse, whose waters formed the return half of the circuit.*

The Abbé Nollet, whose name is famous in the annals of this period, had meanwhile taken up the subject in France. He first, April 1746, transmitted the shock of a Leyden jar through a chain of 180 of the Royal Guards at Paris, and soon after performed a grander experiment of the same kind at the Carthusian convent. By means of iron wires stretched between every two of the monks he formed a large circle of 5400 feet, through which he discharged his jars, with the result in every case that, at the moment of discharge, all the persons in the circuit gave a sudden spring, showing that the shock was felt by each at the same instant and to the same degree of intensity.

---

* Winkler had previously, in 1744, ascertained that the rapidity of an electric discharge was exceedingly great and comparable with the speed of lightning. He also, as the result of his experiments, concluded "that electricity could be transmitted to the ends of the earth, if a conducting body covered, or insulated, with silk be laid so far, it being only necessary to consider that there may be a certain amount of resistance to the transmission."—*Thoughts on the Properties, Operations, and Causes of Electricity*, Leipsic, 1744, pp. 146, 149.

Lemonnier, the younger, also of Paris, employed still longer circuits composed of 2000 toises (12,780 feet) of iron wire laid along the ground, and, although some of the wire dragged upon wet grass, through hedges, and over newly-ploughed fields, the shock was in no way diminished, a fact which was then thought very surprising.   In other experiments he made use of two large basins of water in the gardens of the Tuileries. In April 1746, in the court of the Carthusians, he so laid out two parallel wires of 5700 feet each, that all four ends were close together.   Between one pair he placed a jar, and grasped the other extremities himself; then on causing the circuit to be completed, he could not distinguish any interval (so short was it) between the spark at the jar, and the shock through his arms.*

Upon receiving an account of these performances from Lemonnier, our own distinguished countryman, Watson, took up the inquiry, and pursued it so successfully as not only to eclipse the achievements of his neighbours, but to gain for himself in after years the credit of being the first to propose an electric telegraph —an idea which, as we shall presently see, is quite erroneous.†   Watson's experiments were very numerous, and were carried out on a grand scale, under the auspices of a committee of the Royal Society, con-

* Priestley, pp. 92–5.

† The suggestion has been claimed for Franklin and Cavendish, and with as little reason.   It is time that writers on the telegraph ceased to bandy pretensions for which there is no foundation whatever.

sisting of Mr. Folkes, Lord C. Cavendish, Dr. Bevis, and others. As preparing the way surely, though un-suspectedly, for the first suggestions of an electric telegraph, these investigations must ever possess a peculiar interest for telegraphists, and we therefore make no apology for presenting to our readers the following detailed account of them, for which we are indebted to Dr. Priestley's work, pp 95-102.

Dr. Watson, who wrote a full account* of the labours of the Committee for the Royal Society, begins with observing (which was verified in all their experiments) that the electric shock is not, strictly speaking, con-ducted in the shortest manner possible, unless the bodies through which it passes conduct equally well. The circuit, he says, is always formed through the best conductors, though the length be ever so great—a most sagacious observation for the man and the time.

The first trials took place on the 14th and 18th July, 1747, on a wire carried from one side of the Thames to the other over old Westminster Bridge. One end of this wire communicated with the interior of a charged Leyden jar, the other was held by a person on the opposite bank of the river, who also held in his other hand an iron rod which he dipped into the water. Near the jar stood another person holding in one hand a wire communicating with the exterior

* *An Account of the Experiments made by some Gentlemen of the Royal Society*, &c., 8vo., London, 1748.

coating of the jar, and in the other an iron rod. On dipping this into the water and thus completing the circuit for the discharge, the shock was instantly felt by both persons, but more strongly by him who stood near to the jar—because, as Watson rightly stated, part of the electricity went from the wire down the moist stonework of the bridge, thereby making several shorter circuits to the jar, but still all passing through the observer who stood near it.

The next attempt was to force the shock through a circuit of two miles at the New River, near London. This was accomplished on the 24th July at two places, at one of which the distance by land was 800 feet, and by water 2000; and at the other, 2800 feet of land and 8000 feet of water.

The disposition of the apparatus was similar to that at Westminster Bridge, and the results were equally satisfactory. On repeating the experiments, however, the rods, instead of being dipped into the water, were merely thrust into the ground about twenty feet from the water's edge. The effect was the same, as it was found that the shock was equally well transmitted. This occasioned a doubt whether in the former case the shock might not have been conveyed through the ground between the two rods, instead of passing through all the windings of the river, and subsequent experiments showed that such was the case. Other experiments followed at the same place, on the 28th July, when for the first time the wire was supported in its whole length by dry sticks, and on the 5th August, at Highbury Barn, when it was found that dry ground conducted the electric virtue quite as well as water.

Finally, on the 14th August at Shooter's Hill, an experiment was made " to try whether the electric shock was perceptible at twice the distance to which it had yet been carried, in ground perfectly dry, and where no water was near ; and also to distinguish if possible its velocity as compared with that of sound." The circuit consisted of two miles of wire, and two miles of perfectly dry ground, but one shower of rain having fallen in the previous five weeks. The wire from the inner coating of the jar was 6732 feet long, and was supported all the way upon baked sticks, and that which communicated with the outer coating was similarly insulated, and was 3868 feet long. The observers placed at the ends of these wires, two miles apart, were provided with stop watches with which to note the moment that they felt the shock. The result of a series of careful observations was that " as far as could be distinguished the time in which the electric matter performed its circuit might have been instantaneous."

Not satisfied, apparently, with this result, the inquiry was resumed in the following year, when a series of trials was performed after the manner of Lemonnier's Carthusian experiment of 1746. On the 5th August, 1748, a circuit of two miles was formed at Shooter's

Hill by several turnings of wire in the same field. The middle of this wire was led into the same room as the Leyden jar, and there Watson placed himself in the centre of the line, taking in each hand the ends of the wire, and noting the spark with his eye while he felt the shock in his arms.   Under these circumstances the jar was discharged several times, but in no instance could the observer distinguish the slightest interval between the moments at which the spark was seen and the shock felt ; whereupon it was decided that the time occupied by the passage of electricity along 6138 feet of wire was altogether inappreciable.

In 1748 Benjamin Franklin performed his celebrated experiments across the Schuylkill at Philadelphia, and De Luc some months later (1749) across the Lake of Geneva.   Franklin thus playfully refers to his experiments at the end of a letter to his friend and correspondent, Peter Collinson, of London, dated Philadelphia, 1748 :—

" Chagrined a little that we have hitherto been able to produce nothing in this way of use to mankind, and the hot weather coming on, when electrical experiments are not so agreeable, 'tis proposed to put an end to them for this season, somewhat humorously, in a party of pleasure on the banks of the Skuylkil. Spirits at the same time are to be fired by a spark sent from side to side through the river, *without any other conductor than the water*—an experiment which we some time since performed, to the amazement of many. A turkey is to be killed for our dinner by the electrical shock, and roasted by the electrical jack, before a fire kindled by the electrified bottle, when the healths of all the famous electricians in England, Holland, France, and Germany are to be drank in electrified bumpers, under the discharge of guns from the electrical battery." *

As the words that we have italicised in this extract are apt to mislead, and indeed have misled, some writers into supposing that Franklin here describes an experiment akin to that of telegraphing without wires, from which so much was expected forty years ago, we quote the following details from vol. i. p. 202, of *Franklin's Complete Works*, London, 1806:—"Two iron rods, about three feet long, were planted just within the margin of the river, on the opposite sides. A thick piece of wire, with a small round knob at its end, was fixed on the top of one of the rods, bending downwards, so as to deliver commodiously the spark upon the surface of the spirit. A small wire, fastened by one end to the handle of the spoon containing the spirit, was carried across the river, and supported in the air by

* " An electric battery, famous because it was once owned and operated by Benjamin Franklin, and other distinguished scientific men, has been in constant use at Dartmouth College for years, and is now employed almost daily for class-room experiments in physics. It was at one time in the hands of the celebrated Dr. Priestley, the discoverer of hydrogen."—American newspaper. Another interesting relic— Faraday's first electrical machine—is still in vigorous action at the Royal Institution, and was used by Dr. Gladstone to illustrate his Christmas Lectures in 1874-5.

the rope commonly used to hold by, in drawing ferry-boats over. The other end of this wire was tied round the coating of the bottle, which, being charged, the spark was delivered from the hook to the top of the rod standing in the water on that side. At the same instant the rod on the other side delivered a spark to the spoon and fired the spirit, the electric fire returning to the coating of the bottle, through the handle of the spoon, and the supported wire connected with them." The experiment was, therefore, precisely the same as that of Watson across the Thames, the only difference being in the words used to describe it. In the one case the discharge is said to go out by the water and return by the wire, and in the other to go out by the wire and return by the water.

Notwithstanding the singular suggestiveness of all these experiments, no one up to this time appears to have entertained the faintest suspicion of their applicability to telegraphic purposes ; or, indeed, to any useful purpose whatever. Thus Watson, in a letter to the Royal Society, says :—"If it should be asked to what useful purposes the effects of electricity can be applied, it may be answered that we are not yet so far advanced in these discoveries as to render them conducive to the service of mankind," but, he adds, "future philosophers may deduce from them uses extremely beneficial to society in general." This was in 1746, and with reference to his then recent ignition of spirits by the spark ; but even after his brilliant experiments in the following years, of which we have just given an account, he does not appear to have formed any more hopeful view. We also find the great Franklin, who was always in search of the practical in science, positively expressing his disappointment in the letter just quoted at being unable to find any useful application of electricity.*

* His suggestion of the lightning-conductor was not made until towards the end of July 1750. For this he was indebted to an experiment of his friend, Thomas Hopkinson. This philosopher electrified a small iron ball, to which he fixed a needle, in the hope that from the point, as from a focus, he would draw a stronger spark. Greatly surprised at finding that, instead of increasing the spark, the point dissipated it altogether, he mentioned his failure to Franklin. On repeating the experiment, the latter ascertained, not only that the ball could not be electrified when a needle was fastened to it, but that, when the needle was removed and the ball charged, the charge was silently and speedily withdrawn, when a point connected with the earth was presented to it. Reflecting on this, Franklin conceived the idea that pointed rods of iron fixed in the air might draw down the lightning without noise or danger.—*Franklin's Complete Works*, vol. i. p. 172, London, 1806.

CHAPTER III.

TELEGRAPHS BASED ON STATIC, OR FRICTIONAL, ELECTRICITY.

"Canst thou send lightnings, that they may go, and say unto thee, Here we are?"—*Job* xxxviii. 35.

1753.—*C. M.'s Telegraph.*

THE first distinct proposal to employ electricity for the transmission of intelligence, of which we have any record, is that contained in a letter printed in the number of the *Scots' Magazine*, Edinburgh, for February 17, 1753. As this is one of the most interesting documents to be found in the whole history of telegraphy, we will quote it *in extenso* for the benefit of our readers :—

To the Author of the *Scots' Magazine.*

"Renfrew, Feb. 1, 1753.

" Sir,—It is well known to all who are conversant in electrical experiments, that the electric power may be propagated along a small wire, from one place to another, without being sensibly abated by the length of its progress. Let, then, a set of wires, equal in number to the letters of the alphabet, be extended horizontally between two given places, parallel to one another, and each of them about an inch distant from that next to it. At every twenty yards' end, let them be fixed in glass, or jeweller's cement, to some firm body, both to prevent them from touching the earth, or any other non-electric, and from breaking by their own gravity. Let the electric gun-barrel be placed at right angles with the extremities of the wires, and about an inch below them. Also let the wires be fixed in a solid piece of glass, at six inches from the end; and let that part of them which reaches from the glass to the machine have sufficient spring and stiffness to recover its situation after having been brought in contact with the barrel. Close by the supporting glass, let a ball be suspended from every wire; and about a sixth or an eighth of an inch below the balls, place the letters of the alphabet, marked on bits of paper, or any other substance that may be light enough to rise to the electrified ball; and at the same time let it be so contrived, that each of them may re-assume its proper place when dropt.*

"All things constructed as above, and the minute previously fixed, I begin the conversation with my distant friend in this manner. Having set the electrical machine a-going as in ordinary experiments, suppose I am to pronounce the word *Sir*; with a piece of glass, or any other *electric per se*, I strike the wire S,

---

* It will be observed that in this and most other systems based upon common, or frictional, electricity, the authors constantly, although often unknowingly, used the earth circuit.

so as to bring it in contact with the barrel, then *i*, then
r, all in the same way ; and my correspondent, almost
in the same instant, observes these several characters
rise in order to the electrified balls at his end of the
wires.   Thus I spell away as long as I think fit ; and
my correspondent, for the sake of memory, writes the
characters as they rise, and may join and read them
afterwards as often as he inclines.   Upon a signal
given, or from choice, I stop the machine; and, taking
up the pen in my turn, I write down whatever my
friend at the other end strikes out.

" If anybody should think this way tiresome, let
him, instead of the balls, suspend a range of bells
from the roof, equal in number to the letters of the
alphabet ; gradually decreasing in size from the bell
A to Z ; and from the horizontal wires, let there be
another set reaching to the several bells ; one, *viz.*,
from the horizontal wire A to the bell A, another from
the horizontal wire B to the bell B, &c.   Then let him
who begins the discourse bring the wires in contact
with the barrel, as before ; and the electrical spark,
breaking on bells of different size, will inform his
correspondent by the sound what wires have been
touched.   And thus, by some practice, they may come
to understand the language of the chimes in whole
words, without being put to the trouble of noting
down every letter.

" The same thing may be otherwise effected.   Let
the balls be suspended over the characters as before,

but instead of bringing the ends of the horizontal wires in contact with the barrel, let a second set reach from the electrified cake, so as to be in contact with the horizontal ones ; and let it be so contrived, at the same time, that any of them may be removed from its corresponding horizontal by the slightest touch, and may bring itself again into contact when left at liberty. This may be done by the help of a small spring and slider, or twenty other methods, which the least ingenuity will discover. In this way the characters will always adhere to the balls, excepting when any one of the secondaries is removed from contact with its horizontal ; and then the letter at the other end of the horizontal will immediately drop from its ball. But I mention this only by way of variety.

"Some may, perhaps, think that although the electric fire has not been observed to diminish sensibly in its progress through any length of wire that has been tried hitherto, yet as that has never exceeded some thirty, or forty, yards, it may be reasonably supposed that in a far greater length it would be remarkably diminished, and probably would be entirely drained off in a few miles by the surrounding air. To prevent the objection, and save longer argument, lay over the wires from one end to the other with a thin coat of jeweller's cement. This may be done for a trifle of additional expense, and, as it is an *electric per se*, will effectually secure any part of the fire from mixing with the atmosphere.—I am, &c., "C. M."

From the concluding paragraph it is evident that the writer was not acquainted with Watson's experiments, as detailed in our last chapter, else he would not have suggested insulating the wires, from end to end, with jeweller's cement, and, probably, not even have noticed the objection at all. His suggestions of reading by sound of differently-toned bells, and of keeping his wires charged with electricity, and indicating the signals by discharge, are very ingenious, and deserve to be remembered to his credit in these days of their realisation. The former plan is familiar to us in Bright's Acoustic, or Bell, telegraph of 1855, while the latter was, as we shall presently see, employed by Ronalds in 1816, and is realised to perfection in the method now used in signalling through all long cables.

Unfortunately, little, or nothing, is known of C. M. An inquiry as to his identity was first started by " Inquirendo," Glasgow, in *Notes and Queries*, for October 15, 1853; then by George Blair, also of Glasgow, in the Glasgow *Reformers' Gazette*, for November 1853, in which he, for the first time, republished C. M.'s letter ; and, lastly, by Sir David Brewster, in the Glasgow *Commonwealth*, for January 21, 1854. Nothing, however, came of the inquiry for a long time, and all hopes of solving the question were abandoned, when, on December 8, 1858, the following letter appeared :—

"To the Editor of the *Commonwealth.*

"145, Great Eastern Road.

" Sir,—I have not heard that a name has yet been proposed for the C. M. that wrote to the *Scots' Magazine* last century from Renfrew, giving some hints about the electric telegraph.

" I send you what follows, as I think it gives some probability to C. M. being Charles Marshall.

" In our house was a copy of Knox's *History of the Reformation*, published in Paisley, in 1791. My uncle James's name is in the list of subscribers in Renfrew. Anent this my mother spoke as follows :—' There was a very clever man living in Paisley at that time, that had formerly lived in Renfrew. He asked my uncle, as they were acquainted, to canvass for subscribers in Renfrew. The said clever man could light a room with coal reek, and make lightning speak and write upon the wall,' &c.

" That this was the C. M. of the electric telegraph there can, I think, be no doubt.

" Now, it is probable that the man that solicited my uncle to canvass for subscribers subscribed himself; and in Well Meadow, Paisley, I find the name Charles Marshall, and this is the only name in the list of 1000 names that answers the initials C. M. My list, however, is not complete for Glasgow.

" Peradventure some one belonging to Paisley may have somewhat to say of Charles Marshall.

" ALEX. DICK."

To this letter were appended the following remarks by Sir David Brewster, to whom the editor appears to have submitted it prior to publication :—" That Charles Marshall might have been the inventor, had we known nothing more than that he was a resident in Renfrew about the time when the letter was sent to the *Scots' Magazine,* was very probable ; but when we add to this probability the fact that Charles Marshall was a clever man, and that he was known as a person who could make lightning speak and write upon the wall, and who could also light a room with coal reek (smoke), we can hardly doubt that he was the C. M. who invented the electric telegraph, and that he is entitled to the additional honour of having first invented and used gas from coal." *

Commenting on this correspondence, in *Notes and Queries,* for July 14, 1860, George Blair says :—" That the Charles Marshall who resided at Well Meadows, Paisley, in 1791, was not the C. M. of the *Scots' Magazine,* and, therefore, not the inventor of the electric telegraph, I succeeded in ascertaining positively about a year ago, on the highest possible authority. Through the kindness of a venerable friend in Paisley, I traced out the fact that a Charles Marshall, who once resided in the Well Meadows, had come from Aberdeen ; and that a son of his, a clergyman, was still living. Discovering the address of this gentleman, I applied

---

* These letters, copied from the *Commonwealth,* are reprinted in the *Engineer,* for Dec. 24, 1858, p. 484.

to him for information ; and he states in his reply that
he had no doubt his father was the Charles Marshall
who appears in Mr. Dick's list ; but that he could not
be the C. M. of the *Scots' Magazine.*

*   *   *   *   *   *

"At the time when C. M.'s letter was first dis-
interred, the most diligent search was made by the
schoolmaster of Renfrew, who is also session-clerk,
not only in the records of the kirk-session, but also
among the old people of the parish, without a shadow
of success ; and, strange as it may appear, the name
of C. M. remains at the present moment as great a
mystery as that of Junius."

Whether Sir David Brewster was aware of these
fresh facts we cannot say, but certain it is that, in
October 1859, he accepted the evidence in favour of
C. M. being a Charles Morrison, with as much warmth,
and, we fear, as much haste, as he had done that for
Charles Marshall in the previous December. At
p. 207 of *The Home Life of Sir David Brewster*
(Edinburgh, 1869), Mrs. Gordon says :—"After a
good deal of correspondence on the subject, Sir David
Brewster gave up all hope of discovering the name of
the inventor, and it was not until 1859 that he had
the great pleasure of solving the mystery in the following manner :—He received from Mr. Loudon, of Port
Glasgow, a letter, dated 31st October, 1859, stating
that, while reading the article in the *North British*

Review, his attention was arrested by the letter of C. M., and having mentioned the fact to Mr. Forman, a friend then living with him, he told him that he could solve the mystery regarding these initials. Mr. Forman recollects distinctly having read a letter, dated 1750, and addressed by his grandfather, a farmer, near Stirling, to Miss Margaret Winsgate, residing at Craigengilt, near Denny (to whom he was subsequently married), referring to a gentleman in Renfrew of the name of Charles Morrison, who transmitted messages along wires by means of electricity, and who was a native of Greenock, and bred a surgeon. Mr. Forman also states that he was connected with the tobacco trade in Glasgow, that he was regarded by the people in Renfrew as a sort of wizard, and that he was obliged, or found it convenient, to leave Renfrew and settle in Virginia, where he died. Mr. Forman also recollects reading a letter in the handwriting of Charles Morrison, addressed to Mr. Forman, his grandfather, and dated 25th September, 1752, giving an account of his experiments, and stating that he had sent an account of them to Sir Hans Sloane, the President of the Royal Society of London, who had encouraged him to perfect his experiments, and to whom he had promised to publish an account of what he had done. In this letter Mr. Morrison stated that, as he was likely to be ridiculed by many of his acquaintances, he would publish his paper in the *Scots' Magazine* only with his initials."

How far this statement may be credited we will not undertake to say ; we would, however, just point out that Sir Hans Sloane resigned the presidentship of the Royal Society in 1741, and lived in strict retirement at Chelsea until his death, which occurred on January 11, 1752, at the advanced age of ninety-two years. It is not likely, therefore, that he would have received, or written, any letters of the above-mentioned nature in the last days of his life. At any rate, a careful search through his papers, which we have instituted in the British Museum and the Royal Society, has failed to discover any.

1767.—*Bozolus's Telegraph.*

Joseph Bozolus, a Jesuit and lecturer on natural philosophy in the College at Rome, was the next to suggest an electric telegraph, and one in which the spark was the active principle. This must have been some time anterior to 1767, as we find it familiarly described in a Latin poem,* published in that year.

His proposition was to lay underground two (? in-sulated) wires between the communicating stations, which may be any distance apart. At both stations the ends of the wires were to be brought close together, without touching, so as to facilitate the passage of a spark. When, under these circumstances, at one end, the inner coating of a charged plate, or jar, was con-

* *Electricorum*, by Josephus Marianus Parthenius (*i. e.*, G. M. Mazzo-lari), libri vi., 8vo., Romæ, 1767.

nected to one wire, and the outer coating to the other, the discharge would take place through the wires, and manifest itself, at the break, at the distant end, in the form of a spark.   An alphabet of such sparks, Bozolus says, could be arranged with a friend without any difficulty, and a means of communication be thus contrived, which, as tolerably easy, he leaves to each one's judgment to devise and settle in detail.

Bozolus appears to have been a man of varied acquirements.   As a sort of diversion from more serious studies, he undertook an Italian translation of the *Iliad* and *Odyssey* of Homer, which Mazzolari, himself no mean poet, praises very highly.

As the *Electricorum* is very scarce, and, therefore, not easily accessible, we present our readers with a faithful transcript of the verses descriptive of the telegraph, which we have extracted from a copy of the work in the British Museum.

> " Quid dicam, extrema pendentis parte catenæ,
>   Qui palmam objecit, confestim flamma reluxit,
>   Tenviaque arguto strepuerunt sibila vento ?
>   Et qui continuos secum prius ordine longo
>   Disposuit globulos ; tum flammam excivit, et ignem
>   A primo insinuans sollers traduxit ad imum ?
>   Atque hic arte quidem multa omniginæque Minervæ
>   Instructus studiis vitro impiger instat, et usque
>   Extundit visenda novis spectacula formis.
>   Quid ? quod et elicitas vario discrimine flammas
>   Nunquam tentatos idem detorquet ad usus ;
>   Insuetisque notis absentem affatur amicum.
>   Quippe duo a nexa in longum deducta catena
>   Aenea fila trahit ; spatium distantia amici
>   Definit certum, verum, quo lumina fallat

Spectantum, et miram quo callidus occulat artem,
Fila solo condit penitus defossa sub imo,
Sic tamen; ut capita emergant tum denique ; signa
Conscius opperiens condicta ubi servat amicus.
Ipse autem interea vitri revolubilis orbem
De more exagitans fluctum derivat ; et inde,
Qua duo se extrema respectant aenea parte,
Attactum citra et præscripto limite, fila ;
Composito tot scintillas educit, ad usum
Quot talem elicitis opus est ; quæ singula nempe
Designent elementa; quibus in verba coactis
Sensa animi pateant, certa et sententia constet.
Atque his indiciis, fidaque interprete flamma
Absens absentem dictis compellat amicum."

Lib. i. pp. 32-35.

1773.—*Odier's Telegraph.*

The idea of an electric telegraph appears next to have occurred to Louis Odier, a distinguished physician of Geneva, who thus wrote, in 1773, to a lady of his acquaintance :—

"I shall amuse you, perhaps, in telling you that I have in my head certain experiments by which to enter into conversation with the emperor of Mogol, or of China, the English, the French, or any other people of Europe, in a way that, without inconveniencing yourself, you may intercommunicate all that you wish, at a distance of four or five thousand leagues in less than half an hour! Will that suffice you for glory ? There is nothing more real. Whatever be the course of those experiments, they must necessarily lead to some grand discovery; but I have not the courage to undertake them this winter. What gave me the idea was a word which I heard spoken casually the other day at Sir John Pringle's table, where I had the pleasure of dining with Franklin, Priestley, and other great geniuses." *

Although, according to Professor Maunoir, Odier was about this time devoting much attention to electricity, we do not find that he ever attempted to carry out his telegraphic idea.

1777.—*Volta's (so-called) Telegraph.*

At p. 243, vol. i., of *The Journal of the Society of Telegraph Engineers*, we find the following letter :—

"To the Secretary of the Society of Telegraph Engineers.

"Battle, Sussex, July 4th, 1872.

"Sir,—I have not met with any statement in English histories, or other English treatises, on the Electric Telegraph, relative to *Volta's* proposed Electric Telegraph.

"Professor L. Magrini, member of a committee appointed to examine and report upon Volta's library, manuscripts, and instruments, published a paper in the *Atti del Reale Istituto Lombardo*, vol. ii., entitled, *Notizie, Biografiche e Scientifiche su Alessandro Volta.*

---

* *Chambers's Papers for the People*, 1851, Art. *Electric Communications*, p. 6. Also Dodd's *Railways, Steamers, and Telegraphs*, London and Edinburgh, 1867, p. 226. Odier took out his degrees at Edinburgh, where he might well have read, or heard, of C. M.'s letter in the *Scots' Magazine* of 1753.

This paper was read at various times, in 1861, at the said institute. It contains a paragraph of which the following is a literal translation :—

" ' An autograph manuscript, dated Como, 15th April, 1777, which is suspected (and the suspicion was confirmed by one of the sons of Volta) to have been addressed to Professor Barletti, contains various experiments on his pistols, and the singular proposition, very remarkable for that time, of transmitting signals by means of ordinary electricity. Besides the figure, there are particulars conducive to its practical application.

" ' This letter is of the greatest interest for the history of the science, inasmuch as it indicates the first bold and certain step in the invention and institution of the electric telegraph.'

" Although our Charles Marshall, of Renfrew, in 1753, and others, forestalled this proposition, it is interesting, as proving that the *in re electrica Princeps* believed in the efficiency of frictional electricity for the purpose.

" I am, Sir, your obedient servant,

" FRANCIS RONALDS.

" G. E. Preece, Esq."

Now, although, as Ronalds says, and as we here see, Volta was not the first to propose an electric telegraph, still we were delighted to learn, on such apparently good authority, that the great Italian

G

philosopher had turned his mind to telegraphy at all, and we eagerly sought for some particulars of his plan. After much trouble we succeeded in getting a copy of the letter referred to by Professor Magrini, and great was our disappointment to find that it contained nothing more than the *suggestion* of an experiment which was *carried out* (though on a lesser scale) thirty years before by Lemonnier, Watson, Franklin, De Luc, and others. In order that the reader may be able to form his own opinion on this point, we give Volta's original letter, as well as a translation, which we have made from the French of César Cantu, the distinguished Italian historian.* Volta says :—

"Quante belle idee di sperienze sorprendenti mi van ribollendo in testa, eseguibili con questo stratagemma di mandare la scintilla elettrica a far lo sbaro della pistola a qualsivoglia distanza e in qualsivoglia direzione e posizione ! Invece del colombino che va ad appiccar l'incendio alla macchina di fuochi artificiali, io vi mandero du qualunque sito anche non diretto la scintilla elettrica, che col mezzo della pistola aggiustata al sito della pianta artifiziale, vi metterà fuoco. Sentite. Io non so a quanti miglia un fil di ferro, tirato sul suolo dei campi o della strada, che in fine si ripiegasse indietro, o incontrasse un canal d'acqua di ritorno, condurebbe giusta il sentier segnato la scintilla commovente. Ma prevegga che un lunghissimo

* See *Le Correspondant*, a French scientific periodical, for August 1867, p. 1059, also *Les Mondes*, for December 5, 1867, p. 561.

viaggio, dè tratti di terra molto bagnati, o delle acque scorrenti stabili rebbero troppo presso una communi-cazione e quioi devierebbe il corso del fuoco elettrico, spiccato dall uncino della caraffa per ricondursi al fondo. Ma se il fil di ferro fosse sostenuto alto da terra da pali di legno qua e là piantati, ex. gr., da Como fine a Milane; e quivi interrotto solamente dalla una pistola, continuasse e venisse in fine a pescare nel canale naviglio, continua col mio lago di Comó; non credo impossibile de far lo sbaro della pistola a Milano con una buona boccia di Leyden, da me scaricata in Como."

"The more I reflect, the more I see the beautiful experiments that can be made by means of the spark in exploding the electric pistol at any distance. An iron wire, stretched along the fields, or roads, for I know not how many miles, could conduct the spark. As, however, in long distances moist earth and water-courses would be encountered, which would draw off the electric fire, the wire may be supported on posts placed at regular intervals, say from Como to Milan. At the latter place its continuity would be interrupted only by my electric pistol, from which it would pass into the canal, which communicates with my lake at Como. In this case I do not believe it impossible to explode my pistol at Milan when I discharge a powerful Leyden jar at Como " [through the wire].

"According to this document," says Cantu, " it is incontestable that Volta had in mind an electric

telegraph, half a century before those [alluding to
Ampère] who have been proclaimed its inventors. The
basis of this astonishing discovery lies in the possibility
of transmitting to a great distance the electric virtue,
and there causing it to manifest itself in signs. Now
this is what Volta had clearly perceived, and, further,
he indicated a plan, which is to-day universal, of
insulating the conducting wire on posts." * Perceiving
a fact, or principle, and applying it, are two very
different things. Gray, Dufay, Watson, and all those
who made experiments on the transmission of electricity long before Volta, perceived the same fact as
he did, and, like him, missed its application. To say
then, as Professor Magrini does, that Volta's letter
indicates the first bold and certain step in the invention and institution of the electric telegraph is to

* *Le Correspondant*, p. 1060. In the course of a somewhat effusive
letter on Italy's claim to the discovery of the electric telegraph, Cantu
relates the following interesting particulars. The apartments which
Volta occupied at Como, were, for a time, preserved in the state in
which he left them at his death (March 5, 1827). There one could see
his books, papers, machines, even his tobacco pouch, spectacles, decora-
tions, and cane ; in short, everything that becomes a sacred relic when
death has removed him who used it. Amongst the pieces of apparatus,
were all those which he had himself invented, including the first pile,
and that which he took to Paris, in 1801, when invited by Napoleon to
repeat his experiments before the Institute.

In consequence of the pecuniary embarrassments of Volta's sons,
these precious relics were in danger of being dispersed, when the
Academy of Sciences of Lombardy stepped in, and, while it assisted the
sons, honoured the father The whole collection was purchased for
100,000 livres, and lodged in a chamber of the palace of Brera at Milan,
where, under the appellation of *Cimeli di Volta*, it is preserved with
reverent care.

assign to it a meaning which it was never, we believe, intended to convey ; and we are the more confirmed in this opinion by the fact that, although Volta lived to the year 1827, and must have heard of the numerous telegraphic proposals made up to that time, he never claimed to have done anything in that way himself.

1782.—*Anonymous Telegraph.*

The next proposal, which is an exceedingly interesting one, is contained in an anonymous letter to the *Journal de Paris*, No. 150, for May 30, 1782, a translation of which we append :—

" To the Authors of the Journal.

" A way of establishing a communication between two very distant places has been proposed to me, and those of your readers who care for this kind of scientific amusement will not, perhaps, be angry with me for telling them what it is.

" Let there be two gilt iron wires put underground in separate wooden tubes filled in with resin, and let each wire terminate in a knob. Between one pair of knobs, connect a letter formed of metallic [tin-foil] strips after the fashion of those electrical toys, called ' spangled panes ' ; if, now, at the other end we touch the inside of a Leyden jar to one knob, and the outside to the other, so as to discharge the jar through the wires, the letter will be at the same instant illuminated.

" Thus, with twenty-four such pairs, one could quickly spell all that was desired, it being only requisite to have a sufficient number of charged Leyden jars always ready.

" As it would not be necessary to make the letters very luminous, a slight indication being sufficient, complete darkness would not be required for the perception of the characters, and feebly charged jars would, therefore, suffice, which would greatly facilitate matters. The letters may even be suppressed, and then there would be one instrument common to the twenty-four systems (pairs) of wires [*sic*].

" These means could be simplified by having only five pairs of wires, and attaching a character, or letter, to each of their combinations, $1\ 1°$, $2\ 2°$, $*\ *\ 5\ 5°$; $1\ 1°$, and $2\ 2°$, $1\ 1°$, and $3\ 3°$; $*\ *\ 1\ 1°$, $2\ 2°$, and $3\ 3°$; and so on, which would make thirty-one characters ; six pairs of wires would, in the same way, yield sixty-three, and thus one could arrive at a sort of tachygraphy, or fast writing, one character (or signal) sufficing for a whole word, or phrase, as may be previously agreed upon. There would be some difficulty, however, in discharging at exactly the same instant several (separate) jars through as many separate pairs. One might also use successive combinations of these pairs, 2 to 2, 3 to 3, and so on, in which way five pairs would give 125 signals, and six 216, which would be very fast writing indeed.

" The wooden tubes might, very probably, be unnecessary ; but in view of accidents, such as fractures, it would always be safer to employ them.

" One could use *simple* electricity [*i. e.,* direct from the machine], and so greatly simplify the apparatus, but as the superficial area of a great length of wire, even when a very fine one was used, would be considerable, this plan would necessitate very powerful machines. In either method, however, the object could easily be obtained by using very large *electrophoroi.*

" It would be necessary to give each correspondent a means of notifying that he wished to communicate, to prevent constant watching and cross signalling. For this an electric bell would suffice, and by agreeing beforehand that one stroke shall mean ' I will call you up in 15 minutes,' two strokes ' I am all attention,' &c., all confusion would be avoided.

"As this letter is only intended for those who amuse themselves with physics, they can easily supply for themselves all the details that I have omitted.

" I have the honour to be, &c."

This letter is copied, almost *verbatim,* in *Le Mercure de France,* for June 8, 1782, and is also embodied in a letter, dated June 5, 1782,* where the writer

---

* In Metra's *Correspondance Secrète,* &c., Londres, 1788, vol. xiii. p. 84. Mr. Aylmer, to whom we are indebted for the copy of this letter which appeared in *Le Mercure de France,* tells us that the Comte du Moncel attributes it to Le Sage, but we shall presently see reasons for doubting this.

prefaces it with the following remarks : "We have Linguet once more installed in the career in which his labours have been so disagreeably interrupted. His project of an easy communication between two very distant places appears to be only the dream of some pleasant trifler. It is, however, not new, and would only imperfectly accomplish its object ; but still there may be some good in it."

In these remarks Metra somewhat mixes his facts. There is no more authority for the statement that Linguet was the writer than that he, at this time, was engaged on experiments on some kind of a *luminous* telegraph, which he planned while a prisoner in the Bastille, and in exchange for which he is popularly, though erroneously, supposed to have received his liberty. On the other hand, we have positive proof that he was not the writer, firstly, in the opening sentence of the letter itself, and secondly, in the following passage from his *Mémoires sur la Bastille:*— "I will one day make known my ideas on this subject [of signalling by means of light]. The invention will certainly admit of being greatly improved, as I have no doubt it will be. I am persuaded that in time it will become the most useful instrument of commerce, and all correspondence of that kind ; just as electricity will be the most powerful agent of medicine ; and as the fire-pump will be the principle of all mechanic processes which require, or are to communicate, great force" (Note 13).

1782.—*Le Sage's Telegraph.*

On seeing these accounts, George Louis Le Sage,* a *savant* of French extraction, residing at Geneva, published a method somewhat similar to C. M.'s, in a letter, dated June 22, 1782, and addressed to his friend, M. Prévost, at Berlin.

He writes: " I am going to entertain you with one of my old discoveries, which I see has just been found out by others, at least, up to a certain point. It is a ready and swift method of correspondence between two distant places by means of electricity, which occurred to me thirty, or thirty-five, years ago, and which I then reduced to a simple system, far more practicable than the form with which the new inventor has endowed it.

" I have often spoken of it to one or two persons,† but I see no reason for supposing that the new inventor has drawn his ideas from these conversations. The thing is so natural that, to discover it, it is only necessary that one should be in search of some means of very rapid correspondence ; and people have, on

---

* " Upon the present venerable and learned M. le Sage of Geneva devolved, in a great measure, the education of Lord Mahon, who is frequently heard to mention the name of his preceptor with considerable respect. He even goes so far as to pronounce M. le Sage the most learned man in Europe." *Vide* Life of Earl Stanhope, in *Public Characters of* 1800–1801, London, 1801, p. 88.

† In *Le Journal des Sçavans*, 4to., Paris, 1782 (for Sept., p. 637), this extract is prefaced thus :—" Il y a trente ans qu'il en parla, et une personne à qui il en fit part, offre de l'attester ; mais ceux qui connoissent la sagacité et la candeur de ce digne citoyen, ne formeront à cet égard aucun doute."

occasion, turned their minds to this subject * * * *, as, for example, Mr. Linguet.

"But it is time to tell you briefly in what my plan consisted. One can imagine a subterranean tube, of glazed earthenware, the inside of which is divided, at every fathom's length, by diaphragms, or partitions, of glazed earthenware, or of glass, pierced by twenty-four holes, so as to give passage to as many brass wires, which could in this way be supported and kept apart. At each of the extremities of this tube, the twenty-four wires are arranged horizontally, like the keys of a harpsichord, each wire having suspended above it a letter of the alphabet, while immediately underneath, on a table, are pieces of gold leaf, or other bodies that can be as easily attracted, and are, at the same time, easily visible.

"He, who wishes to signal anything, shall touch the ends of the wires with an excited glass tube, according to the order of the letters composing the words ; while his correspondent writes down the characters under which he sees the little gold leaves play. The other details are easily supplied."

Le Sage had an idea of offering his invention to Frederick the Great, and drew up an introductory note as follows :—

"To the King of Prussia.

"Sire,—My little fortune is not only sufficient for all my wants, but even for all my tastes—except one, *viz.,* that of contributing to the wants and tastes of others ; and this desire all the monarchs of the world, united, could not enable me to fully satisfy. It is not, then, to a patron who can give much, that I take the liberty of dedicating the following discovery, but to a patron who can do much with it, and who can judge for himself of its utility without having to refer it to his advisers." *

Whether he ever carried out this idea or not is difficult to say, but it is certain that his plan was never practically tried, and, like so many of its class, was soon forgotten.

1787.—*Lomond's Telegraph.*

The next plan that we have to notice was a decided improvement, and had an actual existence, though on a very small scale. Seeing, no doubt, the difficulty and expense of using many wires, Lomond of Paris reduced, at one sweep, the number to one, and thus produced a really serviceable telegraph. Arthur Young, the diligent writer on natural and industrial resources, saw this apparatus in action during his first visit to Paris, and thus describes it in his journal, under date October 16, 1787 :—

---

* See *Notice de la vie et des écrits de George-Louis Le Sage de Genève,* &c., par Pierre Prévost, 8vo., Genève, 1805, pp. 176–7. All writers on the Electric Telegraph, copying Moigno (*Traité de Télégraphie Électrique,* Paris, 1849 and 1852), say that Le Sage actually established his telegraph at Geneva in 1774—an assertion for which we have not been able to find any authority.

" In the evening to M. Lomond, a very ingenious and inventive mechanic, who has made an improvement of the jenny for spinning cotton; common machines are said to make too hard a thread for certain fabrics, but this forms it loose and spongy. In electricity he has made a remarkable discovery. You write two or three words on a paper; he takes it with him into a room and turns a machine enclosed in a cylindrical case, at the top of which is an electrometer, a small fine pith-ball* ; a wire connects with a similar cylinder and electrometer in a distant apartment, and his wife, by remarking the corresponding motions of the ball, writes down the words they indicate, from which it appears that he has formed an alphabet of motions. As the length of the wire makes no difference in the effect, a correspondence might be carried on at any distance; within and without a besieged town for instance, or for a purpose much more worthy, and a thousand times more harmless—between two lovers prohibited, or prevented, from any better connection. Whatever the use may

---

* Soon after the discovery of the Leyden jar the necessity of some sufficient indicator of the presence and power of electricity began to be felt, and after some clumsy attempts at an electrometer by Gralath, Ellicott, and others, the Abbé Nollet adopted the simple expedient of suspending two threads, which when electrified would separate by their mutual repulsion. Waitz hung little leaden pellets from the threads for greater steadiness, and Canton, in 1753, improved upon this by substituting two pith balls suspended in contact by fine wires—a contrivance which is used to this day. The electrometer mentioned in the text was of the kind known as the quadrant electrometer, introduced by Henley in 1772.

be, the invention is beautiful. Mons. Lomond has made many other curious machines, all the entire work of his own hands. Mechanical invention seems to be in him a natural propensity."*

As in all systems where the signals were indicated by electroscopes, or electrometers, their action would continue so long as the charge communicated to the wires lasted, and, as during this time it would not be possible to make another signal, the authors must in some way have discharged the wires after every signal, so as to allow the balls, gold leaves, or other indicators, to resume their normal position. This they might have done, either by touching the wires with the finger after the signal had been noted, or by making the indicators themselves strike against some body that would convey their charges to earth. But, probably, there was no need for any such stratagem, as the insulation of the wires would be so imperfect, and the speed of signalling so slow, that the inconvenience would not have been felt.

1790.—*Chappe's Telegraph.*

Most of our readers have, doubtless, heard of Claude Chappe's Semaphore, or Optico-mechanical Telegraph, which, in one form or another (for, like all successful inventions, it had many imitators), did such good service in the first half of this century. Few, however,

* *Travels during the years* 1787, 1788, *and* 1789, *&c., in the Kingdom of France*, Dublin, 1793, vol. i. p. 135

are aware that, before deciding on this form of instrument, he essayed the employment of electricity for
telegraphic purposes.

Reserving a full account of Claude Chappe's life
and works for its proper place in our General History
of Telegraphy, which we hope soon to publish, we
need only concern ourselves here with a brief reference to his early experiments with electricity.

In 1790, he conceived the idea of a telegraph.   He
first employed two clocks, marking seconds, in combination with sound signals, which were produced by
striking on that homely utensil, a stewpan (*casserole*).
Round the seconds dials were marked off equal spaces
corresponding to the numerals 1 to 9, and the cipher 0.
The clocks being so regulated that the second hands
moved in unison, pointing to the same figures at the
same instant, it is clear that, in order to indicate
any particular figure, Chappe had only to strike the
stewpan the moment the hand of his dial entered the
space occupied by that figure ; his correspondent,
hearing the sound, must necessarily note the same
symbol ; and so, successive figures, or groups of
figures, answering to words and phrases in a vocabu-
lary, could be indicated with great ease and rapidity.

But as sound travels so comparatively slowly, it
would in long distances lag behind, and indicate, it
may be, only an A, or B, when an E, or G, was
intended.   Under these circumstances it was but
natural that Chappe should bethink himself of elec-

tricity, of which he was a diligent student, and on which he had just communicated a series of papers to the *Journal de Physique* (which, by the way, obtained his election as a member of the Philomathic Society).

He erected insulated wires for a certain distance,* and arranged that the discharge of a Leyden jar should indicate the precise moment for noting the position of the hands ; but while he was thus removing one difficulty he found himself introducing another, *viz.*, one of electrical insulation. The more he extended his wires, the greater, of course, his difficulty became, until in despair he abandoned the use of electricity, and took to that of optico-mechanics.

In the actual state of telegraphy this circumstance becomes an interesting one, for Chappe held in his hands a power which was destined soon, under another form, to demolish the grand structure on which he was about to spend so much time and labour. Happily, perhaps, he did not live to experience this mortification, for he died January 23, 1805, at the early age of forty-two.

1790.—*Réveroni-Saint-Cyr's Telegraph.*

This gallant officer is said to have proposed in this year an electric telegraph for announcing the result of the lottery drawings, so as to frustrate the knaveries

---

* Gerspach's *Histoire Administrative de la Télégraphie Aérienne en France,* Paris, 1861, p. 7.

of certain individuals; but, apparently, details are wanting.[*]

1794.—*Reusser's Telegraph.*

The next proposal of which we have to speak, and which, in comparison with Lomond's, or Chappe's, was a very clumsy one, is thus described by its author :[†]—

"I have lately contrived a species of electric letter post, by means of which a letter may be sent in one moment to a great distance. I sit at home before my electric machine, and I dictate to some one, on the other side of the street, an entire letter, which he himself writes down. On an ordinary table is fixed, in an upright position, a square board to which a glass plate is fastened. On this plate are glued little squares of tin-foil, cut after the fashion of luminous panes, and each standing for a letter of the alphabet. From one side of these little squares extend long wires, enclosed in glass tubes, which go, underground, to the place whither the despatch is to be transmitted. The distant ends are there connected to tin-foil strips similar in all respects to the first, and, like them, each marked by a letter of the alphabet ; the free ends of all the strips are connected to one return-wire, which goes to the transmitting table. If, now, one touches the outer coating of a Leyden jar with the return-wire,

* Etenaud's *La Télégraphie Électrique*, &c., Montpellier, 1872, vol. i. p. 27.

† Voigt's *Magazin für das Neueste aus der Physik*, vol. ix. part i. p. 183.

and connects the inner coating with the free end of that piece of tin-foil which corresponds to the letter required to be indicated, sparks will be produced, as well at the near, as at the distant tin-foil, and the correspondent there watching will write down the letter."

Reusser concludes : "Will the execution of this plan, on a large scale, ever take place ? That is not the question. It is possible, though it would cost a good deal, but post horses from St. Petersburg to Lisbon are also very expensive. At any rate, whenever the idea is realised I will claim a recompense."

The editor, Johann Heinrich Voigt, appends to the above communication the suggestion of an alarum, which is usually credited to Reusser himself. Voigt says : "Mr. Reusser ought to have proposed to add to his arrangement a flask of some detonating gas, which one could explode by means of the electric spark, and so attract the attention of the distant correspondent to his tin-foil squares."

In comparing the accounts of Reusser's telegraph usually given with our own, many inaccuracies will be observed. Thus, most writers affirm that each piece of tin-foil was cut into the form of a letter of the alphabet, which, on the passage of the spark, became luminous, as in the French telegraph of 1782, or in that of Salvá, which will presently be described. The German text does not admit of this interpretation, for, if such were the case, it would have been unnecessary to affix letters to the squares of tin-foil. Neither is

H

there any authority for the statement that thirty-six circuits for letters and numerals were proposed, which, according to some writers, were entirely metallic, and, therefore, consisted of seventy-two wires, while others assert that there were only thirty-six wires, and that the earth was employed to complete the circuits. Again, it is always said that Reusser, or rather Voigt, was the first to propose an alarum, whereas we have seen that this was done, twelve years before, by the anonymous correspondent of the *Journal de Paris*, 1782.

1794-5.—*Böckmann's, Lullin's, and Cavallo's Telegraphs.*

Böckmann, Lullin, and Cavallo, all about this time, proposed various modifications of Reusser's plan, all requiring but one or two wires, and differing only in their methods of combining the sparks and intervals into a code. Böckmann's, which is a mere suggestion, is to be found at p. 17 of his *Versuch über Telegraphie und Telegraphen*, published at Carlsruhe in 1794 ;* Lullin's we have not been able to trace further back than Reid's *The Telegraph in America*, New York, 1879, p. 69 ; while Cavallo's is described, at length, in his *Complete Treatise on Electricity*, &c., †
from which we condense the following account :—

"The attempts recently made," says Cavallo, "to convey intelligence from one place to another at a great distance, with the utmost quickness, have in-

* Also Zetzsche's *Geschichte der Elektrischen Telegraphie*, p. 32.
† Fourth edition, London, 1795, vol. iii. pp. 285-96.

duced me to publish the following experiments, which I made some years ago, and of which I should not have taken any further notice, had it not been for the above-mentioned circumstance, which shows that they may possibly be of use for that and other purposes."

The object for which those experiments were performed was to fire gunpowder, or other combustible matter, from a great distance, by means of electricity. At first a circuit was made with a very long brass wire, the two ends of which returned to the same place, whilst the middle was at a great distance. At this (middle) point an interruption was made, in which a cartridge of gunpowder, mixed with steel filings, was placed. Then, by applying a charged Leyden phial to the two extremities of the wire (in the usual way) the cartridge was fired.

It proving very troublesome to keep the wires from touching, the experiment was tried with one wire only. A brass wire, one-fiftieth of an inch diameter, and two hundred feet long, was laid on the ground, and one end was inserted in the cartridge of gunpowder and steel filings. Another piece of the same wire had, likewise, one end inserted in the cartridge, whilst the other was thrust into the ground. The distant end of the wire was then connected to the inner coating of a charged jar, while the outer coating was touched with a ground wire. That the discharge took place as before, was proved by the powder being sometimes fired.

Phosphorus and other combustible substances were

next tried, but nothing was found to succeed so well as a mixture of inflammable and common air, confined in specially prepared flasks.

Having made this discovery, Cavallo next directed his attention to the best means of insulating the communicating wire, and at last so contrived that it might be laid indifferently on wet or dry ground, or even through water.

"A piece of annealed copper or brass wire," he says, "being stretched from one side of a room to the other, heat it by means of a flame of a candle, or of a red-hot piece of iron, and, as you proceed, rub a lump of pitch over the part just heated. When the wire has been thus covered, a slip of linen rag must be put round it, which can be easily made to adhere, and over this rag another coat of melted pitch must be laid with a brush. This second layer must be covered with a slip of woollen cloth, which must be fastened by means of a needle and thread. Lastly, the cloth must be covered with a thick coat of oil paint. In this manner many pieces of wire, each of about twenty or thirty feet in length, may be prepared, which may afterwards be joined together, so as to form one continued metallic communication ; but care must be taken to secure the places where the pieces are joined, which is most readily done by wrapping a piece of oil-silk over the painted cloth, and binding it with thread. When a long wire has been thus made out of the various short pieces, let one end of it be formed into a ring, and to the other adapt a small brass ball.

"Through the wires so prepared the flask of inflammable air was always exploded, and whenever the discharge was passed through a flask of common air a spark was seen, and by sending a number of such sparks at different intervals of time according to a settled plan, any sort of intelligence might be conveyed instantaneously from the place in which the operator stands to the other place in which the flask is situated." *

"With respect to the greatest distance to which such communication might be extended," concludes Cavallo, "I can only say that I never tried the experiment with a wire of communication longer than about two hundred and fifty feet; but from the results of those experiments, and from the analogy of other facts, I am led to believe that the above-mentioned sort of communication might be extended to two or three miles, and probably to a much greater distance."

1795–8.—*Salvá's Telegraph.*

Of all the pioneers of the electric telegraph in the last century, Don Francisco Salvá, of Barcelona,†

---

* Moigno (*Télégraphie Électrique*, p. 61) says Cavallo proposed to express signals by the explosion, by the spark, of such substances as gunpowder, phosphorus, phosphuretted hydrogen, &c., but this is an error.

† Don Francisco Salvá y Campillo was born at Barcelona, July 12, 1751. After graduating, with honours, in the universities of his native place, of Valencia, of Huesca, and of Tolosa, he travelled in Italy, France, and other parts of the Continent, and made the acquaintance

deserves the most honourable mention, as well for the extent and completeness of his designs, as for the zeal and intelligence with which he carried them out. His proposals are described with great clearness in a memoir which he read before the Academy of Sciences, Barcelona, December 16, 1795, and from which we cannot do better than make some extracts :*—

" If," he says, " there were a wire from this city to Mataro, and another from Mataro back, and a man were there to take hold of the ends, we might, with a Leyden jar, give him a shock from this end, and so advise him of any matter previously agreed on, such as a friend's death.   But this is not enough, as, if electricity is to be of any use in telegraphy, it must be capable of communicating every kind of information whatsoever ; it must, in a word, be able to speak. This is happily of no great difficulty.

" With twenty-two letters, or even with eighteen, we can express, with sufficient precision, every word in the language, and thus, with forty-four wires from Mataro to Barcelona, twenty-two men there, each to take hold of a pair of wires, and twenty-two charged Leyden jars here, we could speak with Mataro, each man there representing a letter of the alphabet, and giving

---

of many of its learned men, including Le Sage, Reusser, and other well-known electricians. Besides being an able electrician, Salvá was a distinguished physician, and ardently promoted the cause of vaccination in Spain. He died February 13, 1828. See Saavedra's Biography in the *Revista de Telegrafos* for 1876.

* Translated from Saavedra's *Tratado de Telegrafia*, 2nd ed., pp. 119–24 of vol. i.

notice when he felt the shock. Let us suppose that those receive shocks who represent the letters p, e, d, r, o, we shall then have transmitted the word Pedro. All this is within the limits of possibility; but let us see if it cannot be simplified.*

"It is not necessary to keep twenty-two men at Mataro, nor twenty-two Leyden jars at Barcelona, if we fix the ends of each pair of the wires in such a way that one or two men may be able to discriminate the signals. In this way six or eight jars at each end would suffice for intercommunication, for, of course,

---

* Zetzsche (*Geschichte der Elektrischen Telegraphie*, p. 21) says no attempt had been made to construct a telegraph with the physiological effects of static electricity for its basis. Salvá's is an early example; here is another, though of a negative kind. The Rev. J. Gamble, in his excellent treatise on Semaphoric Telegraphs, says, in reviewing the different modes of communication that had been proposed up to his time :—

"Full as many, if not greater, objections will probably operate against every contrivance where electricity shall be used as the vehicle of information. The velocity with which this fluid passes, where the conductors are tolerably perfect, and also that it may be made to pass through water to a very great distance, when it forms part of the circuit, are properties which appear to have given rise to the idea of using it as a means of correspondence. I have never ['even] heard it mentioned, that an alarm may be given to a very great distance, by firing a pistol charged with inflammable air, which explodes by the smallest spark of electricity ; but the further communication could only be maintained by a certain number of shocks being the preconcerted signal of each letter, and requires that the man who receives the intelligence should remain constantly in the circuit of the electric fluid. The whole success of the experiment would likewise depend on an apparatus liable to an infinite number of accidents, scarce in the power of human foresight to guard against."—*Essay on the Different Modes of Communication by Signals*, London, 1797, p. 73. We shall meet with other examples further on.

Mataro can as easily speak with Barcelona, as Barcelona with Mataro.

"It appears, however, little short of impossible to erect and maintain so many wires, for, even with the loftiest and most inaccessible supports, boys would manage to injure them; but as it is not necessary to keep them very far apart, they can be rolled together in one strong cable, and placed at a great height.* In the first trials made with a cable of this kind I covered each wire with paper, coated with pitch, or some other idio-electric substance, then, tying them together, I bound the whole with more paper, which effectually prevented any lateral escape of the electricity. In practice the wire cable could be laid in subterranean tubes, which, for greater insulation, should be covered with one or two coats of resin."

In selecting Barcelona and Mataro, distant about thirteen miles, Salvá did not imply that this was the limit at which his telegraph would be practicable; on the contrary, he thought it very probable that the distance at which the electric discharge would be effective was proportional to the number of jars, and, therefore, that with a large battery telegraphic communication may be established between Barcelona and Madrid, and even between places one hundred, or more, leagues apart.

After showing the superiority of an electric telegraph over the optical (semaphore) system then in

* As is done in London at the present day.

use, he lays special stress on the advantages of the former as regards communication between places separated by the sea, and adds :—

" In no place can the electric telegraph [wires] be better deposited. It is not impossible to construct, or protect, the cables with their twenty-two [pairs of] wires, so as to render them impervious to the water. At the bottom of the sea their bed would be ready made for them, and it would be an extraordinary casualty indeed that should disturb them. * * *

" In 1747, Watson, Bevis, and others, in England, showed how the water of the Thames may be made to form part of the circuit of a Leyden jar, and this makes us consider whether it would not suffice for our telegraph to lay a cable of twenty-two wires only across the sea, and to use the water of the latter in place of the twenty-two return wires." *

In the experiments with which Salvá illustrated his paper, he indicated the letters in a way which, by some strange mistake, has always been ascribed to Reusser. The seventeen essential letters of the

---

* Because Baron Schilling, of St. Petersburg, used a " subaqueous galvanic conducting cord " across the river Neva in 1812, and, in 1837, proposed to unite Cronstadt with the capital by means of a submarine cable, he has been called the Father of submarine telegraphy (Hamel's *Historical Account*, &c., pp. 16 and 67, of W F. Cooke's reprint). But Salvá was, as we here see, at least seventeen years before him with the suggestion, and to Salvá therefore ought to belong the honour which has hitherto been accorded to the Russian philosopher. As we shall see in a future chapter, this is not the only case in which honours justly due to Salvá are unjustly heaped on another.

alphabet (for he omitted those little used, or whose power could be represented by others) were cut out of parallel strips of tin-foil, pasted on bits of glass, after the fashion of spangled panes, and to the ends of each piece of tin-foil were attached the extremities of the corresponding pair of wires. All the wires were bound up in two cables, which were prepared in the way before described, the out-going wires being collected in one cable, and the return wires in the other.

To indicate a letter, A, for example, it was only necessary to take the ends of the corresponding pair of wires, and connect one end with the outer, and the other with the inner coating of a charged jar. Immediately on thus completing the circuit, the observer, at the other end of the cable, heard the noise of the spark, and saw it illuminate the letter A, in its passage across the breaks in the tin-foil.*

From 1796 to 1799 Salvá resided at Madrid, having been invited by the Academy of Sciences of that capital to engage in some experiments of great public interest. There he had the *entrée* of all the *salons*, and was courted by everybody of consideration—amongst the rest, by the Infante Don Antonio, who appears to have assisted him in perfecting his tele-

---

* "The late Dr. Balcells, professor in the Industrial School of Barcelona, whose acquaintance I made towards the latter years of his long life, and who, in his turn, had known the celebrated physicist, Salvá, has often assured me that the apparatus just described was tried by its inventor from the Academy of Sciences to the Fort of Atarazanas, across the Ramblas, a distance of about a kilometre."—Saavedra, *Tratado de Telegrafia,* 2nd ed., vol. i. p. 122

graph. The favourite Godoy, Prince of Peace, was another good friend, to whom Salvá was indebted for an introduction to the King, Charles IV., as we learn from the following paragraph in the *Gaceta de Madrid*, November 29, 1796 :*—" The Prince of Peace, who testifies the most laudable zeal for the progress of the sciences, understanding that Dr. Francisco Salvá had read at the Academy of Sciences, at Barcelona, a memoir on the application of electricity to the telegraph, and presented at the same time an electrical telegraph of his own invention, requested to examine the apparatus himself. Satisfied with the exactness and celerity with which communications may be made by its means, he introduced the doctor to the King of Spain. The Prince of Peace afterwards, in the presence of their Majesties and the whole court, made some communications with this telegraph, completely to their satisfaction. The Infante Don Antonio proposes to have one of them of the most complete construction, which shall possess power sufficient to communicate between the greatest distances, by land

* First translated into English in *The Monthly Magazine*, for February 1797, p. 148. Also noticed in Voigt's *Magazin*, for 1798, vol xi. part iv. p. 61. As a curiosity of bookmaking, we may observe that, in every account of Salvá's telegraph that we have seen, the extracts from the Madrid *Gaceta* and Voigt's *Magazin* are given as if they referred to two entirely different affairs, the latter being usually rendered as follows :—Voigt's *Magazin*, in reference to these experiments, announced two years afterwards that Don Antonio *constructed* a telegraph upon a very grand scale, and to a very great extent. *It also states that the same young Prince was informed at night, by means of this telegraph, of news that highly interested him!* See Highton's *Electric Telegraph : its History and Progress*, London, 1852, p. 43, as a case in point.

or sea. With this view, His Highness has ordered the construction of an electrical machine, the cylinder of which is to be more than forty inches in diameter. He intends, as soon as it is finished, to undertake a series of curious and useful experiments, in conjunction with Dr. Salvá. This is an employment worthy of a great prince. An account of the results will be given to the public in due course."

Notwithstanding this promise, the subject is not again referred to in any succeeding number of the *Gaceta;* but according to Dr. Balcells, the friend of Salvá, a modification of his telegraph which required only one wire was actually constructed in 1798 between Madrid and Aranjuez, a distance of about twenty-six miles. At p. 14 of Gauss and Weber's *Resultate*, &c., for 1837, there is a note of Humboldt's in which he refers to this line, but credits it to Bétancourt, a French engineer. This is clearly a mistake, into which the great traveller might have been led by the probable fact that an engineer of that name was employed to superintend the work—a supposition which is likely enough seeing the greatness of the undertaking.

Dr. Balcells, whose evidence as just quoted should be conclusive on this point, says, further, that the remains of Salvá's telegraph, which, at first, were destined for Don Antonio's museum, were presented, in 1824, to the College of Pharmacy of San Fernando, of which he (Balcells) was then the Adjutant.*

* Saavedra, vol. i. p. 124.

CHAPTER IV.

TELEGRAPHS BASED ON STATIC, OR FRICTIONAL, ELECTRICITY (*continued*).

1802.—*Alexandre's Telegraph.*

TWENTY-FIVE years ago, in the course of a research amongst the imperial archives at Paris, M. Edouard Gerspach, of the French Telegraph Administration, discovered some documents which, in our eyes, are of exceeding value, as establishing for La Belle France the honour of the invention of the first step-by-step, or A.B.C., telegraph. These papers were embodied by M. Gerspach in a memoir for the *Annales Télé-graphiques* for March-April, 1859, pp. 188-99, to which we are indebted for much of what follows in this article.

Jean Alexandre was born at Paris, the natural son, it is said, of Jean-Jacques Rousseau. He had the education of a mechanic, some say of a physician, but his actual career was truly a faithful image of the troublous times in which he lived. In 1787 he was at Poitiers, following the trade of gilder, and, as he had a fine voice, he sang in the churches, which added somewhat to his slender emoluments. But soon the revolution came to Poitiers, and swept away the *clientèle* of the poor gilder and carver. He went to Paris, and there maintained himself for a while by singing in the choir of St. Sulpice ; but the revolutionary tide followed him, and closed the doors of St. Sulpice, as of all the other churches, leaving Alexandre high and dry again, without the means of subsistence.

Feeling there was nothing else to be done, he now took to politics, and, after the manner of the times, soon found himself president of a section of the Luxembourg (club), and, later on, a deputy of the Convention. This latter honour, however, his simple manners made him decline. But greater still were yet in store, and, as he was preparing to return to his workshop at Poitiers, the Government sent him thither, but with the exalted rank of Commissary-General of War. Later on, he was promoted to be chief of the military division of Lyons, where he had to organise an army of 80,000 men. With the title of Chief Agent of the Army of the West, he next went to Angers, where, from the forty-two departments that were under his orders, he had to raise another army of 200,000 men. With all this greatness, he still was not happy ; he yearned for a quiet life — a feeling which seems to have grown daily stronger with him, until, at last becoming irresistible, he quitted honours and politics, and returned to his home at Poitiers, as poor as he had left it—a fact, by the way, which speaks volumes for the integrity of his character.

Here we find him, in 1802, producing his *télégraphe intime*, or secret telegraph. He wrote to Chaptal, Minister of the Interior, acquainting him briefly with the discovery, and asking assistance to enable him to go to Paris, and exhibit his machine to the Government. The Minister asked (and naturally), in the first place, for full particulars and plans of the apparatus, but Alexandre declined to divulge his secret, and addressed himself next to Cochon, Prefect of Vienne, offering to make an experiment before him. The Prefect, agreeably impressed with the conversation of the inventor, whose quick and vigorous imagination he found to contrast singularly with the simplicity of his demeanour, granted his request, and accordingly, on the 13th Brumaire, year X (early in 1802), he went, accompanied by the chief engineer of the department, to Alexandre's house. The experiments were crowned with unhoped-for success, and the Prefect drew up a report for the minister, Chaptal, of which the following is the substance :—

"We were conducted into a room on the ground floor, in the centre of which we found a box nearly 1·5 metre high, and about 30 centimetres broad and deep. This box was surmounted by a dial, around which were traced all the letters of the alphabet. A well-poised needle, or pointer, travelled round the circle at the will of a distant and invisible agent, and stopped over such letters as composed the words that he wished to communicate. The completion of each word and phrase was indicated by an entire revolution of the pointer, which, in its normal state of rest, always occupied a certain determined position [corresponding, no doubt, to our zero].

"A correspondence was established between the [distant] agent and ourselves, and the success was all that we could desire. The dial repeated exactly all the phrases that we had dictated, and the [distant] agent added some from himself, which we had no difficulty in understanding. On asking why the second box was situated in an upper story, about 15 metres distant, instead of being placed on the same level as the first, the inventor replied that it was to show that difference of level had no effect on its action, and that the conductors could in every case go up and down, and adapt themselves to the inequalities of the ground.

"We understood, without, however, his distinctly saying so, that the author derives his power (*usage*) from some fluid, either electric or magnetic. He told us that, in the course of experiment, he had met with a strange matter, or power (of which, until then, he had been ignorant) which, he was almost tempted to believe, is generally diffused, and forms, in some sort, the soul of the universe; that he had discovered the means of utilising the effects of this power, so as to make them conduce to the success of his machine;

and that he was certain of being able to propagate them with the celerity of light, and to any distance that may be required."

In concluding this report on the invention, which the Prefect characterised as a work of genius, he urged that Alexandre should be called to Paris, at the expense of the State, in order that he may repeat his experiments under the eyes of the Government. The minister, Chaptal, did not, however, regard the discovery at all so favourably, evidently imagining it to be a telegraph of the Chappe, or semaphore, kind, and wrote to the inventor's agent, declining to have anything to do with him. Such a rebuff would have ·acted as a *quietus* to ordinary people ; but inventors are proverbially a tenacious race. Alexandre was an inventor, and, firm in his convictions, he quitted Poitiers, and, in hopes of better fortune, betook himself to Tours.

There, at his invitation, General Pommereul, Prefect of the Department of the Indre and Loire, and the mayor and officers of the city of Tours, assembled at his house to assist at a public trial of the apparatus. As before, one of the machines was on the ground floor, and the other on the first story, separated from the lower room by an antechamber and a small court. The Prefect dictated the phrase, "Genius knows no limits," which was transmitted to the distant end, and thence returned with all the success imaginable. The next phrase, "There are no longer miracles," was

I

repeated with the same result, and many others followed, in which all the words were reproduced by the machines, letter for letter, with the greatest exactness.*

All these experiments, conclusive as they were, had, nevertheless, little effect in advancing Alexandre's interests; they drew on him the commendations of the multitude, made his name known, but contributed nothing towards the attainment of his end, which was Paris, and the patronage of the First Consul, to whom only would he confide his secret. Having no money for the further prosecution of his plans, he now entered into partnership with a M. Beauvais, who was to supply all sums necessary, and to receive in return a quarter of the profits of the enterprise, Alexandre keeping to himself the secret of his invention until he had netted 60,000 francs by its exploitation, after which it was to become joint property. No sooner were these terms concluded than Beauvais, provided with the official accounts of the experiments at Poitiers and Tours, addressed himself to Napoleon, and solicited the honour of a trial in his apartments, and in his presence alone. Napoleon, perhaps smelling gunpowder, declined the meeting, but referred the papers to Delambre, the illustrious academician and

---

* The *English Chronicle* newspaper of June 19-22, 1802, has a short account of these experiments, concluding as follows :—" The art or mechanism by which this is effected is unknown, but the inventor says that he can extend it to the distance of four or five leagues, even though a river should be interposed." There is a copy, probably unique, in Mr. Latimer Clark's library.

astronomer, who, some weeks later, returned a report, of which the following is a free translation :—

"Report of Citizen Delambre on the Secret Telegraph of Citizen Alexandre, submitted to the First Consul by Citizen Beauvais.

"Paris, 10 Fructidor, an X

"The papers which the First Consul has caused me to examine do not contain sufficient details to enable me to form an opinion, nor, after the two interviews that I have had with Citizen Beauvais, am I able to do more than offer the merest conjectures on the advantages and disadvantages of the Secret Telegraph.

"Citizen Beauvais knows the secret of Citizen Alexandre, but he has promised to impart it to no one but the First Consul himself. This circumstance must make any report from me valueless, for how can one judge of a machine which one has neither seen nor understands?

"All that we know is that this telegraph is composed of two similar boxes, each having a dial, round whose face are marked all the letters of the alphabet. By means of a winch, or handle, the pointer of one dial is moved to any desired letter or letters, and, at the same instant, the pointer of the other dial repeats the same movements, and in exactly the same order. When these two boxes are placed in two separate apartments, two persons can write and reply without seeing each other, and without being seen, and in such a way that no one can doubt the correspondence, which, moreover, can be carried on at any time, as neither night nor fogs can intercept the transmission.

"By means of this telegraph the governor of a besieged place could carry on a secret and continuous correspondence with a person four or five leagues distant, or even at any distance, and communication can be established between the two boxes as readily as one can hang a bell (*qu'on poserait un mouvement de sonnette*).

"The inventor carried out two experiments with his machines at Poitiers and at Tours, in presence of the prefects and mayors of the respective places, and the official reports of these functionaries attest that the results were completely successful. Now, the inventor and his associate ask, either that the First Consul will be pleased to permit of one of the boxes being placed in his apartment, and the other in that of the Consul Cambacérès, so as to give to their experiments all the *éclat* and authenticity possible ; or, that he will accord an audience of ten minutes to Citizen Beauvais, who will then communicate to him the secret (of the telegraph), which is so simple that the bare description will be equivalent to a practical demonstration. They add that the idea is so natural as to leave little room to fear that it will ever occur to any *savant* [*sic*]. It is said, however, that Citizen Montgolfier divined it, after some hours' reflection, on a description of the apparatus which was given to him.

"After this statement, which is the substance of my conversations with Citizen Beauvais, a very few remarks must suffice. If, as one would be inclined to believe from the comparison with bell-hanging, the means employed comprised wheels, levers, and such like,* the invention would not be very surprising, and one could easily imagine the practical difficulties that would be encountered as soon as it was attempted to employ it over distances of several leagues.

"If, on the contrary, as the official report from Poitiers seems to show, the means of communication is a fluid (*i. e.*, a natural force), the inventor deserves much more credit for having discovered how to utilise it so as to produce, at any distance, effects so regular and so unfailing. But then, one may demand, what guarantee have we for these effects? Neither the experiments at Poitiers, nor those at Tours, in which the distance was only a few metres, supply it. No more would the proposed experiment between the chambers of the First and Second Consuls. So long as the motive power remains a secret, one can never vouch for more than what one sees, and it will be entirely wrong to conclude, from the success of an experiment on a small scale, that like results will be obtained over more considerable distances. If the effect is only attainable at a distance of some few metres, the machine ought to be sent to the scientific toy shops.

"If Citizen Beauvais, who offers to defray the expenses of an experiment, had proposed to carry it out in presence of commissioners appointed for the

* Forming, in fact, a kind of mechanical telegraph like the railway semaphores of to-day.

purpose, there could be no objection to granting his request; for, although an experiment on a small scale would not be very conclusive, still it would enable us to see what might be hoped from a trial of a grander and more expensive kind. But Citizen Beauvais, without expressly declining a commission, desires, in the first place, to secure the testimony and approbation of the First Consul. It only remains, then, for the First Consul to say whether, in view of the little chance of success attaching to an invention so little proved, and announced as so marvellous, he will spare a few moments for the examination of a discovery of an artist, who is described as one as full of genius as he is devoid of scientific learning and of fortune.

"He makes a secret of his discovery, and I ought to judge it with severity, and according to the laws of probability; but the limits of the probable are not those of the possible, and Citizen Alexandre must be sure of his facts, since he offers to expose all to the First Consul. It, therefore, only remains for me to hope that the First Consul will grant him an audience, and that, in the sequel, he will have reason to welcome the inventor, and recompense worthily the author.

" DELAMBRE."

With this most interesting document ends the story of the *Secret Telegraph.* In 1806 Alexandre was at Bordeaux, taking out a patent for a machine for filtering the water of the Garonne for supplying the city ; but, although the authorities seem to have afforded him every facility towards the accomplishment of his scheme, it was never carried out, through want of money. We next hear of him in 1831, when he submitted to the King, Louis Philippe, a project for steering balloons. He died soon after at Angoulême, leaving a widow, who died in 1833, at Poitiers, in extreme want.

Such is the sad story, as told by M. Gerspach, of one who must be regarded as a veritable pioneer in electric telegraphy ; for, although Alexandre chose to surround his invention with an air of mystery, and preserved only too faithfully the secret of its action, we believe that he had, in effect, constructed a telegraph of the A, B, C, sort, with static electricity as his motive power.

Some writers, however, regard his apparatus, like that of Comus, as only another instance of the sympathetic needle telegraph, and seek to explain its action somewhat after the manner figured and described by Guyot.* But there seems to us to be two very good reasons against this theory : first, the impossibility of carrying out any such deception in the apartments of the two consuls ; and second, the character of

---

* *Nouvelles Récréations Physiques et Mathématiques*, Paris, 1769, vol. ı. p. 134. M. Aug. Guerout is the latest exponent of this theory. See *La Lumière Électrique*, for March 3, 1883.

Napoleon, who, as all the world knows, was not a man to be trifled with.

The suspicion of Delambre, that it partook of the nature of a mechanical telegraph, we consider equally disproved by the words of the *procès-verbal* from Poitiers. "He told us that, in the course of experiment, he had met with a strange matter, or power (of which, until then, he had been ignorant), which, he was almost tempted to believe, is generally diffused, and forms in some sort the soul of the universe; that he had discovered the means of utilising the effects of this power, so as to make them conduce to the success of his machine; and that he was certain of being able to propagate them with the celerity of light, and to any distance that may be required." Surely a mechanician would not speak thus of a combination of ropes, wheels, and pulleys. Although, once upon a time, Archimedes glorified the power of the lever, when he said that by its means he could move the world, no Archimedes of our day would be so extravagant as to call the same power, mighty as it is, the soul of the universe.

On the other hand, the language just quoted would apply very well to electricity. Thales called it a spirit, Otto Guericke thought it controlled the revolution of the moon round the earth, and Stephen Gray that of the planets round the sun; Franklin showed its identity with lightning; John Wesley regarded it as an universal healer; and Galvani had just confounded it with life.   Well, then, might Alexandre be excused for calling it the soul of the universe.

Again, let us recollect that while he was still a young man the invention of the Chappe semaphore, and its wonderful performances, were the theme of daily conversation; and that rival plans were being frequently started—some, semaphores more or less like Chappe's, and for night as well as day service; some, based on the properties of acoustics, as those of Gauthey and Count Rumford; and some again, as we have seen in these pages, on those of electricity.*
What more natural, then, than that Alexandre, a clever mechanician, and a man of a quick and vigorous imagination, should invent an electric telegraph.

Now, let us regard the apparatus as described by M. Cochon, in connection with the half admission that electricity was its basis, and that it was operated by a winch, or handle, as mentioned by Delambre.   Do not this handle, the box, the dial on the top, and the conductor recall the telegraph of Lomond, which was the wonder of Paris in 1787, and which has been already described in these pages.   The dial of Alexandre, it is true, is an immense improvement on the

---

* We may here refer to a remark of Amyot's, for which we have not been able to find room before, to the effect that, somewhere about 1798, Henry Monton Berton, the distinguished French composer, conceived the idea of an electric telegraph (*Note historique sur le Télégraphe Électrique*, in the *Comptes Rendus*, for July 9, 1838).   This *note* is reprinted *in extenso* in Julia de Fontenelle's *Manuel de l'Électricité*, but in neither case are any details given.

pith-ball indicator of Lomond, but that (the dial), too, had its prototype in the synchronous clockwork dial with which Chappe essayed an electric telegraph in 1790, and which, no doubt, was equally well known as the machine of Lomond. Indeed, the inference to us seems irresistible, that Alexandre took Lomond's and Chappe's contrivances as his basis, and built upon them his own improvements.

The only point that remains for consideration is, how did the working (? revolving) of the handle actuate the pointers? The explanation to our mind is not far to seek. Given an electrical machine inside the box, and a train of wheels behind the dial, and in gear with the pointer, and it would be easy for a clever mechanician to make the repulsion of a sort of pith-ball electrometer (acting also as a pawl) against a discharging surface, and its subsequent collapse, give motion of a step-by-step character to the wheels, and, through them, to the pointer. The prime conductors of both machines would, under our supposition, be connected by a wire (probably concealed from view), and thus the movements of one pointer would be synchronous with those of the other.

Some writers, as Cézanne[*] and Berio,[†] think it likely that Alexandre used the electricity of the pile, then newly discovered by Volta; but the use of a handle is as fatal to such an assumption, as it is favourable to that of an electrical machine being the *primum mobile.*

[*] *Le Cable Transatlantique,* Paris, 1867, p. 32.
[†] *Ephemerides of the Lecture Society,* Genoa, 1872, p. 645.

1806–14.—*Ralph Wedgwood's Telegraph.*

The next proposal of a telegraph based, presumably, on static, or frictional, electricity, is due to a member of the Wedgwood family. Ralph Wedgwood was born in 1766, and was brought up by his father at Etruria, where he received much valuable aid in chemistry, &c., from his distinguished relative Josiah. He afterwards carried on business, as a potter, under the style of "Wedgwood and Co.," at the Hill Works, Burslem ; but was ruined through losses during the war. After a short and unsatisfactory partnership with Messrs. Tomlinson and Co., of Ferrybridge, Yorkshire, he removed to Bransford, near Worcester, where he issued prospectuses for teaching chemistry at schools. Thence, in 1803, he moved to London, travelling in a carriage of his own constructing, which he described as "a long coach to get out behind, and on grasshopper springs, now used by all the mails."

He appears to have early shown a genius for inventing, and while yet at Bransford had perfected many useful contrivances—amongst them, a "Pennapolygraph," for writing with a number of pens attached to one handle ; and a "Pocket-secretary," since better known as the "Manifold-writer." On coming to London he found that the first-mentioned apparatus had already been invented by another person, but the second, proving to be new, he patented as "an apparatus for producing duplicates of writing."

In 1806, he established himself at Charing Cross, and soon after turned his attention to the construction of an electric telegraph, the first suggestions of which he seems to have obtained from his father.* In 1814, having perfected his plans, he submitted them to Lord Castlereagh, at the Admiralty; and after a proper interval his son, Ralph, waited on his lordship to learn his views with regard to the new invention. He was dismissed with the assurance that "the war being at an end, and money scarce, the old system [of shutter-semaphores] was sufficient for the country."

These chilling words appear to have been stereotyped, ready for use, for, as we shall see in the course of our history, they were the identical missiles with which a wearied and, perhaps, worried bureaucracy repulsed other telegraph inventors, as Sharpe and Ronalds, Porter and Alexander,† and goodness knows how many others besides. They certainly were the death of Wedgwood's telegraph, for he dropped it in disgust, leaving on record only a few words as to its uses and advantages—precisely such as we find them to-day. These show such an appreciation of the value of the electric telegraph, that we feel certain his

* According to Llewellynn Jewitt, whose *Life of Josiah Wedgwood*, &c., London, 1865, we follow in this volume; see chap. ix. pp. 178–81. See also Jewitt's *Ceramic Art in Great Britain*, London, 1878, vol. 1. pp. 489–92.

† The writer of the article "Fifty Years' Progress" in *The Times*, January 5, 1875, says that Alexander could not hear the word "telegraph" without a shudder!

own invention was of no mean order, and we must ever regret, therefore, that he has left us nothing as to its construction or mode of action. His remarks are contained in a pamphlet,* dated May 29, 1815 ; and as they are all that we have on our subject we shall quote them entire :—

"A modification of the stylographic principle proposed for the adoption of Parliament, in lieu of telegraphs, *viz.* :—

"The Fulguri-Polygraph, which admits of writing in several distant places at one and the same time, and by the agency of two persons only.

"This invention is founded on the capacity of electricity to produce motion in the act of acquiring an equilibrium ; which motion, by the aid of machinery, is made to distribute matter at the extremities of any given course. And the matter so distributed being variously modified in correspondence with the letters of the alphabet, and communicable in rapid succession at the will of the operator, it is obvious that writing at immense distances hereby becomes practicable ; and, further, as lines of communication can be multiplied from any given point, and those lines affected by one

* Entitled *An Address to the Public on the Advantages of a proposed introduction of the Stylographic Principle of writing into general use ; and also of an improved species of Telegraphy, calculated for the use of the Public, as well as for the Government.* It will be found at the end of his *Book of Remembrance,* which was published in London, 1814. Wedgwood was an exceedingly reticent man, and, it is feared, carried with him to the grave other scientific secrets, as well as that of the telegraph. He died at Chelsea in 1837.

and the same application of the electric matter, it is evident from hence also that fac-similes of a despatch, written, as for instance, in London, may, with facility, be written also in Plymouth, Dover, Hull, Leith, Liverpool, and Bristol, or any other place, by the same person, and by one and the same act. Whilst this invention proposes to remove the usual imperfections and impediments of telegraphs, it gives the rapidity of lightning to correspondence *when* and *wherever* we wish, and renders *null* the *principal disadvantages* of *distance* to *correspondents.*

" Independent of the advantages which this invention offers to Government, it is also susceptible of much utility to the public at large, inasmuch as the offices which might be constructed for the purposes of this invention might be let to individuals by the hour, for private uses, by which means the machinery might be at all times fully occupied ; and the private uses which could thus be made of this invention might be applied towards refunding the expenses of the institution and also for increasing the revenue. Innumerable are the instances wherein such an invention may be beneficially applied in this country, more especially at a time when her distinguished situation in the political, commercial, and moral world, has made her the central point of nations and the great bond of their union. To the seat of her Government, therefore, it must be highly desirable to effect the *most speedy and certain communication* from every quarter of the world, whilst it would at any moment there concentrate instantaneous intelligence of the situation of each and every principal part of the nation, as well as of each and every branch of its various departments."

In communicating the above extract to *The Commercial Magazine*, for December 1846 (pp. 257-60), Mr. W. R. Wedgwood thus urges the claims of his father to a share in the discovery of the electric telegraph :—" It may be asked, why did not Mr. Ralph Wedgwood carry his invention into practical application ? The answer is very obvious. Railways were not then in existence, and the connecting medium required an uninterrupted course such as railways alone afford. Such an invention also required the assistance either of Government or a powerful company, the scheme being too gigantic for an individual to work. The inventor, then, it will be perceived, did all that was possible to bring the discovery into practical use ; for, in the first instance, he offered it to the Government, who refused it ; and, as it was for the benefit of the nation, he then made public his scheme of an electric telegraph in the manner quoted from his pamphlet."

1816.—*Ronalds' Telegraph.*

This ingenious contrivance belongs to the synchronous class of telegraphs, of which we have already seen two other examples, *viz.*, those of Chappe, 1790, and Alexandre, 1802. It is, in fact, only the realisation of Chappe's idea.   Sir Francis (then Mr.) Ronalds took up the subject of telegraphy in 1816, and pursued it very ardently for some years, until, like Wedgwood, disgusted with the apathetic conduct of the Government, he dropped the matter, and, more in sorrow than in anger, took leave of a science which, as he says, was up to that time a favourite source of amusement.   Fortunately for the science, he returned to his old love in later years, and, dying August 8, 1873, left us a grand legacy in the Ronalds' Library.*

In 1823, he published a thin octavo volume, entitled *Descriptions of an Electrical Telegraph, and of some other Electrical Apparatus;* and, in 1871, the original work having become very scarce, he issued a reprint of the part relating to his telegraph.   From this, in accordance with our rule of consulting, when possible, original sources, we extract the following account.

* A magnificent collection of books on electricity, magnetism, and their applications. The catalogue compiled by Sir Francis is a monument of the concentrated and well-directed labour of its indefatigable author. It has been ably edited by Mr. A. J. Frost, and published at an almost nominal price by the Society of Telegraph-Engineers and Electricians. No student of electricity should be without it. A short, alas! too short, biography of Sir Francis by the editor is prefixed to the catalogue, to which we refer our readers for much interesting information. We would here correct an error—the only one, we believe —into which the biographer has fallen. On p. xv. he says, "Wheatstone, then a boy of about 15, was present at many of the principal experiments at Hammersmith." Wheatstone was born at Gloucester in 1802, where he lived until the year 1823, when he came up to London, and opened business as a maker of musical instruments. It seems impossible to us that a poor lad of 14, as Wheatstone was in 1816, could have been present at Ronalds' experiments, even supposing that he was not then living far away in Gloucester.

The drawings with which our subject is illustrated have been reduced on stone from the original copper-plates which were engraved from Ronalds' own sketches, in 1823.

Ronalds begins by saying :—" Some German and American savans first projected galvanic, or voltaic, telegraphs, by the decomposition of water, &c. But the other [or static] form of the fluid appeared to me to afford the most accurate and practicable means of conveying intelligence ; and, in the summer of 1816, I *amused* myself by wasting, I fear, a great deal of time, and no small expenditure, in trying to prove, by experiments on a much more extensive scale than had hitherto been adopted, the validity of a project of this kind."

These experiments were carried out on a lawn, or grass-plot, adjoining his residence at Hammersmith, and as, of course, it was impossible to lay out in a straight line a great length of wire in such a situation, he had recourse to the following expedient : Two strong frames of wood (see Frontispiece) were erected at a distance of twenty yards from each other, and to each were fixed nineteen horizontal bars. To each of the latter, and at a few inches apart, were attached thirty-seven hooks, from which depended as many silken loops. Through these loops was passed a small iron wire, which, going from one frame to the other, and making its inflections at the points of support, formed one continuous length of more than eight miles.

When a Canton's pith-ball electrometer was con-

K

nected with each extremity of this wire, and it was charged by a Leyden jar, the balls of both electrometers appeared to diverge at exactly the same moment ; and when the wire was discharged, by being touched with the hand, they both collapsed as suddenly and, as it appeared, as simultaneously. When any person took a shock through the whole length of wire, and the shock was compelled to pass also through two insulated inflammable air pistols, one connected with each end of the wire, the shock and explosion seemed to occur at the same instant.

When the spark was passed through two gas pistols, and any one closed his eyes, it was impossible for him to distinguish more than one explosion, although both pistols were, of course, fired. Sometimes one, and sometimes both pistols were feebly charged with gas, but nobody, whose back was turned, could tell from the report, except by mere chance, whether one or both were exploded.

Thus, then, three of the senses, *viz.*, sight, feeling, and hearing, seemed to receive absolute conviction of the instantaneous transmission of electric signs through the pistols, the eight miles of wire, and the body of the experimenter (pp. 4 and 5).

Accepting these experiments as conclusive of the practicability of an electric telegraph, Ronalds next sought out the best means of establishing a communication between any two distant points ; and, after trying a number of ways, at last adopted the following,

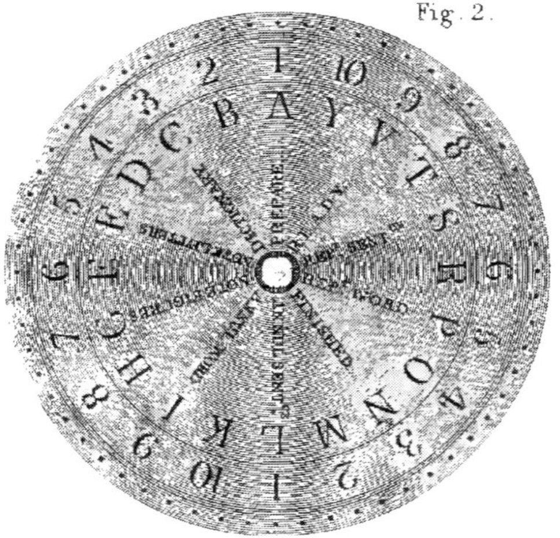

Fig. 2.

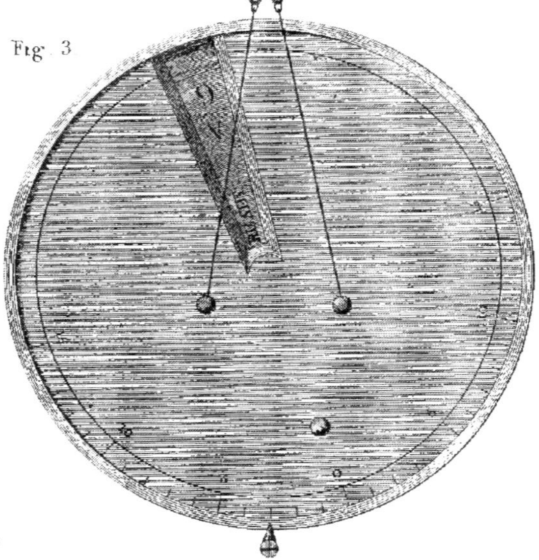

Fig. 3.

Fig 1.

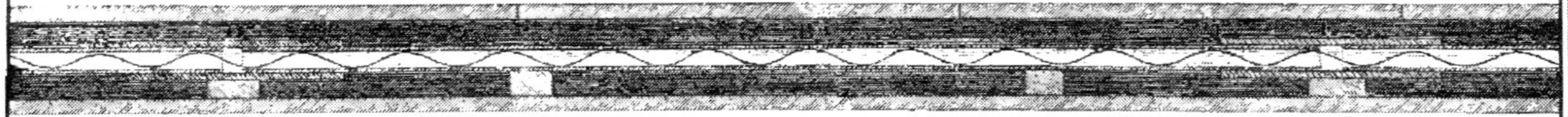

as being the most convenient. A trench was dug in the garden, 525 feet in length, and 4 feet deep. In it was laid a trough of wood, two inches square, and well lined, inside and out, with pitch. In the trough thick glass tubes were placed, through which, finally, the wire (of brass and copper) was drawn. The trough was then covered with strips of wood, previously smeared with hot pitch, and, after painting with the same material the joints so as to make assurance doubly sure, the trench was filled in with earth. Plate I., Fig. 1, represents a section of this trough, tube, and wire. It will be seen that the different lengths of tube A, B, C, did not touch, but that at each joint, or rather interval, other short tubes, or ferrules, D, E, were placed, of just sufficient diameter to admit the ends of the long ones, together with a little *soft* wax. This arrangement effectually excluded any moisture, and yet left the parts free to expand and contract with variations of temperature.

The apparatus for indicating the signals, and its *modus operandi*, are thus described :—A light, circular brass plate, Fig. 2, divided into twenty equal parts, was fixed upon the seconds' arbor of a clock which beat dead seconds. Each division bore a figure, a letter, and a preparatory sign. The figures were divided into two series, from 1 to 10, and the letters were arranged alphabetically, leaving out J, Q, U, W, X, and Z, as of little use. In front of this disc was fixed another brass plate, Fig. 3, capable of being occasionally revolved by the hand round its centre. This plate had an aperture of such dimensions, that, whilst the first disc, Fig. 2, was carried round by the motion of the clock, only one set of letter, figure, and preparatory sign could be seen, as shown in the figure. In front of this pair of plates was suspended a Canton's pith-ball electrometer, B, Fig. 3, from an insulated wire, C, which communicated with a cylindrical machine, D, of only six inches diameter, on one side; and with the line wire, E, insulated and buried in the way just described, on the other.

Another electrical machine and clock, furnished with an electrometer and plates, being connected to the other end of the line in precisely the same way, it is easy to see, that when the wire was charged by the machine at *either* end, the balls of the electrometers at *both* ends diverged; and that when the wire was discharged at either station, both pairs of balls collapsed at the same time. Whenever, therefore, the wire was discharged at the moment that a given letter, figure, and sign of one clock appeared in view through the aperture, the same letter, figure, and sign appeared also in view at the other clock; and thus, by discharging the line at one end, and by noting down whatever appeared in view at every collapse of the pith-balls at the other, any required words could be spelt. By the use of a telegraphic dictionary, the construction of which is explained at pages 8 and 9 of Ronalds' little treatise, words, and even whole sentences, could be intimated by only three discharges, which could be effected, in the shortest time in nine seconds, and in the longest in ninety seconds, making a mean of fifty-four seconds.

Whenever it was necessary to distinguish the preparatory signs from the figures and letters, a higher charge than usual was given to the wire, which made the pith-balls diverge more widely ; and it was always understood that the first sign, *viz.*, *prepare*, was intended when that word, the letter A, and the figure 1, were in view at the sender's clock. Should, therefore, the receiver's clock not exhibit the same sign, in consequence of its having gained, or lost, some seconds, he noted the difference, and turned his outer plate, Fig. 3, through as many spaces, either to the right or left, as the occasion required, the sender all the while repeating the signal, *prepare*. As soon as the receiver had adjusted his apparatus, he intimated the fact by discharging the wire at the moment when the word *ready* appeared through the opening. In order to indicate when letters were meant, when plain figures, and when code figures referring to words and sentences in the dictionary, suitable preparatory signs were made beforehand, as *note letters*, *note figures*, and *dictionary*. Other preparatory signs of frequent use were marked on the dials, and were designated in the same manner whenever required.

The gas pistol F, Plate II., which passed through the side of the clock-case G, was furnished with a kind of discharging-rod, H, by means of which a spark might pass through and explode it when the sender made the sign *prepare*. This obviated the necessity of constant watching on the part of the attendant, which was found so irksome in the semaphores of those days. By a slight turn of the handle, I, the attendant could break the connection between the line wire and the pistol, and so put his apparatus into a condition to "receive."

Midway between the ends of the wire was placed the contrivance, K, K, by which its continuity could be broken at pleasure, for the purpose of ascertaining (in case any accident had happened to injure the insulation of the buried wire) which half had sustained the injury, or if both had. It is seen that the two sides of the wire were led up into the case, and terminated in two clasps, L, and M, which were connected by the metal cross piece, N, carrying a pair of pith-balls. By detaching this wire from the clasp L, whilst it still remained in contact with M, or *vice versâ*, it could at once be seen which half of the wire did not allow the balls to diverge, and, consequently, which half was damaged, or if both were.

One of the stations of this miniature telegraph was in a room over a stable, and the other in a tool-house at the end of the garden, the connecting wire being laid under the gravel walk.*

---

* After a lapse of nearly fifty years, a portion of this line was dug up, in 1862, in the way described in Frost's *Biography*, p. xviii. Some years later the specimen came into the possession of Mr. Latimer Clark, by whom it, as well as the original dial apparatus, was exhibited at

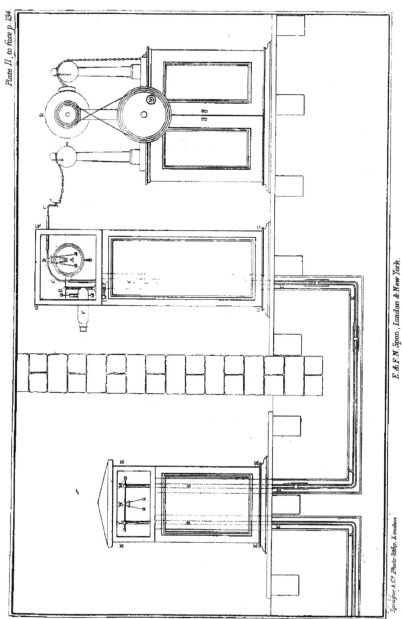

E. & F. N. Spon, London & New York.

Sprague & Co. Photo-lith, London.

Having made a large number of experiments with this line, and having thoroughly proved the practicability of his invention, Ronalds decided upon bringing it to the notice of the Government. This he did on the 11th of July, 1816, in a letter addressed to Lord Melville, the First Lord of the Admiralty, as follows :—

" Upper Mall, Hammersmith,
July 11, 1816.

" Mr. Ronalds presents his respectful compliments to Lord Melville, and takes the liberty of soliciting his lordship's attention to a mode of conveying telegraphic intelligence with great rapidity, accuracy, and certainty, in all states of the atmosphere, either at night or in the day, and at small expense, which has occurred to him whilst pursuing some electrical experiments. Having been at some pains to ascertain the *practicability* of the scheme, it appears to Mr. Ronalds, and to a few gentlemen by whom it has been examined, to possess several important advantages over any species of telegraph hitherto invented, and he would be much gratified by an opportunity of demonstrating those advantages to Lord Melville by an experiment which he has no doubt would be deemed decisive, if it should be perfectly agreeable and consistent with his lordship's engagements to

---

the Special Loan Collection of Scientific Apparatus, South Kensington Museum, 1876, and at the late Electrical Exhibitions in Paris and London (Crystal Palace). They were also shown in the British Section of the Vienna Exhibition, last year.

honour Mr. Ronalds with a call ; or he would be very happy to explain more particularly the nature of the contrivance if Lord Melville could conveniently oblige him by appointing an interview."

Lord Melville was obliging enough, in reply to this communication, to request his private secretary, Mr. Hay, to see Ronalds on the subject, but before an interview could be arranged, and while the nature of the invention was yet a secret, except to Lord Henniker, Dr. Rees, Mr. Brand, and a few particular friends, an intimation was received from Mr. Barrow, the Secretary of the Admiralty, to the effect that telegraphs of any kind were then wholly unnecessary, and that no other than the one then in use (the old Semaphores of Murray and Popham) would be adopted. This much-quoted and now historic *communique* ran as follows : *—

"Admiralty Office, 5th August.

"Mr. Barrow presents his compliments to Mr. Ronalds, and acquaints him, with reference to his note of the 3rd inst., that telegraphs of any kind are now wholly unnecessary, and that no other than the one now in use will be adopted."

In reference to this correspondence Ronalds says :—
"I felt very little disappointment, and not a shadow

* The originals of these important documents, with many other valuable papers, relating to this subject, may be consulted in the Ronalds' Library.   See p. 439 of the Catalogue.

of resentment, on the occasion, because every one knows that telegraphs have long been great bores at the Admiralty. Should they *again* become *necessary*, however, perhaps electricity and electricians may be indulged by his lordship and Mr. Barrow with an opportunity of *proving* what they are capable of in this way. I claim no indulgence for mere chimeras and chimera framers, and I hope to escape the fate of being ranked in that unenviable class " (p. 24).

Ronalds will always occupy a high position in the history of the telegraph, not only on account of the excellence and completeness of his invention, but also for the ardour with which he pursued his experiments, and endeavoured to bring them to the notice of his countrymen.* Had he worked in the days of railways and joint-stock enterprise, there can be no doubt that his energy and skill would have triumphed over every obstacle, and he would have stood forth as the practical introducer of electric telegraphs ; but he was a generation too soon, the world was not yet ready for him.

His little *brochure* of 1823 is the first work ever published on the subject of electric telegraphy, and is

---

* It might have been with a knowledge of Ronalds' telegraphic experiments that Andrew Crosse, in 1816, uttered the prophecy, with which his biographer opens the story of his life :—"I prophesy that by means of the electric agency we shall be enabled to communicate our thoughts instantaneously with the uttermost ends of the earth."— *Memorials Scientific and Literary of Andrew Crosse, the Electrician,* London, 1857.

so marvellously complete, that it might almost serve as a text-book for students at the present day.   In it he proposes the establishment of telegraph offices throughout the kingdom, and points out some of the benefits which the Government and public would derive from their existence.   " Why," he asks, " has no *serious trial* yet been made of the qualifications of so *diligent* a courier ?   And if he should *be* proved competent to the task, why should not our kings hold councils at Brighton with their ministers in London ? Why should not our Government govern at Portsmouth almost as promptly as in Downing-street ? Why should our defaulters escape by default of our foggy climate ? and, since our piteous inamorati are not all Alphei, why should they add to the torments of absence those dilatory tormentors, pens, ink, paper, and posts ?   Let us have *electrical conversazione offices*, communicating with each other all over the kingdom *if we can.*" *

It would hardly be possible at the present day to describe more accurately the progress of electric telegraphy than in these characteristic sentences.   We have " electrical conversazione offices " all over the kingdom. The wires which connect Balmoral, Windsor,

---

* Pp. 2, 3. It is curious to note the similarity of ideas on this subject that occurs in the extract, which we have given, on page 126, from Ralph Wedgwood's pamphlet of 1815. The two trains of thought are perfectly independent, for we believe that Ronalds knew nothing of Wedgwood's invention—a conclusion to which we are led by the absence of the latter's name from the Ronalds' catalogue.

and Osborne with Downing-street, enable Her Majesty to "hold councils with her Ministers in London" at any moment ; while the extensive system of Admiralty and War Office telegraphs enables the Government to "govern at Portsmouth [and many places besides] as promptly as in Downing-street." One of the very first results of the earliest telegraph was the capture of Tawell, the Quaker murderer ; and the curious ramification of police telegraphy in London, if not an absolute protection against our "foggy climate," is, at least, a terror to those who might otherwise elude the grasp of the law. As for our "piteous inamorati," it is perfectly well known that they use the wires as freely as most people, and that "love telegrams" are gradually taking the place of "love letters."

His underground wire was a fair specimen of what exists at the present day. We use iron, or earthenware, pipes in lieu of his wooden trough ; but we are not very far in advance here, for he points out by way of anticipating possible objections to his plan, that "cast-iron troughs might be rendered as tight as gaspipes," should it be deemed desirable to employ them. He did not recommend his glass-tube insulators to the exclusion of all other methods, as, for example, that of Cavallo, by means of pitch and cloth. "No person," says he, "of competent experience in these matters will doubt, that either of them, or several other plans that might be chosen, would be efficient. But

since accident and decay compose the lot of all
inanimate as well as animated nature, let two or more
sets of troughs, tubes, and wires be laid down ; so
that, whilst one may be undergoing repair, the others
may be ready for use " (p. 16).

On the general question of conservancy he says :—
" To protect the wire from mischievously disposed
persons, let the tubes be buried six feet below the
surface of the middle of high roads, and let each tube
take a different route to arrive at the same place.
Could any number of rogues, then, open trenches six
feet deep, in two or more public high roads or streets,
and get through two or more strong cast-iron troughs,
in a less space of time than forty minutes? for we
shall presently see that they would be detected before
the expiration of that time. *If they could,* render
their difficulties greater by cutting the trench deeper :
and should they still succeed in breaking the communication by these means, hang them if you can
catch them, damn them if you cannot, and mend
it immediately in both cases. Should mischievous
devils from the subterranean regions (*viz.,* the cellars)
attack my wire, condemn the *houses* belonging there-
unto, which cannot *easily* escape detection by running
away " (p. 17).

Ronalds, however, proposed to rely upon other
means than Lynch law in maintaining his communications, and here, again, the telegraph engineers of
the present day have followed out his ideas almost to

the letter. He proposed to keep his wire constantly charged with electricity, then to have certain proving stations (like that described on page 134) at frequent intervals along the line; and a staff of persons who would constantly watch the provers, and set out the moment that any indication of an interruption was given. Suitable situations for such proving stations he conceived to be "post-offices in towns and villages, turnpike gates, and the like."

"To put a simple case: we will imagine twenty proving stations established between London and Brighton, or any distance of fifty miles, only four persons employed (but not exclusively) to keep watch over them, and each watchman to have charge of five provers. It is evident that (were he to dwell at the centre one of the five), in order to examine the two on each side of it, he would have to ride only four miles and eight-tenths, which journey even our two-penny post-boy can perform in something less than forty minutes; and he would discover that the defect rested somewhere between two of the provers, a distance of two miles and four-tenths. Let him report his discovery accordingly to the engineer, who may open the trench and the trough at mid-distance of this two miles and four-tenths, make an experiment upon the wire itself similar to that of the provers, and, when he has discovered which half is defective, operate upon that half in the same way. Thus proceeding continually, he must arrive, after ten bisections, within about three yards of the defect" (pp. 19, 20).

Now, what are these innumerable "flush-boxes" which are to be found everywhere in the streets of London and other large cities but "provers" of our underground telegraphic system? Most people are familiar with the snake-like coils of telegraph wires, which are every now and then laid bare in those curious apertures in the pavement, and the little clock-face, with a single handle, which is the invariable companion of the workman engaged in the hole. He is simply "proving" a wire which has been found faulty. Then, again, as regards overhead wires, what are the "linemen" stationed at certain intervals along the route of a trunk line but the "provers" of the section which it is their duty to traverse from time to time, working on either side of their station, precisely as Ronalds would have worked his "sorry little two-penny post cove"?

We must not omit to mention that Ronalds clearly foresaw by the sheer force of reasoning the pheno-menon of retardation of signals in buried wires, such as we find it to-day.* At p. 5 of his *brochure* he says :—

"I do not contend, nor even admit, that an *instanta-neous discharge*, through a wire of *unlimited* extent, would occur in *all* cases" (p. 5). And again, on

---

* Zetzsche tries to combat this assertion at p. 38 of his *Geschichte der Elektrischen Telegraphie*, Berlin, 1867.

p. 12 :—" That objection, which has seemed to most of those with whom I have conversed on the subject the least obvious, appears to me the most important, therefore I begin with *it* ; *viz.*, the probability that the electrical compensation, which would take place in a wire enclosed in glass tubes of many miles in length (the wire acting, as it were, like the interior coating to a battery) *might* amount to the *retention* of a charge, or, at least, might destroy the *suddenness* of a discharge, or, in other words, it might arrive at such a degree as to retain the charge with more or less force, even although the wire were brought into contact with the earth."

Referring to the difficulty that had been urged of keeping the wire charged with electricity, Ronalds says, on p. 21 :—" As to sufficiency (I have no dread of the charge of vanity in borrowing a boast from the great mechanic), give me *materiel* enough, and I will electrify the world. The Harlem machine would probably *in time* electrify, sufficiently for our purpose, a wire circumscribing the half of England : but we want to *save* time ; therefore let us have a small steam engine, to work a sufficient number of plates to charge batteries, or reservoirs, of such capacity as will charge the wire *as suddenly* as it may be discharged when the telegraph is at work ; and when it is *not* at work, let the machine be still kept in gentle motion, to supply the loss of electricity by default of insulation ; which default, perhaps, could not be avoided, because (be the atmosphere ever so dry, and the glass insulators ever so perfect) conductors are, I believe, robbed of their electricity by the same three processes by which Sir Humphrey Davy and Mr. Leslie say that bodies are robbed of their sensible heat, *viz.*, by radiation, by conduction, and by the motion of the particles of air."

While freely admitting that electro-magnetism was much better adapted to the purposes of telegraphy, Ronalds maintained to the last the practicability of his own plans. In a letter to Mr. Latimer Clark, dated Battle, 9th Dec., 1866, he thus writes :—" Had the necessary steps been taken in 1816 to provide a tensional electric telegraph for Government and general purposes, such an instrument might have been constructed and usefully employed, and might have been greatly improved* after the so-called Oersted discovery.   *   *   *

" Do we not all know that an electrophorus (of glass or resin) will remain charged, even when both opposed metallic surfaces are in conducting communication, and can *you* not believe that my difficulty

---

* In the way, for example, suggested in the following extract from Mr. (afterwards Sir) W. F Cooke's letter to Ronalds, of 11th December, 1866 :—" I have often thought what a fortunate thing it would have been if I had known of your labours in 1837. The letters of the alphabet, three letters in a row, might have been distinguished on your clocks by a movement of a needle to the left [for, say, the outer letter], the middle letter by a flourish of the needle right and left, and the inner letter by a movement of the needle to the right." See Ronalds' MSS.

of discharging my wire was greater than that of preserving a charge ? * * *

"I could always supply as much electricity as might be wanted for any length of my telegraphic wire, and it did *not fail*, as many very respectable witnesses well know. I did not (properly speaking) discharge the wire so much as to cause the electro-meter [balls] to collapse, the threads merely vibrated sufficiently to designate a sign when the wire was touched by a rapid stroke." *

* Extracted by kind permission of Mr. Latimer Clark. See also his letter of January 3, 1867, to Mr. (afterwards Sir) W. F. Cooke; and his comments on a letter in *The Reader* of January 5, 1867 ; both preserved in the *Ronalds' MSS. on the Electric Telegraph.*

CHAPTER V.

TELEGRAPHS BASED ON STATIC, OR FRICTIONAL, ELECTRICITY (*continued*).

1824.—*Egerton Smith's Telegraph.*

In *The Kaleidoscope, or Literary and Scientific Mirror* * we find a paragraph in which the editor, Mr Egerton Smith, suggests a telegraph, which resembles that of the Anonymous Frenchman, 1782, or that of Salvá, 1795, in the mode of indicating the signals ; and that of Le Sage, 1782, in the mode of insulating the conducting wires. This paragraph, which was kindly pointed out to us by Mr. Latimer Clark, runs as follows :—

" Amongst the numerous, pleasing, and ingenious, philosophical recreations exhibited by Mr. Charles, at the Theatre of Magic, is the following beautiful electrical experiment :—Mr. Charles presents to any of the company a musical tablet, containing [the names of] twenty-four popular tunes ; any lady or gentleman then privately selects one tune, which is marked with

* Liverpool, October 19, 1824, p. 133.

a silver bodkin. The book, or tablet, is closed without having been seen by Mr. Charles. It is then placed near the stage on a music-stand which communicates with another stand stationed in the orchestra above, at the very extremity of the room, at least thirty yards from the former. On this other stand is fixed a musical tablet corresponding with that below. The connection between the two music-books is made by means of twenty-four stationary wires, being the number of the tunes in each book. The musicians are directed to keep their eyes fixed upon the tablet in the orchestra, until, at Mr. Charles's command, an electrical shock passes from the lower to the upper music-book, illuminating the tune which had been secretly selected. The musicians, at this strange signal, forthwith proceed to play this illuminated air, to the great astonishment of the audience.

"There can be no doubt that most rapid telegraphs might be constructed on this principle, especially to convey intelligence in the night. We will imagine a case which is perfectly practicable, although the trouble and expense attending the project would outbalance all its advantages.

"If by means of pipes underground a communication were formed between Liverpool and London, and throughout the length of this tube twenty-four metal wires [were] stretched [and] supported at intervals by non-conducting substances, one of each of the wires communicating with a letter of the alphabet, formed of metal [foil], stationed at each extremity: If this were done, and it is quite practicable, we have little doubt that an express might be sent from Liverpool to London, and *vice versâ*, in a minute, or perhaps less. It would be necessary to have good chronometers, in order that the parties might be on the look-out at the precise time, or nearly so. The communication on this plan would be letter by letter; the person sending the message would merely have to touch the metallic letters in succession with the electric fluid, which would instantly pass along the wire to the other extremity where it would illuminate the corresponding letter. The communication would thus be made as fast as the operator could impart the shock."

1825.—" *Moderator's " Telegraph.*

The following proposal of what may be called a physiological telegraph * is extracted from the London *Mechanics' Magazine*, for June 11, 1825, p. 148 :—

" Electric Telegraphs.

" Sir,—There is, I think, in one of the numbers of the, *Spectator*, dated about a hundred years ago, a passage tending to ridicule some projector of that day, who had proposed to ' turn smoke into light and light into glory.' This early idea of gas-light, to which it seems plainly to refer, was received as an idle dream,

* See note p. 103, *supra.*

and is only preserved to us, like straws in amber, by the wit and satire of Addison, or Steele. We are to learn, therefore, not too hastily to reject even those hints which are not immediately clear to us.

"Under protection of this remark I venture to propose to you that a telegraphic communication may be held, at whatever distance, without a moment's loss of time in transmission, and equally applicable by day or night, by means of the electric shock.

"An experiment of this kind has been tried on a chain of conductors of three miles in extent, and the shock returned without any perceptible time spent in its going round; and may not the same principle be applicable for 100 or 10,000 miles? Let the conductors be laid down under the centre of the post-roads, imbedded in rosin, or any other the best non-conductor, in pipes of stoneware. The electric shock may be so disposed as to ignite gunpowder; but if this is not sufficient to rouse up a drowsy officer on the night-watch, let the first shock pass through his elbows, then he will be quite awake to attend to the second; and by a series of gradations in the strength and number of shocks, and the interval between each, every variety of signal may be made quite intelligible, without exposure to the public eye, as in the usual telegraph, and without any obstruction from darkness, fogs, &c. It was mentioned before that electricity will fire gunpowder—that is known; we may imagine, therefore, that on any worthy occasion, preparations

having been made for the expected event, as the birth
of a Royal heir, a monarch might at one moment, with
his own hand, discharge the guns of all the batteries
of the land in which he reigns, and receive the con-
gratulations of a whole people by the like return.

" I am, &c.,

"MODERATOR." *

* To the same class belonged the electro-physiological telegraph
proposed by Vorsselmann de Heer, and exhibited by him at a meeting
of the Physical Society of Deventer, on January 31, 1839. In this
system the correspondent received the signals at his fingers' ends, by
placing them upon the ten keys of a finger-board, which, by means of
separate line wires, communicated with corresponding keys at the
distant station. The signals were indicated by sending an induction
current through two of the wires, and the shocks were observed—(*a*)
in one finger of the right and one of the left hand, or (*b*) in two fingers
of the right hand, or (*c*) in two fingers of the left hand. The (*a*) shocks
represented the letters of the alphabet, the (*b*) shocks, the ten numerals,
and the (*c*) shocks, ten code, or conventional signs. See Vorsselmann
de Heer's *Théorie de la Télégraphie Électrique*, &c., Deventer, 1839,
and Moigno's *Traité de Télégraphie Électrique*, Paris, 1852, pp. 90 and
364. Reading by shocks, taken on the tongue, or fingers, has long been
practised as a make-shift by inspectors and line-men all over the world.
Varley mentioned it in the discussion which followed the reading of
the late Sir William Siemens' paper before the Society of Arts, April
23, 1858.

Quite recently (April 1878), yet another form of physiological telegraph has been submitted by M. Mongenot to the French Academy,
in which the transmitter and receiver are the same, and consist of two
ivory plates carrying the disconnected ends of the two line wires. The
sender places this contrivance between his lips, and sends the message
by talking, or by closing the circuit by his lips according to a code
of signals. The receiver, holding the receiving apparatus similarly,
interprets the message by the sensation he feels. This plan was, evidently, suggested by Sulzer's experiment of 1767. See his *Nouvelle
Théorie des Plaisirs*, p. 155; or p. 178, *infra*.

1825.—*R. H.'s Telegraph.*

In reference to the letter which we have just given from the *Mechanics' Magazine,* another correspondent "R. H.," wrote as follows in the number of the journal for June 25, 1825:—

"The present telegraphic communication is effected by means of six shifting boards, in a manner with which your readers are doubtless conversant. Now, if it be practicable to lay down one wire, it will be equally practicable to lay down six ; and the cost of the wire would be nearly all the difference in the expense. Let the wires terminate in a dark room. On one wall let there be the figures 1, 2, 3, 4, 5, 6, pre-pared in *tin-foil,* according to the method practised by electricians, in forming what are called *luminous modes and figures.* Bring the six wires in contact with the six figures separately. With this contrivance, all the signals may be performed, as at present with six shifting boards. A shake of the arm, as 'Moderator' suggests, may call the watch to his duty ; and he could name the signals as they appear, to his assistant, as is the present custom in the established telegraphs. His assistant must, of course, be separated from the dark room by a slight partition, that should be proof against light, but not against the full hearing of the human voice." *

* A further communication on the subject was promised but never made. In the hope of finding some clue to the writers of these letters, we have carefully looked through several succeeding volumes of the *Mechanics' Magazine,* but without success.

1825.—*Porter's Telegraph.*

We copy the following letter from *The Morning Herald*, of September 23, 1837 :—

"The Electric Telegraph.

"16, Somers Place, New Road, St. Pancras,
Sept. 16.

"Mr. Editor,—It now appears that the above application of the electric power is likely to be brought forward for the most useful purposes.

"At Munich, as stated by the *New Wurtsburg Gazette* of the 30th of June, the inhabitants were somewhat astonished by seeing, on the roofs of the loftiest houses, several men employed in passing iron wires, which extended from the towers of the church of Notre Dame to the observatory of Bogenhausen, and back to the church, intended to exemplify a project (so they call it) of Professor Steinheil, for the conveyance of intelligence by means of electric magnetism, whereby they conjecture that, in two seconds, communication may be conveyed from Lisbon to St. Petersburg. It further states there are other candidates beside the above-named Professor in the field, and a little time will decide whether Scotland, France, or Germany, is to carry off the honours for this disputed, and, if practicable, most valuable invention. If, Mr. Editor, you give place in your columns to the above and what follows, I think it will show that not a Scotchman, a Frenchman, or a German, but an Englishman, has the claim.

"On the 8th August, 1825, I requested the Lords

Commissioners of the British Admiralty to afford me an opportunity for bringing under their consideration a method of instantaneous communication with the out-ports, which neither foggy weather nor the darkness of night would obstruct. The next day I received the following answer :—

" ' Admiralty Office, August 9, 1825.

" ' Sir,—In reply to your letter of the 8th inst., I am commanded by my Lords Commissioners of the Admiralty to acquaint you that you may attend here any morning respecting your method of instantaneous communication with the out-ports either in foggy weather or at night.

" ' I am, Sir, your obedient servant,

(Signed)   " ' J. W. Croker.

" ' To Mr. S. Porter.'

"I attended the board, and proposed to their Lordships that electrical machines should be kept ready for use at the Admiralty Office and at each out-port, and that a conducting chain, or wire, of brass, or copper, secured in tubes of glass, be carried under the surface of the most frequented roads, so that any malicious attempt to interrupt the communication would soon be observed by travellers. What, Mr. Editor, under such circumstances, can prevent the electric impulse from proceeding with the utmost velocity to its destination ? The Lords of the Admiralty asked me whether I had prepared a code of signals ? I answered no, but referred them to writings on the subject by Dr. Franklin, which show that more by the power of electricity can be given than by a telegraph of wood.

"The Germans are wrong in using iron wire, a metal most subject to corrosion, particularly when exposed to the changes of the atmosphere. I ask them two questions. How will they carry this wire from Lisbon to St. Petersburg, where lofty buildings on the line are rarely to be found? and how will they secure a poor bird from destruction, which, perching upon the decayed wire, may break it, and, together with a despatch from Lisbon, go into oblivion? The invention has been tried successfully on the London and Birmingham railroad, the conductors being enclosed in hemp, or wood. However, this will not do; both are of a perishable nature; both will absorb damp, and every part of the apparatus employed in electricity should be kept dry. Let the experiment be made with glass to protect the conductor, and it will be found durable; and, as to its effect, I feel confident that if such a method of communication had been prepared from Ramsgate to the Admiralty Office, and continued from thence to Windsor Castle, our most excellent Queen would have been apprised of the arrival of her illustrious relations, the King and Queen of the Belgians, before the last salute gun was fired.

"I am, Sir, your respectful servant,

"SAMUEL PORTER."

1826–7.—*Dyar's Telegraph.*

About this time Harrison Gray Dyar, of New York, constructed a telegraph which was of an entirely different character to any of those hitherto described, as it depended for its action on the power of the spark to effect chemical decompositions. This property of electricity was first observed about the middle of the last century, and, had chemical science attained then to a sufficiently advanced state, it could not have failed to lead to the discovery of electro-chemistry.*

Besides being an *electro-chemical* telegraph (although not the first), Dyar's invention had the great merit of being a (in fact, the first) *recording* telegraph, and a fairly perfect one to boot, and, had he only used

---

* Beccaria, by the electric spark, decomposed the sulphuret of mercury, and recovered the metals, in some instances, from their oxides. Watson found that an electric discharge passing through fine wire rendered it incandescent, and that it was even fused and burned. Canton, repeating these experiments with brass wire, found that, after the fusion by electricity, drops of copper only were found, the zinc having apparently evaporated. Beccaria observed that when the electric spark was transmitted through water, bubbles of gas rose from the liquid, the nature, or origin, of which he was unable to determine. Had he suspected that water was not what it was then supposed to be, a simple elementary substance, the discovery of its composition could scarcely have eluded his sagacity. Franklin found that the frequent application of the electric spark had eaten away iron ; on which Priestley observed that it must be the effect of some acid, and suggested the inquiry whether electricity might not probably *redden vegetable blues ?* Priestley also observed that, in transmitting electricity through a copper chain, a black dust was left on the paper which supported the chain at the points where the links touched it ; and, on examining this dust, he found it to contain copper.—Lardner's *Electricity, Magnetism, and Meteorology*, vol. i. pp. 78–9.

voltaic, instead of static, electricity, the problem of electric telegraphy might have been solved in 1827. And, thus, with a start of several years, there can be little doubt that electro-chemical telegraphs would have made a better stand than they afterwards did in the struggle for existence ; although, perhaps, there can be as little doubt that, in obedience to the inexorable law of the survival of the fittest, they must have eventually yielded to the more practicable electro-magnetic forms of Cooke and Morse.

In connection with one of the many telegraph suits in which Morse was long engaged in America, Dyar gave the following account of his early project, in a letter to Dr. Bell, of Charlestown, dated Paris, March 8, 1848 .—

" Since reading your letter, and when searching for some papers in reference to my connection with this subject, I found a letter of introduction, dated the day before my departure from America, in February 1831, from an old and good friend, Charles Walker, to his brother-in-law, S. F. B. Morse, artist, at that time in Europe. At the sight of this letter, it occurred to me that this Mr. Morse might be the same person as Mr. Morse of the electric telegraph, which I found to be the case. The fact of the patentee of this telegraph, which is so identical with my own, being the brother-in-law of, and living with, my friend and legal counsel, Charles Walker, at the time of, and subsequent to, my experiments on the electric telegraph in 1826 and

1827, has changed my opinion as to my remaining
passive, and allowing another to enjoy the honour of
a discovery, which, by priority, is clearly due to me,
and which, presumptively, is only a continuation
[resumption] of my plans, without any material invention [improvement] on the part of another.

*   *   *   *   *   *

" I invented a plan of a telegraph, which should be
independent of day, or night, or weather, which should
extend from town to town, or city to city, without any
intermediary agency, by means of an insulated wire,
suspended on poles, and through which I intended to
send strokes of electricity, in such a manner as that
the diverse distances of time separating the divers
sparks should represent the different letters of the
alphabet, and stops between the words, &c. This
absolute, or this relative, difference of time between
the several sparks I intended to take off from an
electric machine by a little mechanical contrivance,
regulated by a pendulum ; while the sparks themselves were intended to be recorded upon a moving,
or revolving, sheet of moistened litmus paper, which,
by the formation of nitric acid by the spark in its
passage through the paper, would leave [show] a red
spot for each spark. These so-produced red spots,
with their relative interspaces, were, as I have said,
taken as an equivalent for the letters of the alphabet,
&c., or for other signs intended to be transmitted,
whereby a correspondence could be kept up through

one wire of any length, either in one direction, or back and forwards, simultaneously or successively. In addition to this use of electricity I considered that I had, if wanted, an auxiliary resource in the power of sending impulses along the same wire, properly suspended, somewhat like the action of a common bell-wire in a house.

" Now you will perceive that this plan is like that known as Morse's telegraph, with the exception that his is inferior to mine, inasmuch as he and others now make use of electro-magnetism, in place of the simple spark, which requires that they should, in order to get dots, or marks, upon paper, make use of mechanical motions, which require time; whereas my dots were produced by chemical action of the spark itself, and would be, for that reason, transmitted and recorded with any required velocity.

*   *   *   *   *   *

" In order to carry out my invention I associated myself with a Mr. Brown, of Providence, who gave me certain sums of money to become my partner. We employed a Mr. Connel, of New York, to aid in getting the capital wanted to carry the wires to Philadelphia. This we considered as accomplished; but, before beginning on the long wire, it was decided that we should try some miles of it on Long Island. Accordingly I obtained some fine card wire, intending to run it several times around the Old Union Race-course. We put up this wire at different lengths, in curves and straight lines, by suspending it [with glass insulators] from stake to stake, and tree to tree, until we concluded that our experiments justified our undertaking to carry it from New York to Philadelphia. At this moment our agent brought a suit, or summons, against me for 20,000 dollars, for agencies and services, which I found was done to extort a concession of a share of the whole project.

"I appeared before Judge Irving, who, on hearing my statement, dismissed the suit as groundless. A few days after this, our patent agent (for, being no longer able to keep our invention a secret, we had applied for a patent) came to Mr. Brown and myself and stated that Mr. Connel had obtained a writ against us, under a charge of conspiracy for carrying on secret communication from city to city, and advised us to leave New York until he could settle the affair for us. As you may suppose, this happening just after the notorious bank-conspiracy trials, we were frightened beyond measure, and the same night slipped off to Providence. There I remained some time, and did not return to New York for many months, and then with much fear of a suit. This is the circumstance which put an end [to our project], killing effectually all desire to engage further on such a dangerous enterprise. I think that, on my return to New York, I consulted Charles Walker, who thought that, however groundless such a charge might be, it might give me infinite trouble to stand a suit. From all this the very name of electric telegraph has given me pain whenever I have heard it mentioned, until I received your last letter, stimulating me to come out with my claims; and even now I cannot overcome the painful association of ideas which the name excites."

To this very interesting statement, Dr. Bell has added the following corroborative testimony :—" I was engaged with Harrison Gray Dyar for many months in 1828. We often conversed upon the subject of his having invented an electric telegraph, and I recollect seeing in his apartment a quantity of iron wire which he had procured for the construction of his telegraph. I recollect his saying he had suspended some of this wire at an elevation around the race-course at Long Island, to a length which satisfied him that there were no practical difficulties in carrying it from New York to Philadelphia, which, he stated, was his intention. I recollect suggesting doubts whether the wire would bear the necessary straightening up between the posts, and his reply, that the trial on Long Island had proved to him that there was no difficulty to be apprehended in this direction. My impression, derived from his conversation, was that the electric spark was to be sent from one end of the wire to the other, where it was to leave its mark upon some chemically prepared paper."*

* In these extracts we have followed *History, Theory, and Practice of the Electric Telegraph*, by George B. Prescott, Boston, 1860, pp. 427–30; *Historical Sketch of the Electric Telegraph*, &c., by Alexander Jones, New York, 1852, pp. 35–7; and *The Telegrapher*, New York, vol. i. pp. 48 and 163.

In 1831 Dyar came to Europe on business connected with some of his mechanical inventions, and resided principally at Paris until 1858, when he returned to the United States for good. His connection with telegraphy is somehow little known to the present generation, although, in 1826–7, he was widely known, at least in America, for his electrical researches. It is satisfactory to learn that his pursuits in other departments of science brought him an ample fortune, which was largely augmented by real estate investments in the city of New York. Dyar was born at Boston, Mass., in 1805, and died at Rhinebeck, N.Y., on the 31st January, 1875.

1828.—*Tribouillet de St. Amand's Telegraph.*

In this year* Victor Tribouillet de St. Amand proposed a single line telegraph between Paris and Brussels. The conducting wire was to be varnished with shellac, wound with silk, coated with resin, and enclosed in lengths of glass tubing carefully luted with resin ; the whole being substantially wrapped and water-proofed, and, finally, buried some feet deep in the earth.

Nothing is known for certain of the signalling arrangements, and it is even doubtful to what class the invention belongs ; as, while a strong voltaic battery was the source of electricity, the receiving

* According to *Journal des Travaux de l'Acad. de l'Industrie Française*, Mar. 1839, p. 43.

M

instrument was to be an electroscope, or electrometer. Vail,[*] Prescott,[†] and American writers, generally, evidently regard it as belonging to the electro-magnetic form ; while Zetzsche [‡] and Guerout [§] class it amongst those based on static electricity.

The author appears to have provided no particular form of alphabet, or code, leaving it to each person to devise his own out of the motions of which the electroscope was susceptible.

1830.—*Recy's Telegraph.*

In a *brochure* of 35 pages, entitled *Télétatodydaxie, ou Télégraphie Électrique*, Hubert Recy describes a crude system of syllabic telegraphy. Although his little book was not published in Paris until 1838, we gather from the text that his plans were laid as early as 1830. At p. 34 he writes :—" I had a thought of offering [my teletatodydax] to civilisation, a thought fixed and durable, because, notwithstanding some respectable opinions, I believe it useful to man in seasonable times ; but I did not wish to make it known in 1830 and during the stormy years that followed."

His telegraphic language is composed of (*a*) four initial vowels, (*b*) fifteen diphthongs, and (*c*) six

---

* *American Electro-Magnetic Telegraph*, 1845, p. 135.

† *History, Theory, and Practice of the Electric Telegraph*, 1860, p. 394.

‡ *Geschichte der Elektrischen Telegraphie*, 1877, sec. 6, para. 11.

§ *La Lumière Électrique*, March 3, 1883, p. 263.

monosyllables, all of which, with their various combinations, are figured in tables at pp. 5 and 6 of his
pamphlet.

The line wires were to be of iron, enveloped in
wax-cloth, then well tarred and enclosed in a leaden
tube to preserve them from moisture, and so prevent
the diminution of the force of the electric spark. They
might be placed at some feet underground along the
high roads like water pipes, and those parts destined
for submersion in water, across the sea, for example,
to England, should be prepared with the greatest
care, so as to entirely exclude the moisture.*

In certain cases, he says, the metals of the railway
could be used as lines of communication for the
conveyance of the electric spark, and nothing would
be easier than to put them into a condition to fulfil
this important function, each rail representing a line.

At the sending station were the electrical machines
for producing the sparks, and electrometers, one on
each wire, for indicating the passage and strength
of the same. At the receiving station the lines terminated in needles, or points, which dipped into
little cups containing some inflammable substance
like alcohol, or even hydrogen gas. A sufficient
number of these cups was always at hand, ready

---

* At p. 25 he repeats:—"To communicate with England, Algeria,
and other places, it would suffice to enclose the iron wire in an imper-
meable cloth, well tarred, and covered with sheet lead. In this way
the electricity would operate with as much freedom as in subterranean
lines, to which rivers would be no obstacles."

charged, to take the place of those exploded in the course of correspondence.

The line wires, which were bound together side by side, were marked, the one, say the right, with the *units*, or vowels, and the other, the left, with the *fives*, or diphthongs. As a general rule in teletatodydaxy that vowel termination which aids most in the expression and comprehension of a word, or phrase, is é (*é fermé*) ; for example—" Méhémet-Ali, vice-roi d'Égypte, fait travailler à la découverte des mines de Syrie " might be transmitted thus—Méhémété élé, vécé ré d'Égépété, fé térévéléré é lé dékévérété dé méné dé Séré, in pronouncing which rapidly, and without dwelling too much on the é, the ear would easily comprehend the sense.*

After showing, pp. 10 to 14, how this sentence should be transmitted, the author says that every conceivable communication could be made in the same way, each syllable being expressible, according to his tables, by vowels alone, or by vowels and diphthongs combined. One class of vowel, or *units*, was represented by one spark in the right line, a second class by two sparks, a third by three, and a fourth by four. Each diphthong (and monosyllable), or *five*,

---

* Recurring to this subject, the author says, on p. 16, Suppose the phrase to be pronounced by a stranger, you listen, and, as in a discourse one single sentence, when well understood, enables one to gather the sense of the whole, so in this case one single word well understood aids to a comprehension of the whole sentence, usage and practice will do the rest.

was similarly represented by one, two, three, or four, sparks in the left wire, either alone, or immediately followed by one, two, three, or four, sparks in the right wire, according as the syllable was in the first, second, third, or fourth, class of *fives*, and in the second, third, fourth, or fifth place of the class. Thus, Ba, which is in the first place of the first class of *fives*, would be represented by one spark in the left wire ; while Pa, which is in the third place of the second class, would be indicated by two sparks in the left wire for the class, followed by three sparks in the right for the place.

In case it would be impossible to establish two wires, on account of the expense, or from any other cause, the author shows how one wire would suffice, the signalling requiring, in this case, only a little more time, and a little more attention. The vowels would be transmitted as before, but the diphthongs and monosyllables would be expressed by two sparks in rapid succession *te te*, the interval between being much less than that between the vowels ; and, for greater clearness, the end of each word would be notified by the signal A, which would be neither the end of the last word, nor the commencement of the following.

If desired, each letter, or character, of the teletatodydaxical tables could represent some conventional phrase, or the sparks could stand for figures which would belong to words and phrases in a dictionary, or code.

The author concludes a rhapsody on the uses which the great Napoleon would have made of teletatody-daxy had it been then discovered * in words, which, in these days of their realisation, deserve to be remembered :—" If, in the time of Napoleon, gas-lighting had been as general as it is now, and some one had told him : 'by means of the teletatodydax you can, in less than a second, light all the lamps of the capital at the same time and as one lamp' ; or, as everything sub-lunary has disadvantages as well as advantages, 'you cannot guard yourself against the malefactors who would sow infernal machines under your feet, would fire your ships, arsenals, powder - magazines, and monuments'—enemies all the more difficult to discover, since they can perpetrate their crimes from afar by means of the wire ; would Napoleon have shut his eyes and ears to these facts ?   No, such advantages and disadvantages combined would certainly have fixed his attention, and, not being able to annihilate a power of which he would wish only himself to know the force, he would so control it as to draw for himself all the advantages, and, at the same time, prevent others from putting it to wrongful ends " (p. 34).

1837.—*Du Jardin's Telegraph.*

Du Jardin, of Lille, whose fast-speed type-writer was used, for a short time, in 1866, on the late

* *We* think he would have made short work of it.

Electric and International Telegraph Company's lines, was occupied with the telegraph as far back as 1837.

His first ideas on the subject were, in that year, communicated to the Paris Academy of Sciences; but, except the bare title of the paper in the *Compte Rendu*, for July 10, 1837, nothing appears to have been published. We learn, however, from Professor Magrini* that he proposed to erect a single wire between the Tuileries and the Arc de l'Etoile, and to employ an electric machine and a sensitive electroscope for the signalling apparatus.

If none of the contrivances that we have described in the foregoing pages ever passed the stage of experiment, it is because they, one and all, laboured under two heavy disadvantages—the one, that they were in advance of the age, and the other, the intractable nature of the force employed, rendering its transmission to any distance impossible in the open air, and exceedingly difficult through buried wires.

Of course, if no other form of electricity had been discovered, some of these inventions—notably those of Alexandre, Ronalds, and Dyar—could be improved, so that we should have at this day electric telegraphs, not so simple, nor with so many resources as those at present in use, but yet instruments that would fulfil the grand object of communicating at a distance with lightning speed. Many practical difficulties

* *Telegrafo Elettro-Magnetico*, Venezia, 1838, p. 23.

would, however, remain, which, even with our present extended knowledge, we could not entirely obviate, and which would, therefore, have hindered their complete success.

If, then, none of their authors, though through no fault of his own, deserves the title of inventor of a really practicable and commercially successful telegraph, we must, at least, give one and all the credit of having fully appreciated its importance, and of having dedicated their energies to the accomplishment of the task they set themselves in the face of many difficulties and disappointments.*

* Since 1837 the following telegraphs have been proposed in which static electricity was to be employed :—By the Rev. H. Highton, in 1844 (Patent No. 10,257 of 10th July); by Isham Baggs, in 1856 (Patent No. 1775 of 25th July)—a most interesting document, which will repay perusal in these days of multiplex and fast-speed apparatus , by C. F. Varley, in 1860 (Patent No. 206 of 27th January); and by Wenckebach, a Dutch electrician, in 1873 (*Journal Télégraphique de Berne*, for March 25, 1873).

CHAPTER VI.

DYNAMIC ELECTRICITY—HISTORY IN RELATION TO TELEGRAPHY.

> " The hooked torpedo, with instinctive force,
> Calls all his magic from its secret source ;
> Quick through the slender line and polished wand
> It darts, and tingles in the offending hand."
>
> Pennant's Oppian.

THE discoveries of the Italian philosophers, Galvani and Volta, at the close of the last century, marked a new era in the history of telegraphy, by furnishing a form of electricity as tractable and copious, as that derived from friction was volatile and small.

Before entering into this subject it may be well to say a few words on the early history of what has been called *Animal Electricity*—a force which is identical with, and whose early manifestations in certain fishes led up to, Galvanism.

Although this power is now known to exist in many fishes, and even in some of the lower animals,*

---

* With the aid of a microscope sparks have been seen to issue from the *annelides* and *infusoria*, and the luminosity of the glow-worm and other shining insects is thought to be due to the same cause. Margrave describes an insect, a native of Brazil, which, on being touched, gives a very perceptible shock ; and specimens of the *Sepia* and *Polypi* have also been observed to do the same.—Kirby and Spence's *Introduction to Entomology*, London, 1856, 7th ed., p. 56.

the torpedo was the only instance known to the ancients.*

Aristotle says :—" This fish hides itself in the sand, or mud, and catches those that swim over it by benumbing them, of which some persons have been eye-witnesses. The same fish has also the power of benumbing men."† Pliny writes :—" From a considerable distance even, and if only touched with the end of a spear, or staff, this fish has the property of benumbing the most vigorous arm, and of rivetting the feet of the runner, however swift he may be in the race." ‡ Plutarch declares that the torpedo affects fishermen through the drag net, and that, were water to be poured on a living one, the person pouring it would be affected, the sensation being communicated through the water to the hand. Claudian and Galen have much to the same effect, and Oppian is even more explicit, for he describes the organs by which the fish exerts its extraordinary power. " It is," he says, "attributable to two organs of a radiated texture, which are situated one on each side of the fish." §

The ancients knew something also of what we

---

* The name of the torpedo in the Arabian language is *ra'ad*, which means *lightning*.

† *History of Animals*, ix. 37.

‡ *Natural History*, xxxii. 2.

§ Lib. ii. v. 62. In the *Phil. Trans.*, for 1773, p. 481, the celebrated Hunter published the anatomical structure of the torpedo, showing the position of the electric organs. In a fish eighteen inches long it was found that the number of columns composing each organ amounted to 470.

would now call *Medical Electricity*. Thus, we read that Dioscorides, the physician of Anthony and Cleopatra, used to cure inveterate headaches by applying a live torpedo to the head;* and that (as related by Scribonius Largus†) Anthero, a freedman of Tiberius, was cured of the gout by the same means. The patient in such cases had to stand on the sea-shore with a live torpedo under foot, until not only the feet but the legs as far as the knees became numb.

The *Gymnotus electricus* was first made known in Europe in 1671 by Richer, one of a party sent out by the French Academy for astronomical observations at Cayenne. The accounts which he brought home of its *shocking* powers were, however, received with much scepticism, and it was not until towards the middle of the last century that the observations of Condamine, Fermin, Bancroft, and others had fully established their credibility.

The gymnotus, which inhabits the warmer regions of Africa and South America, delivers far stronger shocks than the torpedo, the strokes of the larger ones being, according to Bancroft, instantly fatal. When one of average dimensions is touched with one hand a smart shock is felt in the hand and forearm; and when both are applied it affects the whole frame, striking, apparently, to the very heart. Thus, Humboldt mentions that, treading upon an ordinary

* Lib. ii., Art. Torpedo.
† *De Compositione Medicamentorum Medicæ*, cap. i. and xli.

specimen, he experienced a more dreadful shock than he ever received from a Leyden jar, and that he felt severe pain in his knees, and other parts of his body, which continued for several hours. According to Bryant, a discharge sometimes occasions such strong cramps of the muscles which grasp the fish that they cannot let it go.*

On the river Old Calabar, the electrical properties of the gymnotus are used by the natives to cure their sick children ; a small specimen of the fish is put into a dish containing water, and the child is made to play with it, or the child is put into a tub of water and the fish put in beside it.

Of the remaining electrical fishes, the Silurus, introduced by Adanson, in 1751, is an inhabitant of the Nile and Senegal ; the Trichiurus inhabits the Indian Seas ; and the Tetraodon is found near the Canary Islands and along the American coast.

Although Redi, 1678, Kempfer, 1702, and others had made many and accurate observations on the torpedo, the electrical nature of the phenomena exhibited by this and the other fishes that we have named was not known, nor even suspected, up to the middle of the last century. The idea first occurred to Professor Musschenbröck of Leyden in reference to the torpedo, and nearly at the same time (1751), Adanson formed a similar notion regarding the

---

* *Transactions of the American Society,* vol. ii. See also *Mechanics' Magazine,* for August 6, 1825.

Silurus; but it was not till the years 1772–4 that the fact was clearly established by the experiments of Walsh, S'Gravesande, Hunter, Ingenhousz, and others.*

Walsh, in transmitting to Benjamin Franklin, then in London, the results of his researches for communication to the Royal Society, says:—"It is with peculiar interest that I make to you my first communication, that the effect of the torpedo appears to be absolutely electrical," and he concludes, after going fully over the details, "He, who predicted and showed that electricity wings the formidable bolt of the atmosphere, will hear with attention that in the deep it speeds a humbler bolt, silent and invisible ; he, who analysed the electric phial, will hear with pleasure that its laws prevail in *animated* phials ; he, who by reason became an electrician, will hear with reverence of an instinctive electrician gifted at its birth with a wonderful apparatus, and with skill to use it.†

It is singular that, while the examination of the torpedo was going on in Europe, similar investigations were taking place in America with respect to the gymnotus. These were made in Philadelphia and Charleston by Drs. Williamson and Garden, and the same conclusions, grounded on the same *data*, were arrived at. These are thus summed up by their authors :—"As the fluid discharged by the eel affects the same parts that are affected by the electric fluid;

* *Phil. Trans.*, 1773 and 1775.   † Ibid., 1773, pp. 461–72.

as it excites sensations perfectly similar; as it kills and stuns animals in the same manner; as it is conveyed by the same bodies which convey the electric fluid, and refuses to be conveyed by others that refuse to convey the electric fluid, it must itself be the electric fluid, and the shock given by the eel must be the electric shock." *

Though these early experiments thus led to the strong presumption that this peculiar animal power was precisely of the same nature with common electricity, yet they were very far from affording that absolute demonstration which alone satisfies the requirements of modern science ; and, hence, naturalists have ever been on the watch to seize every opportunity which could supply additional evidence. The science of electricity, likewise, has since those days been prosecuted with the greatest success, and the phenomena of the respective subjects have mutually thrown light upon each other.   As regards electricity, there are now a number of palpable effects which are considered as demonstrative of its presence and operation, chief amongst which are the shock, the electric spark, heat, magnetic virtue, and chemical agency. These positive proofs of the operation of electricity were soon desiderated in connection with the animals we have named, and one after another, by the ingenuity of experimenters, have been at last obtained.†

Now to resume our subject.   In the hundred years

* *Phil. Trans.*, 1775, pp. 94 and 102.
† Faraday's *Exper. Researches*, series iii. and xv.

preceding the discoveries of Galvani and Volta, we find record of many observations of a character closely resembling the fundamental ones, which, in their hands, led to the grand discovery of dynamic electricity. Thus, in 1671, Richter noticed that the gymnotus was able to produce by its shocks a sort of sympathetic quivering in dead fishes lying around it. In 1678, Swammerdam, in some experiments before his friend and patron, the Grand Duke of Tuscany, produced convulsions in the muscle of a frog, by holding it against a brass ring from which it hung by a silver wire—an experiment which, as we shall presently see, exactly resembles that by which Galvani became so famous more than a hundred years later.

This celebrated experiment is thus described in Swammerdam's *Biblia Naturæ*, vol. ii. p. 839 :—" Let there be a cylindrical glass tube, in the interior of which is placed a muscle, whence proceeds a nerve that has been enveloped in its course with a small silver wire, so as to give us the power of raising it without pressing it too much, or wounding it. This wire is made to pass through a ring bored in the extremity of a small copper support and soldered to a sort of piston, or partition ; but the little silver wire is so arranged that, on passing between the glass and the piston, the nerve may be drawn by the hand and so touch the copper. The muscle is immediately seen to contract."

Du Verney, in 1700, made a similar observation, and Caldani, 1757, described what he called "the revival of frogs by electric discharges." Du Verney's experiment is thus described :— "M. Du Verney showed a frog just dead, which, in taking the nerves of the belly that go to the thighs and legs, and irritating them a little with a scalpel, trembled and suffered a sort of convulsion. Afterwards he cut the nerves, and, holding them a little stretched with his hand, he made them tremble again by the same motion of the scalpel." *

The experiments described in the following extract from the *Philosophical Transactions*, for 1732, are of an exactly similar kind. We copy from a paper headed " Experiments to prove the existence of a fluid in the nerves," by Alexander Stuart, M.D. :—

"The existence of a fluid in the nerves (commonly called the *animal spirits*) has been doubted of by many ; and, notwithstanding experiments made by ligatures upon the nerves, &c., continues to be controverted by some. This induced me to make the following experiments, which I hope may help to set that doctrine, which is of so much consequence in the animal economy and practice of physic, in a clearer light than I think it has hitherto appeared in.

" *Experiment I.*—I suspended a frog by the forelegs in a frame leaving the inferior parts loose ; then,

* Martyn and Chambers' *The Phil. Hist. and Mems. of the Royal Acad. of Sciences at Paris*, London, 1742, vol. i. p. 187. Du Verney was a celebrated anatomist, for whom the use of vaccine as early as 1705 is claimed with a great show of reason. See Fournier's *Le Vieux-Neuf*, Paris, 1859, vol. ii. p. 385.

the head being cut off with a pair of scissors, I made a slight push perpendicularly downwards, upon the uppermost extremity of the *medulla spinalis*, in the upper vertebra, with the button-end of the probe, filed flat and smooth for that purpose; by which all the inferior parts were instantaneously brought into the fullest and strongest contraction; and this I repeated several times, on the same frog, with equal success, intermitting a few seconds of time between the pushes, which, if repeated too quick, made the contractions much slighter.

"*Experiment II.*—With the same flat button-end of the probe, I pushed slightly towards the brain in the head, upon that end of the *medulla oblongata* appearing in the occipital hole of the skull; upon which the eyes were convulsed. This also I repeated several times on the same head with the same effect.

"These two experiments show that the brain and nerves contribute to muscular motion, and that to a very high degree."*

In their results these experiments were precisely the same as those with which the name of Galvani is associated. Nor was the mode of operating very different, even in the use of only one kind of metal. In Galvani's experiments, excitation was produced by contact, or communication, of nerves and muscles. In Stuart's the convulsions were produced by exciting the spinal marrow.

* Vol. xxxvii. p. 327.

N

Sulzer, in his *Nouvelle Théorie des Plaisirs*, published at Berlin in 1767, described the peculiar taste occasioned by pieces of silver and lead in contact with each other and with the tongue. He, however, had no suspicion of the electrical nature of this effect, but thought it "not improbable that, by the combination of the two metals, a solution of either of them may have taken place, in consequence of which the dissolved particles penetrate into the tongue ; or we may conjecture that the combination of these metals occasions a trembling motion in their respective particles, which, exciting the nerves of the tongue, causes that peculiar sensation." *

The next person to whom chance afforded an opportunity of making the discovery of galvanism, but who let it pass with as little profit as Sulzer and his predecessors had done, was Domenico Cotugno, professor of anatomy at Naples. His observations are contained in the following letter,† dated Naples, October 2, 1784, and addressed to the Chevalier Vivenzio :—

"Sir,—The observation which I mentioned some days ago, when we were discoursing together of the electrical animals upon which I said that I believed the mouse to be one of the number, is the following:—

"Towards the latter end of March I was sitting with

---

* Note to text on p. 155. The date of this experiment is variously stated as 1752, and 1760. See note under Sulzer in *Ronalds' Catalogue*.

† Extracted from Cavallo's *Complete Treatise on Electricity*, 4th ed., London, 1795, vol. iii. p. 6.

a table before me ; and observing something to move about my foot, which drew my attention, looking towards the floor, I saw a small domestic mouse, which, as its coat indicated, must have been very young. As the little animal could not move very quick, I easily laid hold of it by the skin of the back, and turned it upside down ; then with a small knife that laid by me, I intended to dissect it. When I first made the incision into the epigastric region, the mouse was situated between the thumb and first finger of my left hand, and its tail was got between the two last fingers. I had hardly cut through part of the skin of that region, when the mouse vibrated its tail between the fingers, and was so violently agitated against the third finger, that, to my great astonishment, I felt a shock through my left arm as far as the neck, attended with an internal tremor, a painful sensation in the muscles of the arm, and such giddiness of the head, that, being affrighted, I dropped the mouse.

The stupor of the arm lasted upwards of a quarter of an hour, nor could I afterwards think of the accident without emotion. I had no idea that such an animal was electrical ; but in this I had the positive proof of experience." *

* Volta, in telling this story in after years, used to say that Cotugno was a pupil of Galvani, and that it was his drawing his master's attention to the phenomenon that put Galvani on the trail of his great discovery—Robertson's *Mémoires Récréatifs Scientifiques et Anecdotiques*, Paris, 1840, vol. i. p. 233.

Galvani's great discovery is popularly supposed to have resulted from an accidental observation on frogs made in 1790 ; but as early, at least, as 1780, he was engaged, as we learn from Gherardi, his biographer, in experiments on the muscular contractions of these animals under the influence of electricity.*

One day in that year (November 6), while preparing "in the usual manner" a frog in the vicinity of an electrical machine with which some friends were amusing themselves, he observed the animal's body to be suddenly convulsed.   Astonished at this pheno-menon, and supposing that it might be owing to his having wounded the nerve, Galvani pricked it with the point of his knife to assure himself whether or not this was the case, but no convulsion ensued.   He again touched the nerve with his knife, and, directing a spark to be taken at the same time from the machine, had the pleasure of seeing the contortions renewed.   Upon a third trial the animal's body remained motionless, but observing that he held the knife by its ivory handle, he grasped the metal, and immediately the convulsions took place each time that a spark appeared.†

---

* From two papers in the *Bolognese Transactions*, one, *On the Muscular Movement of Frogs*, dated April 22, 1773 ; and the other, *On the Action of Opium on the Nerves of Frogs*, dated January 20, 1774. it is evident that Galvani's acquaintance with frogs was long anterior even to the year 1780.   We follow mainly, in our account of Galvani's researches, Professor Forbes' Dissertation (Sixth), chap. vii., in the *Encyclopædia Britannica*, 8th ed.

† These experiments are similar to, and are explained by, the

After a number of similar experiments with the machine, Galvani resolved to try the effect of atmospheric electricity, and with this object erected a lightning conductor on the roof of his house to which he attached metallic rods leading into his laboratory. These he connected with the nerves of frogs and other animals, and fastened to their legs wires which reached to the ground. As was anticipated, the animals were greatly convulsed whenever lightning appeared, and even when any storm-cloud passed over the apparatus. These experiments were continued in 1781 and 1782, and were afterwards embodied in a paper (not published) *On the Nervous Force and its Relation to Electricity*. In 1786, Galvani resumed the inquiry with the aid of his nephew, Camillo, and it was in the course of these studies that certain facts were observed which led immediately to the discovery of galvanism.

One day (the 20th) in September 1786, Camillo Galvani had prepared some frogs for experiment, and

phenomenon of the *lateral shock*, or *return stroke*, first observed by Wilson, of Dublin, in 1746, but first explained by Lord Mahon in 1779. In Galvani's experiment the frog, while it merely lay on the table, so being insulated, had its electricities separated by induction at every turn of the machine, and on the passage of every spark their reunion took place, but with so small effect that it escaped notice. When, however, the animal was placed in connection with the ground, through the knife and body of the professor, one of the separated electricities freely escaped, thus rendering a greater inductive charge possible, and raising the *return stroke* to a sufficient strength to convulse the dead limbs. It is but fair to add that Galvani himself suggested this explanation some years later.

had hung them, by an iron hook, from the top of an iron rail of the balcony outside Galvani's laboratory to be ready for use. Soon he noticed that when, by accident, a frog was pressed, or blown, against the rail, the legs contracted as they were wont to do when excited by the electricity of the machine, or of the atmosphere. Surprised at this effect where there was apparently no exciting cause, he called his uncle to witness it, but Galvani dismissed it on the easy assumption that the movements were connected with some unseen changes in the electrical state of the atmosphere. He soon, however, found that this was not the case, and, after varying in many ways the circumstances in which the frogs were placed, at length discovered that the convulsions were the result of the simultaneous contact of the iron with the nerves and muscles, and that the effect was increased by using a combination of different metals—such as iron and silver, or iron and copper.

Galvani, who was an anatomist first and an electrician afterwards, accounted for these effects by supposing that in the animal economy there exists a natural source of electricity; that at the junction of the nerves and muscles this electricity is decomposed, the positive fluid going to the nerve, and the negative to the muscle; that these are, therefore, analogous to the internal and external coatings of a charged Leyden jar; that the metallic connection made between the nerve and the muscle serves as a conductor for these opposite electricities; and that, on establishing the connection, the same discharge takes place as in the Leyden experiment. Galvani's re-searches were not made public until the year 1791, when they were embodied in his celebrated paper printed in the Bolognese *Transactions* of that year.

It will be evident from this account, which is based upon the researches of Gherardi, Galvani's biographer, supported by original documents, how absurd is the popular story, first invented by Alibert in his *Éloges historiques de Galvani* (Paris, 1802), and constantly repeated since, that "this immortal discovery arose, in the most immediate and direct way, from a slight cold with which Madame Galvani was attacked in 1790, and for which her physician prescribed the use of frog-broth." As if frog-broth were usually prepared in the laboratory!

Luigi Galvani was born at Bologna on the 9th of September, 1737, and died there December 4, 1798. From his youth he was remarkable for the ardour with which he prosecuted his studies in anatomy and physiology, and at the early age of twenty-five he was appointed professor of these sciences in the University of his native place.

The closing years of his life form a sad contrast to those of his great contemporary, Volta, who died, in 1827, covered with honours.* At the moment when

* Alessandro Volta was born at Como, February 19, 1745. Soon after his discovery of the pile, in 1801, he was invited to Paris, and

Galvani was immortalising his name, he was obliged to undergo the most cruel blows of destiny; for he lost his dearly loved wife, Lucia Galeazzi, and, a short time afterwards, had the misfortune to be ordered by the Cisalpine Republic to take an oath which was entirely opposed to his political and religious convictions. He did not hesitate a moment, but promptly refused, and permitted himself to be stripped of his position and titles. Reduced nearly to poverty, he retired to his brother's house, and soon fell into a state of lethargy from which he could be aroused, neither by medicine, nor by the decree of the government, which, out of respect for his celebrity, reinstated him in his position as professor of anatomy in the University of Bologna. The great physicist died without having again occupied the chair which he had rendered so illustrious.

---

was honoured with the presence of the First Consul while repeating his experiments before the Institute Bonaparte conferred upon him the orders of the Legion of Honour, and of the Iron Crown, and he was afterwards nominated a count, and senator of the kingdom of Italy. At the formation of the Italian Institute, a meeting was held, at which Bonaparte presided, for the purpose of nominating the principal members When they were considering whether or not they should draw up a list of the members in an alphabetical order, Bonaparte wrote at the head of a sheet of paper the name of Volta, and, delivering it to the secretary, said, "Do as you please at present, provided that name is the first." At his death, on March 5, 1827, his fellow-citizens struck a medal, and erected a monument to his memory; and a niche in the façade of the public schools of Como, which had been left empty for him between the busts of Pliny and Giovio, natives of the town, was filled by his bust. See note on p. 84.

In 1879, the city of Bologna erected a statue in his honour, from the chisel of Adalbert Cincetti, the eminent Roman sculptor. It represents him at the moment when the muscles of the frog are revealing to him the effects of electricity on the animal organism.

Galvani's theory fascinated for a time the physiologists. The phenomena of animal life had hitherto been ascribed to an hypothetical agent, called the *nervous fluid,* which now the new discovery had consigned to oblivion. Electricity was, henceforth, the great vital force, by which the decrees of the understanding, and the dictates of the will, were conveyed from the organs of the brain to the obedient members of the body.

CHAPTER VII.

DYNAMIC ELECTRICITY—HISTORY IN RELATION TO TELEGRAPHY (*continued*).

ALEXANDER Volta, then Professor of Physics at Pavia, and already well-known for his researches in electricity, had naturally his attention directed, in common with other philosophers, to the Bolognese experiments, and, although at first he warmly espoused Galvani's opinions, his superior sagacity soon enabled him to detect their want of basis.   He first ascertained that the contractions of the frog ensued on simply touching, with the extremities of the metallic arc, two points of the same nervous filament ; he next found that it was possible with the metallic arc to produce, either the sensation of light, or that of taste, by applying it to the nerves of the eye and tongue respectively.*   In short, he ended by showing that the exciting cause was nothing more nor less than ordinary electricity, produced by the *contact* of the two metals, the convulsion of the frog being simply due

---

* These observations were independently made in England about the same time (1793) ; the one by Fowler, and the other by Professor Robison, of Edinburgh.

to the passage of the electricity so developed along the nerves and muscles.*

The first analogy which Volta produced in support of his *theory of contact* was derived from the well-known experiment of Sulzer, which we have just described in these pages. From that it is seen that if two pieces of dissimilar metal, such as lead and silver, be placed one above, and the other below, the tongue, no particular effect will be perceived so long as they are not in contact with each other; but if their outer edges be brought together, a peculiar taste will be felt. If the metals be applied in one order, the taste will be acidulous. If the order be inverted, it will be alkaline. Now, if the tongue be applied to the conductor of a common electrical machine, an acidulous, or alkaline, taste will be perceived, according as the conductor is electrified positively, or negatively. Volta contended, therefore, that the identity of the cause should be inferred from the identity of the effects; that, as positive electricity produced an acid savour, and negative electricity an alkaline, on the conductor of

* Volta first broached his contact theory in two letters, in French, to Cavallo, dated September 13 and October 25, 1792. See *Phil. Trans.*, 1793, pp. 10–44; also chaps. x. and xiii. vol. i. of Robertson's *Mémoires Récréatifs*, Paris, 1840, for much interesting information on the early history of galvanism. Robertson was a celebrated aeronaut, a friend of Volta, and one of the founders of the Galvanic Society of Paris early in the present century. Dubois Reymond, in his *Untersuchungen uber thierische Elektricität*, Berlin, 1848, gives a good account of this celebrated dispute, from a physiologist's point of view. See pp. 3–19 of the English translation, edited by Dr. Bence Jones, London, 1852.

the machine, so the same effects on the organs of taste produced by the metals ought to be ascribed to the same cause.

In August 1796, Volta arranged an experiment which, by eliminating the physiological element, afforded, as he thought, a direct and unequivocal proof of the correctness of his hypothesis. He took two discs, one of copper and the other of zinc, and, by means of their insulating handles, carefully brought them into contact and suddenly separated them without friction; then, on presenting them to a delicate condensing electroscope, the usual indications of electricity were obtained, the zinc being found to be feebly charged with positive, and the copper with negative, electricity.

Of the numerous philosophers in every part of Europe who took part in the discussions, and varied and repeated the experiments connected with these questions, one to whom attention is more especially due was Fabroni, who, in the year 1792, communicated his researches to the Florentine Academy. In this paper is found the first suggestion of the chemical origin of galvanic electricity.

Fabroni supposed that, in the experiments of Galvani and Volta, a chemical change was made by the contact of one of the metals with the liquid matter always found on the parts of the animal body; and that the immediate cause of the convulsions was not, as supposed by Galvani, due to animal electricity, nor,

as assumed by Volta, to a current of electricity emanating from the surface of contact of the two metals,
but to the decomposition of the fluid upon the animal
substance, and the transition of oxygen from a state
of combination with it to combination with the metal.
The electricity produced in the experiments Fabroni
ascribed entirely to these chemical changes, it being
then known that chemical processes were generally
attended with sensible signs of electricity.*

Galvani's theory was soon rejected on all hands,
but a bitter war raged for a long time between the
partisans of the contact theory of Volta and of the
chemical theory of Fabroni. Now, however, it is generally conceded that both contact between dissimilar
substances and chemical action are necessary to produce the effect. "Perhaps," says Fleeming Jenkin, in
his excellent little text-book, "it is strictly accurate
to say that difference of potential is produced by contact, and that the current which is maintained by it is
produced by chemical action." †

In pursuing his inquiries on galvanic electricity,
Volta felt the necessity of collecting it in much greater
quantities than could be obtained from the combination of a single pair of copper and zinc plates as
above described, and he, therefore, sought for some

---

* *Journal de Physique*, xlix. p. 348.

† This theory was, we believe, first propounded in England by Sir
Humphry Davy in 1806. See Lardner's *Electricity, Magnetism, and
Meteorology*, vol. i. p. 164.

means by which he could combine, and, as it were, superpose two or more currents, and thus multiply the effect. With this object he conceived the idea of placing alternately, one over the other, an equal number of discs of copper and zinc; but he found that the effect produced was no greater than that of a single pair, and for reasons which he was not slow to perceive. With such an arrangement as that described, there would proceed, according to Volta's own theory, from the first surface a negative downward and a positive upward current, from the second a positive downward and a negative upward, and from the third a negative downward and a positive upward, and so on. The downward currents would be alternately positive and negative, and the same would be the case with the upward currents; and since the surfaces of contact were equal, all the *intermediate* currents would neutralise each other, and the effect of the pile would simply be that of the two extreme discs.

Volta, therefore, saw the necessity of adopting some expedient by which all the currents in the same direction should be of the same kind, and, if this could be accomplished, it was easy to see that the resulting currents, negative at the bottom and positive at the top, would be as many times more intense as there were surfaces of contact. To effect this it was necessary to destroy the galvanic action at all those surfaces from which descending positive and ascending negative currents would proceed; but while this was being done it was also essential that the progress of the descending negative and ascending positive currents should still be uninterrupted. The interposition of any substance, which would have no sensible galvanic action on either of the metals between each disc of copper and the disc of zinc immediately below it, would attain one of these ends, but in order to allow the free progress of the remaining currents in each direction, such substance should be a sufficiently good conductor of electricity. Volta selected, as the fittest means of fulfilling these conditions, discs of wet cloth, which would give rise to no galvanic action, while their moisture would endow them with sufficient conducting power.

Although the principle of the pile was thus evolved as early as the middle of 1796, Volta does not appear to have actually constructed the instrument with which his name has become so imperishably associated until some three years later, and it was not until the 20th March, 1800, that he published a description of it in a letter to Sir Joseph Banks, President of the Royal Society. In this letter he thus writes :—" I took some dozens of discs of copper, brass, or better, of silver, one inch in diameter (coins for instance), and an equal number of plates of tin, or, what is much better, of zinc. I prepared also a sufficient number of discs of card-board, leather, or some other spongy matter capable of imbibing and retaining water, or, what is much better, brine. I placed on a table a disc of silver, and on it a disc of zinc, then one of the moist discs ; then another disc of silver, followed by one of zinc, and one of card-board. I continued to form of these several stages a column as high as could sustain itself without falling."

FIG. 3.

Fig. 3 represents one of the earliest forms of the pile. It consists of discs of silver, S, zinc, Z, and some bibulous substance soaked in brine, W. The rods R, R, R, are of glass, or baked wood, and with the piece O, which slides freely up and down, serve to keep the discs in position.

The invention of the pile had been scarcely more than hinted at, when the compound nature of water was discovered by its means. The first four pages only of the letter of Volta to Sir Joseph Banks were despatched on the 20th of March, 1800 ; and as these were not produced in public till the receipt of the remainder, the letter was not read at the Royal Society, or published, until the 26th of June following. This first portion, in which was described, generally, the formation of the pile, was, however, shown in the latter end of April to some scientific men, and, among others, to Sir Anthony (then Mr.) Carlisle, who was engaged at the time in certain physiological inquiries. Mr. W. Nicholson, the conductor of the scientific journal known as *Nicholson's Journal*, and Carlisle constructed a pile of seventeen silver half-crown pieces alternated with equal discs of copper and cloth soaked in a weak solution of common salt, with which, on the 30th of April, they commenced their experiments.

It happened that a drop of water was used to make good the contact of the conducting wire with a plate to which the electricity was to be transmitted ; Carlisle observed a disengagement of gas in this water, and Nicholson recognised the odour of hydrogen proceeding from it. In order to observe this effect with more advantage, a small glass tube, open at both ends, was stopped at one end by a cork, and being then filled with water was similarly stopped at the other end. Through both corks pieces of brass wire were inserted, the points of which were adjusted at a distance of an inch and three-quarters asunder in the water. When these wires were put in communication with the opposite ends of the pile, bubbles of gas were evolved from the point of the negative wire, and the end of the positive wire became tarnished. The gas evolved appeared on examination to be hydrogen, and the tarnish was found to proceed from the oxydation of the positive wire. Thus was inaugurated on the

o

2nd of May, 1800, a new line of research, the limits of which, even now, it is impossible to foresee.[*]

Nicholson observed that the same process of the decomposition of water was carried on in the body of the pile, as between the two ends of the wire, the side of the zinc next the fluid being covered with oxide in two, or three, days, and the apparatus then ceasing to act. He also observed that the common salt, which had been dissolved in the water, was precipitated, for, gradually, an efflorescence of soda appeared round the margin of the discs.

Nicholson made the further important observation that, by employing discs of considerably more extensive surface, no greater effect was produced in the decomposition of water, or in the strength of the shock ; whence he concluded that the repetition of the series is of more consequence to these actions than the enlargement of the surface.[†]

Cruickshank, of Woolwich, confirmed the observations of Nicholson respecting the appearance of sparks and the decomposition of water. This last pheno-

[*] Nicholson's *Journal of Natural Philosophy*, for 1800, vol. iv. p. 179. For the composition and decomposition of water by the electric spark, see Lord Brougham's paper in the *Mechanics' Magazine*, for November 9, 1839.

[†] In some experiments on the combustion of metals, Fourcroy, Thénard, and Vauquelin made the same observation in connection with the trough form of the pile   They found that the energy of the shock and the power of decomposition were not increased by the size of the plates, but by the number of the repetitions ; while the same extent of surface, arranged in the form of a few large plates, readily consumed metallic leaves.—*Annales de Chimie*, xxxix. 103.

menon he varied in different ways. By employing silver terminals, or *electrodes*, and passing the current through water tinged with litmus, he found that the wire connected with the *zinc* end of the pile imparted a red tinge to the fluid contiguous to it ; and that, by using water tinged with Brazil wood, the wire connected with the *silver* end of the pile produced a deeper shade of colour in the surrounding fluid ; whence it appeared that an acid was formed in the former, and an alkali in the latter, case.

He next tried the effects of the wires on solutions of acetate of lead, sulphate of copper, and nitrate of silver, with the result that, in each case, the metallic base was deposited at the negative, and the acid at the positive pole. In the latter case he observes, "the metal shot into fine needles, like crystals articulated, or jointed, to each other, as in the *Arbor Dianæ*." Muriate of ammonia and nitrate of magnesia were next decomposed, the acid, as before, going to the positive, and the alkali to the negative, pole.

These experiments were made as early as June 1800 ; and in the September following, Cruickshank published a second memoir, in which he directed his attention to the nature of the gases emitted at the electrodes ; to the effects of different kinds of electrodes ; and to the influence of the fluid medium. The following are the most important of his conclusions :—

(1). From the wire connected with the silver, or copper, end of the pile, whatever be its composition, if it terminate in water, the gas evolved is, chiefly, hydrogen; if it terminate in a metallic solution, the metal is reduced and is deposited upon the wire.

(2). When the wire connected with the zinc end is composed of a non-oxydable metal, nearly pure oxygen is set free; when of an oxydable metal, it is partly oxydated, and partly dissolved, and only a small quantity of oxygen is set free.

(3). When fluids contain no oxygen they appear to be incapable of transmitting the voltaic current; while, on the contrary, it would seem that it may be transmitted by every one which contains this element.[*]

These views were confirmed by some experiments that were performed, about the same time, by Colonel Haldane. He found that the pile ceased to act when immersed in water, or when placed in the vacuum of an air-pump; that it acted more powerfully in oxygen gas than when confined in an equal bulk of atmospheric air,[†] and that azote had the same effect as a vacuum. These circumstances led him to conceive that its action depended essentially upon the consumption of oxygen, which it derives from the atmosphere. Haldane also made some experiments on the

[*] Nicholson's *Journal*, iv. pp. 187 and 254.

[†] Biot and Cuvier observed the converse of this. When the pile was enclosed in a limited quantity of air, they found that, after some time, the air was sensibly deoxydated.—*Annales de Chimie*, xxxix. 242.

series of metals which are the best adapted for producing the voltaic effects, and the relative power
which they possess in this respect.*

While these investigations were proceeding, Ritter,
afterwards so distinguished for his experimental researches, but then young and unknown, made various
experiments at Jena on the effects of the pile ; and,
apparently without knowing what had been done in
England, discovered its property of decomposing water
and saline compounds, and of collecting oxygen and
the acids at the positive, and hydrogen and the bases
at the negative, pole. He also showed that the
decomposing power in the case of water could be
transmitted through sulphuric acid, the oxygen being
evolved from a portion of water on one side of the
acid, while the hydrogen was produced from another
separate portion on the other side of it.†

When the chemical powers of the pile became
known in England, Sir Humphry (then Mr.) Davy
was commencing those labours in chemical science
which subsequently surrounded his name with so
much lustre, and have left traces of his genius in the
history of scientific discovery, which must remain as
long as the knowledge of the laws of nature is valued
by mankind. The circumstance attending the decompositions effected between the poles of the pile
which caused the greatest surprise was the production
of one element of the compound at one pole, and the

* Nicholson's *Journal*, iv. pp. 247, 313.　　† Ibid., iv. 511.

other element at the other pole, without any discoverable transfer of either of the disengaged elements between the wires.   If the decomposition was conceived to take place at the positive wire, the constituent appearing at the negative wire must be presumed to travel through the fluid in the separated state from the positive to the negative point ; and if it was conceived to take place at the negative wire, a similar transfer must be imagined in the opposite direction.

Thus, if water be decomposed, and the decomposition be conceived to take place at the positive wire where the oxygen is visibly evolved, the hydrogen from which that oxygen is separated must be supposed to travel through the water to the negative wire, and only to become visible when it meets the point of that wire ; and if, on the other hand, the decomposition be imagined to take place at the negative wire where the hydrogen is visibly evolved, the oxygen must be supposed to pass invisibly through the water to the point of the positive wire, and there become visible.   But what appeared still more unaccountable was, that in the experiment of Ritter it would seem that one, or other, of the elements of the water must have passed through the intervening sulphuric acid.   So impossible did such an invisible transfer appear to Ritter, that at that time he regarded his experiment as proving that one portion of the water acted on was wholly converted into oxygen, and the other portion into hydrogen.

This point was the first to attract the attention of Davy,* and it occurred to him to try if decomposition could be produced in quantities of water contained in separate vessels united by a conducting substance, placing the positive wire in one vessel and the negative in the other. For this purpose, the positive and negative wires were immersed in two separate glasses of pure water. So long as the glasses remained unconnected, no effect was produced; but when Davy put a finger of the right hand in one glass, and of the left hand in the other, decomposition was immediately manifested. The same experiment was afterwards repeated, making the communication between the two glasses by a chain of three persons. If any substance passed between the wires in these cases, it must have been transmitted through the bodies of the persons forming the line of communication between the glasses.

The use of the living animal body as a line of communication being inconvenient where experiments of long continuance were desired, Davy substituted fresh muscular fibre, the conducting power of which, though inferior to that of the living animal, was sufficient. When the two glasses were connected by this substance, decomposition went on as before, but more slowly.

To ascertain whether metallic communication

* In our account of Davy's researches we follow mainly Lardner's *Electricity, Magnetism, and Meteorology*, vol. i. pp. 119-29, which we have carefully collated with Davy's own memoir in the *Phil. Trans.*, for 1801.

between the liquid decomposed and the pile was essential, he now placed lines of muscular fibre between the ends of the pile and the glasses of water respectively, and at the same time connected the two glasses with each other by means of a metallic wire. He was surprised to find oxygen evolved in the *negative*, and hydrogen in the *positive*, glass, contrary to what had occurred when the pile was connected with the glasses by wires. In none of these cases did he observe the disengagement of gas, either from the muscular fibre, or from the living hand immersed in the water.

In October 1800, after many experiments on the chemical effects of the pile, Davy commenced an investigation of the relation which its power had to the chemical action of the liquid conductor on the more oxydable of its metallic elements. He showed that at common temperatures zinc connected with silver suffers no oxydation in water which is well purged of air and free from acids; and that, with such water as a liquid conductor, the pile is incapable of evolving any quantity of electricity which can be rendered sensible, either by the shock, or by the decomposition of water; but that, if the water hold in combination oxygen or acid, then oxydation of the zinc takes place, and electricity is sensibly evolved. In fine, he concluded that the power of the pile appeared to be, in great measure, proportional to the power of the liquid between the plates to oxydate the zinc.

To ascertain whether a liquid solution, capable of conducting the electric current between the positive and negative wires of a voltaic pile, but not capable of producing any chemical action on its metallic elements, would, when used between its plates, evolve electricity, Davy constructed a pile in which the liquid was a solution of sulphuret of strontia.[*] Twenty-five pairs of silver and zinc plates, alternated with cloths moistened in this solution, produced no sensible action, though the moment the sides of the pile were moistened with nitrous acid, the ends gave shocks as powerful as those of a similar pile constructed in the usual manner.

The inventor of the pile maintained that, among the metals, those which held the extreme places in the scale of electromotive power were silver and zinc ; and that, consequently, these metals, paired in a pile, would be more powerful, *cæteris paribus*, than any other. But as he had shown that pure charcoal was a good conductor of the electric current, and that the electromotive virtue seemed also to depend on the different conducting powers of the metallic elements, it was consistent with analogy that charcoal, combined with another substance of different conducting power, would produce voltaic action. Dr. Wells [†] was

[*] When the current from an active pile was transmitted through this liquid, the shock was as sensible as if the communication had been made through water.

[†] *Phil. Trans.*, 1795.

the first to demonstrate this by showing that a combination of charcoal and zinc produced sensible convulsions in the frog ; and, subsequently, Davy constructed a pile, consisting of a series of eight glasses, with small pieces of well-burned charcoal and zinc, using a solution of red sulphate of iron as the liquid conductor.   This series gave sensible shocks, and rapidly decomposed water.   Compared with an equal and similar series of silver and zinc, its effects were much stronger.

In considering the various arrangements and combinations in which voltaic action had been manifested, Davy observed, as a common character, that one of the two metallic elements was oxydated, and the other not. Did the production of the electric current, then, depend merely on the presence of two metallic surfaces, one undergoing oxydation, separated by a conductor of electricity ? and, if so, might not a voltaic arrangement be made by one metal only, if its opposite surfaces were placed in contact with two different liquids, one of which would oxydate it, and the other transmit electricity without producing oxydation ? To reduce these questions to the test of experiment with a single metallic plate would have been easy ; but in constructing a series, or pile, the two liquids, the oxydating and the non-oxydating, must be in contact, and subject to intermixture. To overcome this difficulty, different expedients were resorted to, with more or less success; but the most convenient and effectual method of attaining the desired end was that suggested to Davy by Count Rumford.

Let an oblong trough be formed as a substitute for the pile; and let grooves be made in it such as to allow of the insertion of a number of plates, by which the trough may be divided into a series of water-tight cells. Let plates of the metal of which the apparatus is to be constructed be made to fit these grooves; and let as many plates of glass, or other non-conducting material, of the same form and magnitude, be provided. Let the metallic plates be inserted in alternate grooves of the trough, and the glass plates in the intermediate grooves, so as to divide the trough into a succession of cells, each having on one side metal, and on the other, glass. Let the alternate cells be filled with the oxydating liquid, and the intermediate cells with the liquid which conducts without oxydating. Let slips of moistened cloth be hung over the edge of each of the glass plates, so that their ends shall dip into the liquids in the adjacent cells. This cloth, or rather the liquid it imbibes, will conduct the electric current from cell to cell, without permitting the intermixture of the liquids.

In the first arrangements made on this principle, the most oxydable metals, such as zinc, tin, and some others, were tried. The oxydating liquid was dilute nitric acid, and the other plain water. In a combination consisting of twenty such pairs sensible but weak effects were produced on the organs of sense, and water was decomposed slowly by wires from the extremities. The wire from the end towards which the oxydating surfaces were directed evolved hydrogen, and the other oxygen.

To determine whether the evolution of the electric current was dependent on the production of *oxydation* only, or would attend *other chemical effects* producible by the action of substances in solution upon metal, the oxydating liquid was now replaced by solutions of the sulphurets, and metallic plates were selected on which these solutions would exert a chemical action. Silver, copper, and lead were tried in this way, solution of sulphuret of potash and pure water being the liquids employed. A series of eight metallic plates produced sensible effects. Copper was the most active of the metals tried, and lead the least so. In these cases, the terminal wires effected, in the usual manner, the decomposition of water, the wire from which hydrogen was evolved being that which was connected with the end of the series to which the surface of the metal not chemically acted on was presented.

It will be observed that in this case the direction of the electric current relatively to the surfaces of the metallic plates was the reverse of the former, for when oxydation was produced, the oxydating sides of the plates looked towards the *negative* end of the series. Comparing these two effects, Davy was led by analogy to suspect that if one set of cells was filled with an oxydating solution, while the other set was filled with a solution of sulphuret, or any other which would produce a like chemical action, the combined effects of the currents proceeding from the two distinct chemical processes would be obtained. This was accordingly tried, and the results were as foreseen. A series, consisting of three plates of copper, or silver, arranged in this way, produced sensible effects; and twelve or thirteen decomposed water rapidly.

. As it appeared from former experiments that charcoal possessed, as a voltaic element, the same properties as the metals, the next step in this course of experiments was, naturally, to try whether a voltaic arrangement could not be constructed without any metallic element, by substituting charcoal for the metallic plates in the series above described. This was accomplished by means of an arrangement in the form of the *couronne des tasses*. Pieces of charcoal, made from very dense wood, were formed into arcs; and the liquids were arranged in alternate glasses The charcoal arcs were placed so as to have one end immersed in each liquid, the intermediate glasses being connected by slips of bibulous paper. When the liquids were dilute acid and water, a series consisting of twenty pieces of charcoal gave sensible shocks, and decomposed water. This arrangement also acted, and with increased effect, when the liquids were sulphuric acid and solution of sulphuret of potash.

Soon after the discovery of the pile, Dr. Wollaston

turned his attention to the subject, and in the *Philo-sophical Transactions*, for 1801 (p. 427), records his observations, which are marked by his accustomed sagacity and penetration. He observed, like Davy, that the energy of the pile seemed to be in proportion to the tendency which one of the metals had to be acted upon by the interposed fluid. If, he says, a plate of zinc and a plate of silver be immersed in dilute sulphuric acid, and kept asunder, the silver is not affected, but the zinc begins to decompose the water, and to evolve hydrogen. If the plates be now placed in contact, the silver discharges hydrogen, and the zinc continues, as before, to be dissolved. From these and other analogous facts, he concludes, that whenever a metal is dissolved by an acid, electricity is disengaged.*

Davy's experiments have shown that in all voltaic combinations only one of the metallic elements is attacked by the liquid; but this condition, although desirable, is not essential to the production of electricity. It is sufficient if the chemical action of the liquid upon one of the metals be greater than upon the other ; for then the two metals may be considered to give rise to two separate currents, of which the one proceeding from the metal most attacked is the stronger, the current perceived being the difference

* He extends this principle to the action of the electrical machine, which, he conceives, has its power increased by applying to the cushion an amalgam, into the composition of which enters an easily oxydable metal. Clearly the zinc used by Wollaston was very impure, for, as we now know, *pure* zinc is unaffected so long as it is not joined to the silver.

between the two. If the currents were absolutely equal, a condition, however, practically impossible to realise, we must assume that no electrical effects would be produced.

As a voltaic current, then, is produced whenever two metals are placed in metallic contact in a liquid which acts more powerfully upon one than upon the other, it is easy to see that there must be a great choice in the mode of producing such currents. The following is a list of the principal metals, arranged in what is called an *electromotive series*, and from which any two being taken and placed in contact in, say, dilute sulphuric acid, that metal highest in the list is the one that will suffer oxydation. This is called the *electropositive* metal in contradistinction to its fellow, which is denominated *electronegative* :—

Zinc	Nickel	Silver
Cadmium	Bismuth	Gold
Tin	Antimony	Platinum
Lead	Copper	Graphite
Iron	Mercury	

It will be seen that the electrical deportment of any metal depends upon the metal with which it is associated. Iron, for instance, is electronegative towards zinc, but electropositive towards copper; while copper, in its turn, is electronegative towards iron and zinc, but electropositive towards silver, platinum, or graphite.

The force resulting from the contact of two metals, in a liquid is called the *electromotive force*, and, as may be supposed, is greater in proportion to the distance of the two metals from one another in the above list. Thus, the electromotive force between zinc and platinum is greater than that between zinc and iron, or between zinc and copper. Indeed the law, as established by Poggendorff, is, that the electromotive force between any two metals is equal to the sum of the electromotive forces between all the intervening metals.[*]

The electromotive force is influenced by the condition of the metal; rolled zinc, for example, is negative towards cast zinc. It also depends on the *degree of concentration* of the liquid; thus, in dilute nitric acid zinc is positive towards tin, and mercury positive towards lead; while in concentrated nitric acid, the reverse is the case, mercury and zinc being respectively electronegative towards lead and tin.

The *nature* of the liquid is also of influence, as is seen from the change in the relative position of the metals in the following lists:—

CAUSTIC POTASS.	SULPHIDE OF POTASSIUM.
Zinc	Zinc
Tin	Copper
Cadmium	Cadmium
Antimony	Tin
Lead	Silver
Bismuth	Antimony
Iron	Lead
Copper	Bismuth
Nickel	Nickel
Silver	Iron

In short, anything that affects the energy of the chemical action on the positive plate, or the resultant

[*] Ganot's *Elementary Treatise on Physics*, London, 1881, p. 707.

actions on the two plates, affects to a like degree the electromotive force of the combination.

Of the theories proposed to explain chemical decomposition by the voltaic apparatus, that of Grotthus was the earliest and most plausible.   To simplify the view of this theory, we shall take as an example of its application the decomposition of water.   Each molecule of water being composed of a molecule of oxygen and a molecule of hydrogen, their natural electricities are in equilibrium when not exposed to any disturbing force, each possessing equal quantities of the positive and negative fluids.   The electricity of the positive wire acting on the natural electricities of the contiguous molecule of water, attracts the negative and repels the positive fluid.   It is further assumed in this theory, that oxygen has a natural attraction for negative, and hydrogen for positive electricity ; therefore the positive wire in attracting the negative fluid of the contiguous molecule of water, and repelling its positive fluid, attracts its constituent molecule of oxygen, and repels its molecule of hydrogen.   The particle of water, therefore, places itself with its oxygen next the positive wire, and its hydrogen on the opposite side.

The positive electricity of the first particle of water thus accumulated on its hydrogen molecule, produces the same action on the succeeding molecule of water as the wire did upon the first molecule ; and a similar arrangement of the second molecule of water is the

P

result.   This second molecule acts in like manner on the third, and so on.   All the particles of water between the positive and negative wires thus assume a polar arrangement, and have their natural electricities decomposed; the negative poles and oxygen molecules looking towards the positive wire, and the positive poles and hydrogen molecules looking towards the negative wire.   The electro-positive wire now separates the oxygen molecule of the contiguous particle of water from its hydrogen molecule, neutralises its negative electricity, and either dismisses it (the oxygen) in the gaseous form, or combines with it, according to the degree of its affinity for the metal of the wire.   The hydrogen molecule thus liberated effects in like manner the decomposition of the second particle of water, combining with its oxygen, and thus again forming water, and liberating hydrogen.   The latter acts in the same manner on the next particle of water, and so on.

Thus, a series of decompositions and recompositions is supposed to be carried on through the fluid, until the process reaches the particle of water contiguous to the negative wire.   The molecule of hydrogen there disengaged gives up its positive electricity (by which an equal portion of negative electricity proceeding from the wire is neutralised), and then escapes in the gaseous form.   It is equally compatible with this theory to suppose the series of decompositions and recompositions to commence at the negative and terminate at the positive wire, or to commence simultaneously at both, and terminate at an intermediate point by the union of the last molecule of oxygen disengaged in the one series with the last molecule of hydrogen disengaged in the other.

Grotthus illustrated this ingenious hypothesis by comparing the supposed phenomena with the mechanical effects produced when a number of elastic balls —ivory balls for example—are suspended, so that their centres shall be in the same straight line, and their surfaces mutually touch, and either of the extreme balls of the series is raised and let fall against the adjacent one. The effect is propagated through the series, and although action and reaction are suffered by each ball, and each is instrumental in transmitting the effect, no visible change takes place in any ball except the last, which alone recoils in consequence of the impact.*

The investigations of which the pile became the instrument now began to assume an importance which rendered it necessary to give it greater power, either by increasing its height, or by enlarging the surfaces of the plates. In either case, inconveniences were encountered which imposed a practical limit to the increase of its power. When the number, or magnitude, of the discs was considerable, the incumbent pressure discharged the liquid from the intermediate

---

* Lardner's *Electricity, Magnetism, and Meteorology,* vol. i. pp. 135-37 ; also *Phil. Mag.,* for 1806, vol. xxv. p. 334.

card-board, so that the energy of the pile gradually diminished from the first, and ultimately ceased altogether.* It had then to be taken to pieces, the metals cleaned, and the card-board re-soaked in the solution, every time it was required.

Volta himself, seeing these inconveniences, proposed an arrangement which he called *la couronne des tasses*, and which consisted of a circle, or row, of small cups containing a solution of salt. In each cup were placed a small plate, or bar, of zinc, and another of silver, not touching, the zinc of the first cup being connected metallically to the silver of the second, the zinc of this to the silver of the third, and so on. The silver rod of the first cup and the zinc rod of the last formed the poles of the apparatus. Twenty such combinations were able to decompose water, and thirty gave a distinct shock to the moistened hands.

A still more convenient form was that known as *Cruickshank's Battery*, which was introduced in 1800, within a few weeks of the announcement of Volta's discovery. It consisted of a number of pairs of zinc and copper plates soldered together and cemented into grooves in an oblong trough of wood, the spaces between each being filled with the exciting liquid. On this plan was constructed the great battery of

* To prevent this, Ritter turned up the edges of the lower discs so as to retain the liquid. His piles were thus able to preserve their powers for a fortnight. See the *Phil. Mag.*, vol. xxiii. p. 51.

600 pairs given to the Polytechnic School of Paris by Napoleon I., and with which Gay Lussac and Thénard made their experiments in 1808.*

Dr. Babington's arrangement was a great improvement upon this form. The plates of copper and zinc, four inches square, were united in pairs by soldering at one point only. The trough in which they were immersed was made of porcelain, and divided into ten, or twelve, equal compartments. The plates were attached to a strip of wood, well baked and varnished, and so arranged that each pair should enclose a partition between them when let down into the trough. By this means the whole set could be lifted at once into, or out of, the little cells, and thus, while the exciting fluid remained in the trough, the action of the battery could be suspended at pleasure, and the plates, when corroded, could be easily replaced.

A battery of 2000 pairs, with a surface of 128,000 square inches, was made on this plan for the Royal

* An amusing story anent this battery is told in Dr. Paris's *Life of Sir Humphry Davy* :—When Napoleon heard of the decomposition of the alkalies by an English philosopher, he angrily questioned the *savans* of the Paris Institute why the discovery had not been made in France. The excuse alleged was the want of a battery of sufficient power. He immediately commanded one to be made, and when completed he went to see it. With his usual impetuosity, the Emperor seized the terminal wires, and, before he could be checked by the attendant, applied them to his tongue. His Imperial Majesty was rendered nearly senseless by the shock, and as soon as he recovered from its effects he walked out of the laboratory with as much composure as he could assume, not requiring further experiments to test the power of the battery, nor did he ever afterwards allude to the subject.—Vol. ii. p. 24.

Institution of London, with which Davy and Faraday performed those long-continued and brilliant series of experiments, for which the Royal Institution will ever be celebrated.*   Children in 1809, Wollaston in 1815, Berzelius in 1818, and others, proposed various modifications of the trough battery, all having for their object increase of power, with more cleanliness and less waste.

But all these arrangements of two metals in one fluid, constituting what are now called *single-fluid batteries*, had one great defect : their power, variable from the first, rapidly declined, and, sooner or later, ceased altogether.

This defect was due to two causes, first, the decrease in the chemical action owing to the neutralisation of the sulphuric acid by its combination with the zinc ;

* "When the whole series was put into action, platina, quartz, sapphire, magnesia, and lime, were all rapidly fused ; while diamond, charcoal, and plumbago, in small portions disappeared, and seemed to be completely evaporated.   A singularly beautiful effect was produced by placing pieces of charcoal at the two ends of the wires ; when they were brought within the thirtieth, or fortieth, part of an inch of each other, a bright spark was produced, above half the volume of the charcoal, which was rather more than an inch long, and the points became ignited to whiteness.   By withdrawing them from each other, a constant discharge took place through the heated air, in a space equal to at least four inches, producing a most brilliant arc of light."— Bostock's *History of Galvanism*, p. 95.   This refers to Davy's experiments of 1809, when he for the first time produced a *continuous* arc of light, but long before this the electric light, *as a spark*, had been obtained from charcoal points, as by Davy himself (Nicholson's *Journal*, Oct. 1800, p. 150), by Moyes (*Phil. Mag.*, vol. IX. p. 219), and by Robertson, to whom we referred on p. 187 (*Journal de Paris*, Mar. 12, 1802).   See *The Electrician*, vol. XI. p. 162.

second, *polarisation* of the negative, or copper, plate, giving rise to *secondary currents*. These are currents which are produced in the battery in a contrary direction to the principal one, and which destroy it, either totally, or partially. In a couple of zinc, copper, and sulphuric acid diluted with water, for example, when the circuit is closed sulphate of zinc is formed which dissolves in the liquid, and at the same time hydrogen gas, in what is called its *nascent* state, is gradually deposited on the copper.

Now, it has been found that hydrogen deposited in this manner on metallic surfaces acts far more energetically than ordinary hydrogen. In virtue, therefore, of this increased action it gradually reduces some of the sulphate of zinc dispersed in the liquid, causing a layer of metallic zinc to be formed on the surface of the copper plate ; hence, instead of having two different metals, copper and zinc, we have two metals becoming gradually less different, and, consequently, in the connecting wire there are two currents in opposite directions tending to become equal, and so to neutralise one another. When the copper plate is entirely covered with zinc the action of the couple ceases, for the condition essential to this action no longer exists, *viz.*, two dissimilar metals.

Becquerel, of Brussels, was the first to recognise these causes of the inconstancy of the voltaic battery, and, in 1829, he devised the first double-fluid arrangement, which, while it prevented polarisation, maintained the supply of acid around the positive plate, thus removing both sources of weakness at once.

It was composed of two small glass vessels, one of which contained concentrated nitric acid, and the other a solution of caustic potash, also concentrated. The two vessels communicated with each other by means of a bent glass tube, filled with fine clay, moistened with a solution of sea-salt. In the vessel which contained the alkali was immersed a plate of gold, and in the other a plate of platinum. By connecting the two through a galvanometer a constant and tolerably energetic current was perceived, resulting from the reaction of the acid on the sea-salt and potash.[*]

In 1830, Wach constructed double-fluid batteries on this plan, using animal bladders as the separating medium.

Professor Daniell, of King's College, London, is commonly supposed, in England at all events, to have been the first to make a double-fluid battery; but although he was not the first, as we see, his independent researches and his beautiful memoir on the subject, in the *Philosophical Transactions*, for 1836, were the means of bringing the improvement into general notice, and hence, no doubt, the popular belief.[†]

[*] *Comptes Rendus*, for 1837, No. 2.
[†] For much valuable information on this question, as between Daniell and Becquerel, see Sturgeon's *Annals of Electricity*, vol. ix. pp. 534–49. Zetzsche, at p. 45 of his *Geschichte der Elektrischen Telegraphie*, says that Dobereiner, Privy Councillor at Jena, had, as far back as 1821, constructed such a battery as the constant form of Daniell. See also Dove's *Über Elektricität*, Berlin, 1848, p. 24.

Double-fluid batteries are so simply constructed nowadays that our readers will probably be surprised to learn the pains that Daniell was at in contriving his.　Fig. 4 represents a section of one of his original cells ; *a, b, c, d,* is a cylinder of copper, six inches high and three and a half inches wide ; it is open at the top *a, b,* but closed at the bottom, except for a collar *e, f,* one inch and a half wide, intended for the re-ception of a cork into which a glass syphon tube, *g, h, i, j, k,* is fitted water-tight. On the top *a, b,* a copper collar, corresponding with the one at the bottom, rests by two horizontal arms.　Pre-viously to fixing the cork, a membranous tube, formed of part of the gullet of an ox, is drawn through the lower collar *e, f,* and fast-ened with twine to the upper, *l, m, n, o,* and, when

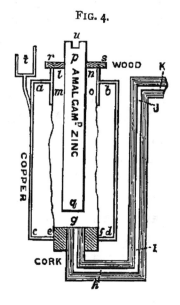

FIG. 4.

tightly fixed by the cork plug below, forms an in-ternal cavity to the cell, communicating with the syphon tube, so that, when filled with any liquid to the level, *m, o,* any addition causes an overflow at the aperture *k.*

The objects which Daniell proposed to himself in

constructing this cell were (1) the removal of the oxide of zinc as fast as formed, and (2) the absorption of the hydrogen evolved upon the copper, without the precipitation thereon of any substance that could impair its action.

The first he effected by suspending the zinc rod, which he took care to be amalgamated,* in the interior membranous cell, into which fresh acidulated water was allowed slowly to drop (from a funnel suspended over it, whose aperture was adjusted to this purpose), whilst the heavier solution of the oxide was withdrawn from the bottom at an equal rate by the syphon tube. The second object was attained by charging the exterior space surrounding the membrane, with a saturated solution of sulphate of copper. When the circuit was completed the current passed freely, no hydrogen was observed to collect on the negative plate, but, instead, a beautiful pink coating of pure copper was deposited upon it, and thus perpetually renewed its surface.

Although this cell was much more steady and permanent in its action than one of the ordinary single-

---

* The first mention of amalgamated zinc in voltaic arrangements occurs in Sir H. Davy's Bakerian lecture, for 1826, in which he simply remarked that "zinc in amalgamation with mercury is positive with respect to pure zinc" (*Phil. Trans.*, 1826, part iii.), without any allusion to the probable beneficial employment of it in the general construction of batteries. Kemp of Edinburgh was the first to employ amalgamated zinc and copper in the regular construction of batteries. See his paper in Jameson's *New Edinburgh Philosophical Journal*, for Dec. 1828.

fluid construction, it still showed a gradual but very slow decline, which Daniell traced to the weakening of the saline solution by the precipitation of its copper and consequent decline of its conducting power. To obviate this defect some crystals of sulphate of copper were suspended in muslin bags, which just dipped below the surface of the solution in the copper cylinder, and which, gradually dissolving as the precipitation proceeded, kept the solution in a state of saturation. This expedient fairly answered the purpose, and its author was delighted to find that "the current was now perfectly steady for six hours together."

Such, in brief, is the evolution of the far-famed Daniell cell. Its further development we need not pursue in these pages, as all the later forms, as well as many other kinds of double-fluid, or so-called *constant*, batteries, are to be found in all the text-books on electricity.

CHAPTER VIII.

TELEGRAPHS (CHEMICAL) BASED ON DYNAMIC ELECTRICITY.

" Awhile forbear,
*   *   *   *   *
Nor scorn man's efforts at a natural growth,
Which in some distant age may hope to find
Maturity, if not perfection."
Household Words, June 14, 1851.

1800–4.—*Salvá's Telegraph.*

IT is generally supposed that Sömmerring was the first to employ the electricity of the pile for telegraphic purposes ; but M. Saavedra * has shown that this honour belongs to his distinguished countryman, Don Francisco Salvá, whose name has already occurred in our pages (pp. 101–8).

At a meeting of the Academy of Sciences of Barcelona, held on the 14th May, 1800, Salvá read a paper, entitled *Galvanism and its application to Telegraphy*, in which, after an elaborate dissertation on the phenomena and theories of the new science, he proceeds to consider its application to telegraphy.

---

* *Tratado de Telegrafía*, Barcelona, 1880, vol. i. pp. 331–35. In our account of Salvá we translate, literally, from this excellent treatise.

He relates the experiments made for this purpose at his residence with line wires, some 310 metres long, stretched across the terrace and garden, and fastened at the ends to varnished glass insulators, and through which he distinctly perceived the convulsions of the frog, notwithstanding the distance.   The fact that the contractions sometimes took place without closing the circuit led to the discovery, that, on account of the wire being uncovered, its extension permitted its taking electricity from the atmosphere, so as to act upon the frog.   The conducting wires, adds Salvá, can act by means of galvanism alone, as he demonstrated by insulating his small line.*   He expressed the conviction that he could obtain a telegraphic communication at a much greater distance.

The memoir does not enter into details as to this new telegraphic proposal, limiting itself to saying that it could be made by a process analogous to that described at the meeting of December 16, 1795,† with the advantages of greater durability and cheapness as compared with the old plan.

Salvá employed, as his motive power, the electricity produced by a great number of frogs.

This illustrious Spanish physician was therefore the first person who attempted to apply electricity dynamically for the purpose of telegraphing.   "It is," says Saavedra, "not without reason I must con-

---

* These passages are obscure in the original.
† See p. 101, *ante*.

fess, notwithstanding my cosmopolitan opinions on scientific questions, that the Catalans hold Salvá to be the inventor of electric telegraphy. With documents as authentic as those which I have seen with my own eyes, in the very handwriting of this distinguished professor (which documents are at this present moment to be found in the library of the Academy of Sciences of Barcelona), it is impossible for any author to henceforth deny, even if others did precede Salvá in telegraphic experiments with static electricity, that no one preceded him in the application of the docile electro-dynamic fluid to distant communications."

On the 22nd February, 1804, when the invention of the voltaic pile had hardly begun to be known in Europe (for in that period there were no telegraphs or railroads), Don Francisco Salvá Campillo read before the Academy of Sciences at Barcelona another paper, called *The Second Treatise on Galvanism applied to Telegraphy*. He therein enumerates the difficulties which optic telegraphy presents in actual practice, and shows its inadequacy to the amount of work required, and its unproductiveness to the State, on account of the great expense attending its erection and maintenance. He relates, referring to two personal friends as witnesses, that Napoleon I., in the midst of a Session of the National Institute of Paris, declared that he had often received news by express sooner than by the optic telegraph, which, he says, is not to be wondered at, considering the fogs and other impediments peculiar to that system.

Salvá says in this paper, that when he read the previous one in 1800, he had not heard of the instrument invented by Volta, called *Volta's Column*, which is not strange, considering its so recent invention, and that it was not made public at all until the middle of the year 1800, when it was published in Nicholson's *Journal*. This, says Salvá, yields more fluid than the electric machine, and could be well applied to telegraphy, as the force can be obtained more simply and steadily than in the static form. He describes what had been done by scientific men towards improving its form ; he observes that the force of the shock is in proportion to the number of pairs, but not to the extent of surface in contact, and relates the experiments made to demonstrate this ; he proposes to avoid the excessive height of the pile (the well-known objection to which is the great weight of the upper discs pressing on those below), by forming a battery of several piles united; he complains of its being so difficult to clean, and concludes with his belief in the eventual obtainment of piles in a much greater state of perfection.

As to the means of indicating the signals, Salvá shows some hesitation, since, although he alludes to the contractions of frogs as adequate to the effect, he manifests an inclination for employing the decomposition of water.

For this last he gives sufficient explanation as to the system that could be adopted. It would suffice for the ends of each pair of wires to be inserted through a cork into a glass tube containing water. As the wire that communicates with the zinc plate of the pile is covered with bubbles of hydrogen gas, and the other is oxydated, these actions would economise, in the diversity of their effects, one half the conductors, since in applying the wires in a certain way to the poles of the voltaic pile, the letter A, for example, could be had, and, effecting the contact in a contrary way, the letter B could be indicated. Six wires would thus be enough for a telegraph, which would greatly reduce the expense and simplify the installation. After reading this paper the author proceeded to the experiments necessary towards perfectly understanding the above statements.

The apparatus constructed by Salvá on this occasion has not been preserved, but from the description of it in the memoir M. Saavedra thinks that it took a form similar to that shown in Fig. 5. On a table was arranged a number of flasks of water, one flask serving for two letters, or signals. Into each dipped two metallic rods, one of which was connected to a corresponding line wire, and the other to a return wire, of which there was only one, common to all. The different line wires and the return wire were similarly connected at the distant end.

When it was desired to transmit a signal, all the

line wire rods at the sending end were removed from
the flasks, or raised so as to be clear of the water, then
one pole, say the positive, of a pile was touched to
the return wire, and the other pole, the negative, to
the wire corresponding to the letter desired to be
signalled. Immediately this was done the water in
the flask, into which the distant ends of the wires

FIG. 5.

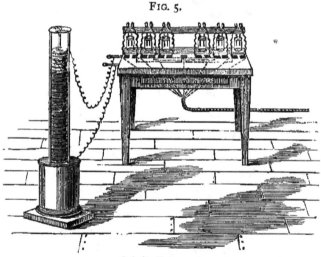

Salvá's Telegraph.

dipped, began to be decomposed, bubbles of oxygen
gas being given off at one rod, and bubbles of hydrogen at the other. By reversing the poles of the
pile at the sending end the bubbles of oxygen and
hydrogen changed places, thus making it possible for
one wire to serve for two signals, for since the bubbles
of hydrogen (being the more numerous) were taken

Q

to represent the signals, their evolution at the line wire might stand for the letter A, and at the return wire for B, and so on. When the communication was ended the rods were let down into the water, and the distant correspondent proceeded in the same manner to transmit his reply.

These notable and interesting memoirs, says Saavedra, have, ever since they were read, slept the sleep of the innocent on the shelves of the modest Scientific Academy of Catalonia; no one took the trouble to publish them at the time—a thing not strange in those days when their transcendent value was not appreciated, and when scientific journals were few in number, and little given to investigation; but it was unpardonable to neglect their publication subsequently, when the glorious realisation of public telegraphy excited general enthusiasm, and all the civilised nations made every effort to allege—through their numerous scientific and literary publications—the part each had taken in the great discovery. If, therefore, neither the author, nor any one else in Barcelona, or even in Spain, took the trouble to publish these trials of an electric telegraph, is it to be wondered at that foreign authors do not mention them, attributing to Sómmerring and to Coxe of 1809 and 1816 the first application of voltaic electricity to telegraphy? Is it strange that this should be the case when even the few Spanish writers who pay any attention to these matters, repeat the same words in chorus, as though the unfortunate country of Cervantes and Balmes were not also the birthplace of Blasco de Garay and of Salvá?

In this connection we would ask our readers to peruse again our account of Salvá's earlier experiments, which will be found at pp. 101–8, *ante.*

We join with M. Saavedra in the hope that his distinguished countryman will in future be better known and appreciated for his early and valuable contributions to the art of telegraphy.

1809–12.—*Sommerring's Telegraph.*

Sömmerring's telegraph was based on the same principle as Salvá's, and was not very dissimilar in detail. There is an interesting account of this contrivance in the *Journal of the Society of Arts,** for 1859, contributed by Dr. Hamel, of St. Petersburg. According to this indefatigable writer, the war between France and Austria, in 1809, gave rise to Sömmerring's discovery. On the 9th April in that year the Austrian troops crossed the river Inn, and on the 16th occupied Munich, whence King Maximilian had fled on hearing of their approach. The Emperor Napoleon, having speedy intelligence of this move by Chappe's sema-

---

* Vol. vii. pp. 595–99 and 605–10, *Historical Account of the Introduction of the Galvanic and Electro-Magnetic Telegraph into England.* Republished in pamphlet form, in Nov. 1859, by W. F. Cooke, with comments. See also *Der Elektrische Telegraph als Deutsche Erfindung S. T. Von Sömmerring's aus dessen Tagebuchern nachgewiesen,* 21 pp., published at Frankfort in 1863, by Sömmerring's only son.

phore, hastened away with some troops, and, so rapid and unexpected were his movements, that in less than a week the Austrians were obliged to retire, and on the 25th Maximilian re-entered his capital.

This event in which Chappe's semaphore played so important a part caused much attention to be directed to the subject of telegraphy, and on the 5th July following we find the Bavarian minister, Montgelas, requesting his friend, Dr. Sommerring, to bring the subject before the Academy of Sciences (of Munich), of which he was a distinguished member.*

Sömmerring at once gave the matter his attention, and soon it occurred to him to try whether the visible evolution of gases from the decomposition of water by the voltaic current might not answer the purpose. He worked at this idea incessantly, and, before three days had elapsed, had constructed his first apparatus, shown in Fig. 6. He took five wires of silver, or copper, and, insulating each with a thick coating of sealing-wax, bound the whole up into a cable. These wires, at one end, terminated in five pins which penetrated a glass vessel containing acidulated water ; and, at the other, were capable of being put in connection with the poles of a pile of fifteen pairs of zinc discs, and Brabant thalers, separated by felt soaked in hydrochloric acid. By touching any two of the wires to the poles of the pile he was able to produce, at their distant ends, a disengagement of gases, and

* Hamel, Cooke's reprint, pp. 5-7.

thereby indicate any of the five letters *a*, *b*, *c*, *d*, *e*.
Having thus shown the feasibility of his project, he set
himself to perfect his apparatus, and worked at it with
such a will that by the 6th of August it was completed. He wrote in his diary on that day :—" I have
tried the entirely finished apparatus which completely
answers my expectations. It works quickly through

Fig. 6.

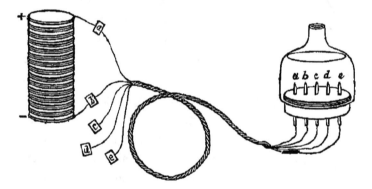

wires of 362 Prussian feet." Two days later he
worked it through 1000 feet, and then through 2000
feet, the wire in each case being wound round a
glass cylinder for greater compactness.*

As there is great diversity amongst writers on the
telegraph not only as regards the date of this invention, but as to the number of wires used, and other
details of its construction, we translate the following

* Hamel, pp. 7, 8. On the 4th February, 1812, he worked through
4000 feet, and on the 15th March following through as much as 10,000
feet.

description from the author's own paper, which was read before the Munich Academy of Sciences, on the 29th August, 1809, on which occasion the telegraph was exhibited in action :—

" In the bottom of a glass reservoir of water, 170 mm. long, 25 mm. broad, and 65 mm. high, of which C, in Fig. 7 is a sectional view, are thirty-five gold points, or pins, passing up through the bottom of the trough, and corresponding with the twenty-five letters of the German alphabet and the ten numerals. The thirty-five pins are each connected to as many insulated copper wires, E, E,* which, extending to the distant station, are there soldered to thirty-five brass terminals arranged on a wooden bar B.   Through the front end of each of these terminals there is a small hole for the reception of brass pegs, one of which is attached to the wire coming from the positive pole, and the other to that from the negative pole of the voltaic pile A. Each of the thirty-five terminals corresponds, through its wire, with a pin in the distant reservoir, and is lettered accordingly.

" When thus arranged, the two pegs from the pile are taken, one in each hand, and, two terminals being selected, are pushed into the holes.   The communication is now established and gas is evolved at the corresponding pins in the distant reservoir, hydrogen

---

* In September 1811, Sòmmerring reduced the number of wires to twenty-seven, of which twenty-five were for the letters, one for the stop, and one for the note of interrogation, or repetition.—Hamel, p. 19.

at the pin in connection with the positive pole of the pile, and oxygen at the other. In this way every letter and numeral may be indicated at pleasure, and,

FIG. 7.

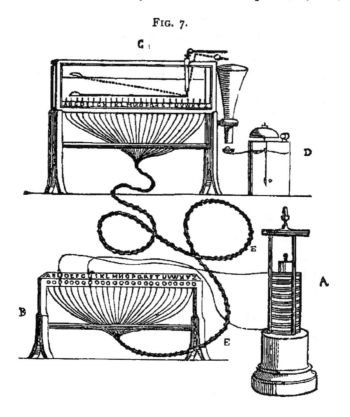

if the following rules be observed, one can communicate as much as, if not more than, is possible by the common (semaphore) telegraph.

"*First Rule.*—As the hydrogen gas evolved is greater in quantity than the oxygen, therefore those letters which the former gas represents are more easily distinguished than those of the latter, and must be so noted. ·For example, in the words containing *ak, ad, em, ie,* we indicate the letters *a, a, e, i,* by the hydrogen; *k, d, m, e,* on the other hand, by the oxygen.*

" *Second Rule.*—To telegraph two letters of the same name, we must use a unit, unless they are separated by the syllable. For example, the word *anna* may be telegraphed without the unit, as the syllable *an* is first indicated and then *na.* The word *nanni,* on the contrary, cannot be telegraphed without the use of the unit, because *na* is first telegraphed, and then comes *nn,* which cannot be indicated in the same vessel. It would, however, be possible to tele-graph even three, or more, letters at the same time by increasing the number of wires from twenty-five to fifty, but this would very much augment the cost of construction and the care of attendance.

" *Third Rule.*—To indicate the conclusion of a word, the unit 1 must be used with the last single letter, being made to follow the letter. It must also be prefixed to the letter commencing a word when that letter follows a word of *two letters* only. For example : *Sie lebt* must be represented *Si, el, le, bt* ; and *Er lebt*

---

* This plan of sending the current down one wire and making it return by any other was employed by Cooke and Wheatstone in their five-needle telegraph of 1837. It proved to be a little complicated in the present instance, and Sommerring subsequently adopted the prac-tice of signalling only one letter at a time, the oxygen signal being neglected.

must be represented *Er*, 1*l*, *eb*, *tl*. Instead of using
the unit, another signal may be introduced, the cross †,
to indicate the separation of syllables.

"Suppose now the decomposing table is situated in
one city, and the peg arrangement in another, connected by thirty-five wires. Then the operator, with
his voltaic pile and pegs at one station, may communicate intelligence to the observer of the gas at
the decomposing table of the other station.

"The metallic plates, or terminals, with which the
wires are connected have conical-shaped holes in their
ends ; and the pegs attached to the two wires of the
voltaic pile are likewise of a conical shape, so that,
when they are put in the holes, there may be a close
fit, preventing oxydation and ensuring a good contact,
as it is well known that slight oxydation of the parts
in contact will interrupt the communication. The
peg arrangement might be so contrived as to use
permanent keys, which for the thirty-five plates would
require seventy pins. The first key might be for
hydrogen A ; the second key for oxygen A ; the third
key for hydrogen B ; the fourth key for oxygen B,
and so on.

"The preparation and management of the voltaic
pile is so well known that little need be said, except
that it should be of that durability as to last more
than a month. It should not be of very broad surfaces,
as I have proved that six of my usual pairs (each one
consisting of a Brabant dollar, felt moistened with a

saturated solution of common salt, and a disc of zinc, weighing 52 grains) would evolve more gas than five pairs of the great battery of our Academy.

"As to the cost of construction, this model, which I have had the honour to exhibit to the Royal Academy, cost 30 florins.   One line, consisting of thirty-five wires, laid in glass, or earthen, pipes, each wire insulated with silk, and measuring 22,827 Prussian feet, or a German mile, might be made for less than 2000 florins, as appears from the cost of my short one."

On the 23rd August, 1810, Sömmerring perfected his apparatus by adding a contrivance D, Fig. 7, for attracting attention at the distant station. He made the gas, rising in small bubbles from two contiguous pins in the water, collect under a sort of inverted glass spoon at the end of a long lever, which, rising, made a second lever bent in the opposite direction on the same axle descend and throw off a little perforated leaden ball, resting lightly on it, and which, falling on an escapement, set the clockwork of an ordinary alarum, D, in motion.   This arrangement, simple as it is, gave Sömmerring much trouble. He writes in his diary, " If the principal part of the telegraph gave me no trouble, and demanded no alteration, but was ready in a few days, this secondary object, the alarum, cost me a great deal of reflection, and many useless trials with wheelwork, till, at last, I hit upon this very simple arrangement." The existence of this alarum is not generally known, and, probably, because it is not represented on the two plates which accompanied the description of the telegraph in the Memoirs of the Munich Academy, which plates, as Dr. Hamel has shown, were engraved before the alarum was invented.*

Sommerring's wires, which were of brass, or copper, were insulated with a coating of gum-lac, then wrapped round with silk thread, and united into a cable, which was also bound with thread, and covered with gum-lac, or with a ribbon, soaked in that material. The cable was then wound on reels. In practice there was no appreciable retardation in the action of the apparatus through the greatest length of wire, the evolution of the gas appearing to begin as quickly as if the current had only to traverse two feet. The only effect of distance (*i. e.*, resistance) was to reduce the quantity of the gases evolved in a given time.†

This telegraph, complex and unpractical as it was, was earnestly prosecuted by its author for several years, and received a large share of attention, princes, statesmen, and philosophers thronging to his lodgings to witness its performances. He had a complete apparatus connected up in his house, which worked through insulated wires carried round on the outside, and which he always took great delight in showing to

---

* Hamel, pp. 13, 14.

† In connection with this evolution Sömmerring noticed a curious phenomenon. Whenever the gases were evolved at two neighbouring gold points, as A, and B, the hydrogen bubbles always ascended vertically, but those of the oxygen inclined towards the hydrogen.

his visitors. At the suggestion of one of the most constant and intimate of these, Baron Schilling (of whom
more hereafter), Sommerring, on the 6th June, 1811,
tried the action of his telegraph when the conducting
wires were cut and the ends separated by an interval
of water in wooden tubs. The result was that the
signals appeared just as well as if the wires had not
been cut, but they ceased as soon as the water in the
two tubs was connected by a wire, the current then
returning by this shorter road. On the two following
days he performed with his friend some other experiments, first across a canal off the river Isar, and then
along the river itself, in which he showed that the
water and earth might be used instead of a return
wire, an experiment which was probably suggested by
the similar one of Aldini, in 1802, across the harbour
of Calais.*

Besides his own apparatus, Sömmerring had prepared other models for exhibition in France, Austria,
Russia, Switzerland, and England. With the latter,
which, by the way, was never forwarded, "fearing
difficulties at the custom-house," he sent a description in English, in which he expressed the hope that
"Sir Humphry Davy would receive it favourably,
perhaps improve it, and further its application in Great
Britain.†

The instruments designed for exhibition in Paris
were entrusted to Baron Larrey (an old friend and

* Hamel, pp. 15–17.   † Ibid., p. 33.

correspondent of Sommerring, and then Inspector-General of the Army Medical Department under the Empire), who happened to be passing through Munich on his way home from the battle-fields of Aspern, Esslingen, and Deutsch Wagram. This was on the 4th November, 1809, and, immediately after Larrey's departure, Sommerring drew up an account of his telegraph in French, under the title *Mémoire sur le Télégraphe*, which, on the 12th November, he forwarded to his friend at Paris, accompanied by a private note as follows :—

"I have the honour to enclose a memoir, which, with the model that you have kindly taken charge of, will, I hope, explain matters clearly and briefly. I am anxious to know what reception His Imperial Majesty will accord to my ideas. The memoir, as you will see, describes the principal results of some experiments as varied as I could make them ; and I hope that they will interest many members of the Institute, for independently of the great interest that attaches to them they are not wanting in novelty."

Replying on the 10th December, 1809, Larrey says :—" His Majesty, being prevented by press of business from occupying himself just now with scientific matters, has informed me that he will inspect your apparatus later on. Meanwhile I have decided to present it to the first class of the Institute, which [on the 5th December] received it with interest, and appointed a commission to report upon it."

Writing again, on the 28th December, to Baron de Kobell, one of the Bavarian Secretaries of State, Larrey says:—"I have the honour to send you the little case for Doctor Sommerring of which I have already spoken to you. Will you kindly send it on to him by the first safe means that you can find. I hope that the contents will please him.

"I shall take care to inform him of the nature of the report of the Institute upon his telegraphic machine as soon as it appears ; but in this there will probably be some delay, as the academicians who have charge of the matter are greatly occupied at present on some pressing Government affairs."

The commission consisted of Biot, Carnot, Charles, and Monge, but, for some unaccountable reason (at least, Biot, in after years, could recollect none), no report was ever made, and full eighteen months later (May 12, 1811) the instruments were sent back to Munich.

Writing to Sòmmerring, on April 19, 1810, Larrey says:—

"My dear and respected friend,—Three months ago I sent to Mr. de Kobell, for transmission to you, a small case containing some remarkably diseased bones, and some brief notices upon them ; your long silence makes me fear that this case has not reached you. Will you, my dear Doctor, kindly enlighten me on this point ?

"I also informed you that the Institute had appointed a commission to report upon your telegraph, but certain persons, no doubt moved by jealousy, do not regard the discovery with the interest that it ought to inspire, and the report is, consequently, not yet made.

"Wishing to meet your desires, at least in part, I have inserted a notice of your instrument in the current number of the *Bulletin de la Société Médicale d'Émulation*, after having submitted it to the Society. If you have not received the Journal I will send you a copy."

In this paper, *On the Origin and Structure of the Encephalic Nerves*, Larrey briefly introduces the telegraph, and then goes on to speak with much detail of the analogy which its many wires offered to the single fibres of the nervous system, an analogy which Sömmerring himself had pointed out in his French memoir, as well as in the original account of his telegraph in the *Transactions* of the Munich Academy.* Larrey's article was republished twenty years later (in November 1829), in his *Clinique Chirurgicale ;* † but, in both publications, it was placed in the midst of

---

* The same analogy between the nerves of the body and the telegraphic system of the world has since been frequently noticed. See Fechner's *Lehrbuch des Galvanismus*, Leipsic, 1829, p. 269 ; *Mechanics' Magazine*, for 1837–8, p. 262 ; and *Notes and Queries*, for August 27, 1870, p. 173.

† Vol. 1. p. 361.

pathological and surgical subjects, where one would never look for an invention for telegraphic purposes.

On the 30th July, 1810, Sömmerring replied in the following curious letter :—

"I have read with great pleasure, sir, you dissertation on my telegraph. Have you received my memoir, which I posted on the 12th November; and have you kindly communicated it to the Institute?

"The old conducting wires are somewhat damaged, and as it was entirely to avoid delay that I did not renew them before despatching the apparatus, I would be glad if they could be replaced by new wires of the sort used in harpsichords, covered with silk thread, as the material of which these are composed is more durable than the old copper wires. Had I imagined, sir, that you would have taken such an interest in my invention as to charge yourself with its transport to Paris, I would certainly not have omitted, beforehand, to effect the necessary changes, which, without counting the time, require only a little care. I am very much afraid that, besides the fragility of the copper wires, the rough usage to which they have been subjected in the course of experiment may have rubbed off in places the silk, and so may cause intermediate contacts of the metal, whence must result a derangement of the whole system.

"Allow me, then, to beg of you not to show the instrument to the Prince de Neufchâtel, or even to His Majesty the Emperor, until the above-mentioned repairs have been effected, either by myself, or, if its return would take too long, by some competent mechanic in Paris."

Regarding the model of the telegraph which Sommerring delivered to Count Jeroslas Potocki, a colonel of Russian Engineers, for exhibition at Vienna and St. Petersburg, the following letter has been preserved by Sommerring's family :*—

" Baaden, near Vienna, July 5, 1811.

" Sir,—I hasten to inform you that, on my return to Vienna, their Majesties, the Emperor and Empress, signified their desire to see the electric telegraph—an invention which does honour to human genius. On the first of the current month I had the pleasure to show your telegraph to their Majesties, and they were enchanted. His Majesty was so pleased that he expressed his desire to have a telegraph from Laxenburg to Vienna (a distance of about nine miles). He did not omit to ask me to whom we are indebted for so ingenious an invention. He knows you by reputation, and says that you are one of the first anatomists living. In fact, I can assure you that their Majesties, and the Archdukes, who were also present, were enchanted.

"Professor Jacquin, of Vienna, wishes to come to

---

* We are indebted for this and the preceding extracts to Mr. Karl Sömmerring, of Frankfort, a grandson of the distinguished physicist of whom we are writing.

R

see me at Baaden, with the view of inspecting the telegraph.

"In fine, your invention has had the greatest success, and I do not doubt for an instant that, especially in Russia, it will be carried out on a grand scale. I shall not fail to acquaint you with the reception that it may there meet with. Meanwhile I pray you accept the assurance of the highest sentiments with which I am, sir, your very humble and very obedient servant,

"JEROSLAS POTOCKI."

The apparatus which Sommerring sent to his son, Wilhelm, at Geneva, where he was then studying, is still preserved by the family of the latter at Frankfort. It was exhibited at Vienna in 1873, at London (South Kensington Museum) in 1876, and at the Paris Exhibition of 1881, at all of which places it elicited, as may be supposed, the liveliest interest.

Sommerring, who was a distinguished anatomist and physiologist, was born at Thorn, West Prussia, on January 28, 1755, and died at Frankfort, on 2nd March, 1830. He was elected a member of the Munich Academy of Sciences in 1805, was made Knight of the Order of St. Anne of Russia in 1818, and, in 1819, was elected an honorary member of the Imperial Academy of Sciences of St. Petersburg. Quite recently, we believe, a monument has been erected to his memory in the city of Frankfort, where he passed the last ten years of his interesting life.

Dr. Hamel, in his *Historical Account*, &c., pays a very just tribute to his worth, with which we will close this account of his telegraph:—" When one studies the life and the labours of Sömmerring, it is impossible not to feel the highest esteem for him, as a man and as a philosopher. Not vanity, not eagerness of gain, but pure love of science and the wish to be useful were the motives of his incessant activity. Nor was Sömmerring too sanguine in his expectation with regard to the application of his invention. He expressed the hope that it might serve to telegraph from Munich to Augsburg, nay, from one end of the kingdom to the other, without intermediate stations " (pp. 34, 35).

1811.—*Schweigger's Telegraph.*

In preparing an account of Sömmerring's telegraph for insertion in his *Journal fur Chemie und Physik,*[*] Schweigger of Nürenberg, and later of Halle—the same who afterwards invented the galvanometer—was struck with the insuperable difficulty there would be in dealing practically with so many wires, and in his paper he suggested a plan which required only two wires, and two piles of different strengths, so that at one time the weaker may be used, and at another time the stronger, or even both combined. In this way the quantity of gas produced in a given time at the distant station would be varied, a small quantity

* Vol. ii. p. 240, for 1811.

denoting one letter, and a larger another. Again, by varying (1) the duration of the evolutions, and (2) the intervals between, other letters might be indicated; and thus, by the combination of these primary elements of quantity and time, all the letters of the alphabet could be expressed through two wires instead of thirty-seven. In ignorance of Sommerring's alarum, Schweigger suggested the firing of Volta's gas-pistols by Leyden jars as a means of drawing attention.*

1813.—*Sharpe's Telegraph.*

All that we know of this invention is contained in the following paragraph, which we copy from the *Repertory of Arts*, 2nd series, June 1816, p. 23 :—

" *On the Electrical Telegraph.* Communicated by Mr. J. R. Sharpe, of Doe Hill, near Alfreton.—In the *Repertory of Arts*, vol. xxiv., 2nd series, p. 188, is an account of an electric telegraph by M. Soemmering. This account I did not see till a few weeks ago. Without the slightest wish to throw a doubt over the originality of M. Soemmering's invention, I beg leave to mention that an experiment, showing the advantages to be obtained from the application of the certain and rapid motion of the electric principle

---

* He also described a sort of manifold short-hand, or sign-printer, like that patented by Wheatstone in 1841 ; but this had nothing to do with electricity, and was mentioned *par parenthèse.*

through an extensive voltaic circuit to the purposes of the ordinary telegraph, was exhibited by me before the Right Hon. the Lords of the Admiralty, in the beginning of February 1813."

My Lords are said to have approved the design, but passed it over with the remark that "As the war was over, and money scarce, they could not carry it into effect." *

A nephew of the inventor, writing in 1861,† says, in reference to the above announcement, that Mr. Sharpe "conveyed signals a distance of seven miles under water." In the hope of getting further information we addressed ourselves to this gentleman, but, unfortunately, he could add nothing to the abovementioned facts.

Mr. J. R. Sharpe was bred in London as a solicitor, but early left the profession, and retired to Doe Hill, which he built in 1801. He was always of a studious turn, and even in advanced age amused himself with mathematical problems. He died November 11, 1859, aged eighty-four years.

It was probably anent Sharpe's proposals that the following squibs were written, which we extract from *The Satirist*, September and October 1813.

"On the report that it is in contemplation to substitute an electrical mode of communication with the

---

* *Saturday Review*, August 21, 1858, p. 190.
† *A Treatise on the Construction and Submersion of Deep-Sea Electric Telegraph Cables*, by Benjamin Sharpe, London, 1861, p. 16.

outposts (by means of wires laid underground) for the existing telegraphic system :—

> "Our telegraphs, just as they are, let us keep,
>   They forward good news from afar,
> And still may send better—that Boney's asleep,
>   And ended oppression and war.
>
> Electrical telegraphs all must deplore,
>   Their service would merely be mocking ;
> Unfit to afford us intelligence more
>   Than such as would really be *shocking!*
>   "TAM GLEN."

"On the Proposed Electrical Telegraph, October 1813 :—

> " When a victory we gain
>   (As we've oft done in Spain)
> It is usual to load well with powder,
>   And discharge 'midst a crowd
>   All the Park guns so loud,
> And the guns of the Tower, which are louder.
>
>   But the guns of the Tower,
>   And the Park guns want power
> To proclaim as they ought what we pride in ;
>   So when now we succeed
>   It is wisely decreed
> To announce it from the *batteries of Leyden.*"

1816.—*Coxe's Telegraph.*

In February 1816, Dr. J. Redman Coxe, professor of chemistry at Philadelphia, published some suggestions for an electro-chemical telegraph on the same principle as those already described. His views are given in the following letter, which we have extracted from Thomson's *Annals of Philosophy*, vol. vii. pp. 162-3, headed "Use of Galvanism as a Telegraph" :—

"I observe in one of the volumes of your *Annals of Philosophy* a proposition to employ galvanism as a solvent for the urinary calculus, which has been very properly, I think, opposed by Mr. Armiger. I merely notice this, as it gives me the opportunity of saying that a similar idea was maintained in a thesis three years ago by a graduate of the University of Pennsylvania.

"I have, however, contemplated this important agent as a probable means of establishing telegraphic communication with as much rapidity, and, perhaps, less expense than any hitherto employed. I do not know how far experiment has determined galvanic action to be communicable by means of wires, but there is no reason to suppose it confined as to limits, certainly not as to time. Now, by means of apparatus fixed at certain distances, as telegraphic stations, by tubes for the decomposition of water and of metallic salts, &c., regularly arranged, such a key might be adopted as would be requisite to communicate words, sentences, or figures, from one station to another, and so on to the end of the line. I will take another opportunity to enlarge upon this, as I think it might serve many useful purposes ; but, like all others, it requires time to mature. As it takes up little room, and may be fixed in private, it might in many cases, of besieged towns, &c., convey useful intelligence with scarcely a chance of detection by the enemy. However fanciful in speculation, I have no doubt that sooner or later it will be rendered useful in practice.

"I have thus, my dear sir, ventured to encroach upon your time with some crude ideas that may serve perhaps to elicit some useful experiments at the hands of others. When we consider what wonderful results have arisen from the first trifling experiments of the junction of a small piece of silver and zinc in so short a period, what may not be expected from the further extension of galvanic electricity? I have no doubt of its being the chiefest agent in the hands of nature in the mighty changes that occur around us. If metals are compound bodies, which I doubt not, will not this active principle combine their constituents in numerous places so as to explain their metallic formation; and if such constituents are in themselves aeriform, may not galvanism reasonably tend to explain the existence of metals in situations in which their specific gravities certainly do not entitle us to look for them?"

Dr. Coxe does not appear to have ever reduced his ideas to practice, but the large faith which he expresses in the capabilities of galvanism deserves to be remembered to his credit. Indeed there can be no doubt that, if electrical science had made no further advances, the early projects that we have been describing in this chapter would have gradually developed themselves into practical electro-chemical telegraphs, such as were afterwards proposed by E. Davy in 1838, by Smith in 1843, by Bain in 1846, and by Morse in 1849.* But the grand discovery of electro-magnetism was at hand, and soon turned the tide of invention into quite another channel.

* Besides all these inventions, other electro-chemical telegraphs have been proposed by Bakewell, Caselli, Bonelli, D'Arlincourt, Sawyer, and others. All are dependent on a fact which, as we have shown in our seventh chapter (p. 195), was first observed by Cruickshank in 1800, very soon after the announcement of the voltaic pile, *viz.*, the power of electricity to discolour litmus paper.

CHAPTER IX.

ELECTRO-MAGNETISM AND MAGNETO-ELECTRICITY —HISTORY IN RELATION TO TELEGRAPHY.

> " Around the magnet, Faraday
> Is sure that Volta's lightnings play,
> But how to draw them from the wire?
> He took a lesson from the heart—
> 'Tis when we meet—'tis when we part
> Breaks forth the electric fire."
>
> *Impromptu*, by Herbert Mayo,
> in *Blackwood's Magazine.*

FROM an early period in the history of electricity philosophers began to point out strong resemblances between the phenomena which it exhibits and those of magnetism. In both sciences there existed two forces of opposite kind, capable, when separate, of acting with great energy, and being, when combined, perfectly neutralised and exhibiting no signs of activity; there was the same attraction and repulsion between the two magnetisms as between the two electricities, and according to the same law of inverse squares; the action of free electricity on a neighbouring body was not unlike that which a magnet exercises upon iron; and, lastly, the distribution of the two forces in the one seemed to differ little from that of the two forces in the other.

These analogies were powerfully supported by several facts. Thus, as early as 1630 Gassendi observed that magnetism was communicated to ferruginous bodies by lightning ; the compass needles of ships were known to have their poles weakened, and even reversed, by a similar cause—a fact first recorded by English navigators in 1675 ; and, in 1750, Professor Wargentin remarked that delicately-suspended magnets were affected by the aurora borealis.*

With such analogies, and supported by such remarkable facts as these, the suspicion was but natural that the two sciences were allied in some close and intimate way, and accordingly we find that, about the middle of the last century, the discovery of this relation became a favourite pursuit.

Swedenborg was the first to boldly express his views on this subject in his *Principia Rerum Naturalium* (Dresden, 1734), in which he argued a close relationship between electricity and magnetism on the ground of their both being polar forces.

In 1748, Beraut, professor of mathematics in the College of Lyons, published at Bordeaux a thin volume of 38 pages,† which is probably the first dis-

---

* *Encys. Brit. and Metropol.*, articles Electricity and Magnetism. A similar observation to that of Wargentin was made by Halley, and afterwards more accurately by Dalton, both of whom likewise found that the beams of the aurora were always parallel to the magnetic meridian.—*Trans. Cambridge Phil. Soc.*, vol. i.

† *Dissertation sur le rapport qui se trouve entre la Cause des effets de l'Aiman, et celles des Phénomènes de l'Électricité.*

tinct treatise on its subject, and which also goes to
show that a true connection exists ; that, in fact, it is
the same force only differently disposed, which produces both electric and magnetic phenomena.

In studying the points of analogy between lightning
and electricity, the great Franklin remarked that the
latter, like the former, had the power not merely of
destroying the magnetism of a needle, but of completely reversing its polarity.   By discharging four
large Leyden jars through a common sewing needle,
he was able to impart to it such a degree of magnetism
that, when floated on water, it placed itself in the
plane of the magnetic meridian.   When the discharge
was sent through a steel wire perpendicular to the
horizon it was permanently magnetised, with its lower
end a North, and its upper end a South pole ; and, on
reversing the position of the wire and again transmitting through it the discharge, the polarity was either
destroyed, or entirely reversed.   Franklin also found
that the polarity of the loadstone could be destroyed
in a similar manner.*

Dalibard, about the same time, imagined that he
had proved that the electric discharge gives a northern
polarity to that point of a steel bar at which it enters,
and a southern polarity to that at which it makes its
exit, while Wilcke, for his part, was equally satisfied
that an invariable connection existed between negative
electricity and northern polarity.

* Priestley's *History of Electricity*, London, 1767, p. 178.

From a review of all these, and other observations by himself made between 1753 and 1758, Beccaria came to the conclusion that the polarity of a needle magnetised by electricity was invariably determined by the direction in which the electric discharge was made to pass through it; and as a consequence he assumed the polarity acquired by ferruginous bodies which had been struck by lightning as a test of the kind of electricity with which the thunder-cloud was charged.

Applying this criterion to the earth itself, Beccaria conjectured that terrestrial magnetism was, like that of the needle magnetised by Franklin and Dalibard, the mere effect of permanent currents of natural electricity, established and maintained upon its surface by various physical causes; that, as a violent current, like that which attends the exhibition of lightning, produces instantaneous and powerful magnetism in substances capable of receiving that quality, so may a more gentle, regular, and constant circulation of the electric fluid upon the earth impress the same virtue on all such bodies as are capable of it. "Of such fluid, thus ever present," observes Beccaria, "I think that some portion is constantly passing through all bodies situate on the earth, especially those which are metallic and ferruginous; *and I imagine it must be those currents which impress on fire-irons, and other similar things, the power which they are known to acquire of directing* *themselves according to the magnetic meridian when they are properly balanced."* *

Diderot, one of the editors of the celebrated "Encyclopædia," and whom the *Revue des deux Mondes* † calls a "Darwinist a century before Darwin," was also, as early as 1762, a firm believer in the identity of electricity and magnetism, and has left in his writings some arguments in support of this hypothesis.

In his essay *On the Interpretation of Nature* he says:—"There is great reason for supposing that magnetism and electricity depend on the same causes. Why may not these be the rotation of the earth, and the energy of the substances composing it, combined with the action of the moon? The ebb and flow of the tides, currents, winds, light, motion of the free particles of the globe, perhaps even of the entire crust round its nucleus, produce, in an infinite number of ways, continual friction. The effect of these causes, acting as they do sensibly and unceasingly, must be, at the end of ages, very considerable. The nucleus or kernel of the earth is a mass of glass, its surface is covered only with remains of glass— sands and vitrifiable substances. Glass is, of all bodies, the one that yields most electricity on being rubbed. Why may not the sum total of terrestrial

* Ampère's theory of electro-magnetism, and likewise his view of terrestrial magnetism, are here distinctly foreshadowed by this most acute and accurate observer. For a full account of Beccaria's researches, see Priestley's *History of Electricity*, London, 1767, pp. 340-352.

† For December 1, 1879, p. 567.

electricity be the result of all these frictions, either at the external surface of the earth, or at that of its internal kernel?

" From this general cause it is presumable that we can deduce, by experiments, a particular cause which shall establish between two grand phenomena, *viz.*, the position of the aurora borealis and the direction of the magnetic needle, a connection similar to that which is proved to exist between magnetism and electricity by the fact that we can magnetise a needle without a magnet and by means only of electricity.

" These notions may be either true or false. They have no existence so far but in my imagination. It is for experience to give them solidity, and it is for the physicist to discover wherein the phenomena differ, or how to establish their identity." *

In the year 1774, the following question was proposed by the Electoral Academy of Bavaria as the subject of a prize essay :—" Is there a real and physical analogy between electric and magnetic forces, and, if such analogy exist, in what manner do these forces act upon the animal body ? " The essays received on that occasion were collected and published ten years later by Professor Van Swinden, of Franeker—the winner of one of the prizes.† Some of

---

* The physicist has been true to the trust. See *Collection Complète des Œuvres Philosophiques, Littéraires, et Dramatiques de Diderot*, 8vo., 5 vols., Londres, 1773, vol. ii. p. 28.

† *Recueil de Mémoires sur l'Analogie de l'Electricité et du Magnétisme*, &c., 3 vols., 8vo., La Haye, 1784.

the essayists, and amongst them Van Swinden, maintained that "the similarity was but apparent, and did not constitute a real physical resemblance;" while, on the other hand, Professors Steiglehner and Hubner contended that "so close an analogy as that exhibited by the two sciences indicated a single agency acting under different circumstances." *

In this unsettled state the subject remained for many years until the discovery of galvanism and the invention of the voltaic pile, which, by furnishing the philosopher with the means of maintaining a continuous current of electricity in large quantity, enabled him to study its effects under the most favourable circumstances.

Early in the present century philosophers thought they saw an analogy between magnetism and galvanism in a phenomenon, which we find thus referred to in Lehot's *Observations sur le Galvanisme et le Magnétisme :* †—"It has long been known that the two wires which terminate a pile attract one another, and, after contact, adhere like two magnets. This attraction between the two wires, one of which receives, and the other loses, the galvanic fluid, differs essentially from electrical attraction, as Ritter observed, since it is not followed by a repulsion after contact, but continues as long as the chain is closed" (note on p. 4).‡

* Noad's *Manual of Electricity*, p. 641.
† Paris, *circa*, 1806, 8vo., 8 pp.
‡ This discovery appears to have been made independently, and about the same time, by Gautherot, in 1801 (*Philosophical Magazine*,

In the same spirit of inquiry Desormes and Hachette, in 1805, tried to ascertain the direction which a voltaic pile, whose poles were not joined, would take when freely suspended horizontally. The pile, "composed of 1480 thin plates of copper tinned with zinc, of the diameter of a five-franc piece," was placed upon a boat, which floated on the water of a large vat; but it assumed no determinate direction, although " a magnetised steel bar, of a weight nearly equal to that of the pile, and placed like it upon the boat, would turn, after some oscillations, into the magnetic meridian." *

The honour of the discovery of the much-sought-for connection between electricity and magnetism has often, in the last fifty years, been claimed for Romagnosi, an Italian writer who is justly esteemed for his works on history, law, and political philosophy. Govi,† however, in 1869, showed in the clearest manner possible the absurdity of this claim; but, notwithstanding, it has been again put forward, and this time by no

---

for 1828, vol. iv. p. 458); by Laplace; and by Biot (*Journal de Physique et de Chimie*, &c., for 1801, vol. liii. p. 266). The latter made the further very acute observation that, if the wires be attached to plates of metal, and these plates be approached by their *edges*, they will attract one another; while if approached by their *faces* no action whatever takes place.

For other interesting experiments of this kind, see Nicholson's *Journal*, for 1804, vol. vii. p. 304.

* *Philosophical Magazine*, for 1821, vol. lvii. p. 43.

† *Romagnosi e l'Elettro-Magnetismo*, Turin, 1869.

S

less an authority than Dr. Tommasi, of Paris, in a recent number of *Cosmos les Mondes* (June 30, 1883).

Dr. Tommasi, in republishing Romagnosi's experiment, asks the following questions, which he submitted, in particular, to the managing committee of the (late) Vienna Exhibition, in the hope that they might have been brought before electricians :—

" Is it to Oersted, or to Romagnosi, that we should ascribe the merit of having first observed the deviation of the magnetic needle by the action of the galvanic current ?

" Had Oersted any knowledge of the experiment of Romagnosi when he published his discovery of electro-magnetism ? *

" Have any other *savants* taken part in this discovery ? "

Now, we should have thought that after the admirable *exposé* of Govi, to which we have just referred, no electrician would seriously put to himself these questions. But it appears that our Paris *confrère* does so, although, if he had only read carefully the facts on which he bases them, he would perceive that they have no relation whatever to electro-magnetic action, but are simply effects of ordinary electrical attraction and repulsion brought about by the static charge which is always accumulated at the poles of a strong voltaic *pile*—the form of battery used in Romagnosi's

---

* Dr. Hamel, for one, thought he had, and tries to prove it in his *Historical Account*, &c., of 1859 (pp. 37–9 of W. F. Cooke's reprint).

experiments, and which, as is well known, exhibits this phenomenon in a far more exalted degree than the ordinary cell arrangement.

We cannot establish better the correctness of our conclusions than by quoting in full the recital of Romagnosi's experiment, as it originally appeared in the *Gazzetta di Trento*, of August 3, 1802 : *—

" Article on Galvanism.

"The Counsellor, Gian Domenico de Romagnosi, of this city, known to the republic of letters by his learned productions, hastens to communicate to the physicists of Europe an experiment showing the action of the galvanic fluid on magnetism.

" Having constructed a voltaic pile, of thin discs of copper and zinc, separated by flannel soaked in a solution of sal-ammoniac, he attached to one of the poles one end of a silver chain, the other end of which passed through a short glass tube, and terminated in a silver knob. This being done, he took an ordinary compass-box, placed it on a glass stand, removed its glass cover, and touched one end of the needle with the silver knob, which he took care to hold by its glass envelope. After a few seconds' contact, the needle was observed to take up a new position, where it remained, even after the removal of the knob. A fresh application of the knob caused a still further

* Our translation is made from the reprint at p. 8 of Govi's *Romagnosi e l' Elettro-Magnetismo.*

deflection of the needle, which was always observed to remain in the position to which it was last deflected, as if its polarity were altogether destroyed.

FIG. 8.

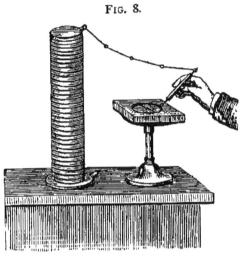

Romagnosi's Experiment, according to Govi.

" In order to restore this polarity, Romagnosi took the compass-box between his fingers and thumbs, and held it steadily for some seconds. The needle then returned to its original position, not all at once, but little by little, advancing like the minute or seconds hand of a clock.

" These experiments were made in the month of May, and repeated in the presence of a few spectators, when the effect was obtained without trouble and at a very sensible distance."

Here it will be seen that Romagnosi uses only *one*

pole of the pile, and *never speaks of the circuit being closed*—facts which show that his experiment has no resemblance to that of Oersted.

The effects which he describes are, moreover, easily explainable on another hypothesis. The compass needle, we may imagine, received a charge of static electricity by contact with the charged pole of the pile. Being insulated, it could not part with this charge, and, consequently, as soon as it had attained the same potential as the voltaic pole, mutual repulsion ensued. As the needle belonged to "an ordinary compass-box," we may assume it was neither strongly magnetised, nor delicately suspended. Friction at the point of support, then, might more than counterbalance the directive force of the earth, and so the needle would always remain in the position to which it had been last repelled.

The "restoration of polarity," or the bringing back of the needle to the magnetic meridian, by merely holding the compass-box steadily between the fingers and thumbs, although savouring of legerdemain, was really due to a "simple turn of the wrist." Romagnosi may have imagined that he held the compass-box *steadily*, but there can be no doubt that his hands suffered a slight and imperceptible tremor, which, aided by the directive force of terrestrial magnetism, sufficed to shake the needle into a north and south position.

Another, and to us convincing, argument against

the supposition that Romagnosi had any share in the discovery of electro-magnetism is that he himself never claimed any, although he lived down to the year 1835, or fifteen years after the announcement of the Danish philosopher.*

To the same category belongs the contrivance of Schweigger, which is described in Gehlen's *Journal für die Chemie und Physik*, for 1808 (pp. 206–8), and on the strength of which a recent writer † says that the celebrated inventor of the galvanometer ought also to be considered as the discoverer of electro-magnetism. There is no ground whatever for this statement. Schweigger's paper, which is headed *On the Employment of the Magnetic Force for Measuring the Electrical*, simply describes an electroscope for indicating the attraction and repulsion of ordinary, or frictional, electricity, and which he used as a substitute for the torsion electrometer of Coulomb. It consisted of a magnetic needle, armed at each end with a brass knob, and mounted on a pivot, as in an ordinary compass.

In fact, these experiments of Romagnosi and Schweigger are but modifications of one which dates back to the very earliest days in the history of electricity, and upon which, we have no doubt, Milner, in 1783, constructed the electrometer now known as

* For some very interesting experiments of this kind, see Van Mons' *Journal de Chimie*, for Jan. 1803, p. 52 ; also Nicholson's *Journal*, vol. vii. p. 304.

† In the *Journal für Math. und Physik*, Berlin, 1873, p. 609.

Peltier's. In our second chapter (p. 31) we have said, when speaking of Gilbert :—"In order to test the condition of the various substances experimented upon, Gilbert made use of a light needle of any metal, balanced, and turning freely on a pivot, like the magnetic needle, to the extremities of which he presented the bodies after excitation." Romagnosi and Schweigger have done no more than this—hardly as much, for Gilbert's contrivance was a valuable instrument of research, while those of the later philosophers were barren of results.

Other instances of this phenomenon, contributed by Robins and Kinnersley respectively, occur in the *Philosophical Transactions*, for 1746 and 1763 ; and a recent example, which is described in the *American Polytechnic Review*, for 1881, is considered so puzzling that "it is given for what it is worth" in the scientific paper in which we find it :*—

"An American surveyor, who had been taking some delicate bearings, was puzzled to find that the magnetic needle did not give the same bearing twice, and he observed that it never *quite* settled. This could not be explained as due to metallic articles in the dress, or pockets, of the observer ; and an examination of the magnifying glass used in reading the needle was made. The magnifier was similar to those now universally used to read the verniers and needle-bearings

* An exactly similar case is recorded at p 280, vol. xxi., of *The Quarterly Journal of Science and the Arts* (Royal Institution), for 1826.

of field instruments, having a black vulcanite frame, highly polished, and in this, it is stated, the whole cause of the trouble lay. It was found that this frame was peculiarly liable to become electrified, that the slightest friction, even the mere carrying in the pocket, was sufficient to charge it, and that, when thus electrified, if brought near the needle of a compass, it had almost the effect of a loadstone in drawing it from its true settling place. On discarding this magnifier and using an ordinary glass lens without a frame, no further trouble was found in the field work done with the compass. This must be taken for what it is worth."

As little value attaches to the observation of Mojon which we find recorded by Aldini, and which seems to us but a repetition of Franklin's experiment (before mentioned, p. 252), with this difference, that a voltaic battery was used instead of one of Leyden jars. Aldini says :—"The following experiment has been quite recently communicated to me by its author Mojon :—

" Having placed horizontally sewing-needles, very fine, and two inches long, he put the two extremities in communication with the two poles of a battery of one hundred cups, and on withdrawing the needles, at the end of twenty days, he found them a little oxidised, but at the same time endowed with a very sensible magnetic polarity. This new property of galvanism has been verified by other observers, and lately by

Romanesi, who has found that galvanism is able to deflect a magnetic needle." *

At p. 120 of his *Manuel du Galvanisme* (Paris, 1805), Joseph Izarn describes Mojon's experiment, and appends an illustration, which shows most conclusively that it had no reference to electro-magnetism. His words are :—

"Apparatus for observing the action of galvanism on the polarity of a magnetised needle :—

FIG. 9.

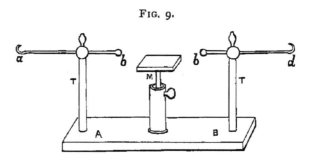

Mojon's Experiment, according to Izarn.

"*Preparation.* Arrange the horizontal rods *a b, b d* (Fig. 9) so that they may approach the magnetic bar shown between them, in place of the knobs *b b*, screw on little pincers which take hold of the magnetic bar, and attach one pole of a pile to *a*, and the other to *d*, thus completing the voltaic circuit through the length of the magnet.

* *Essai Théorique et Expérimental sur le Galvanisme*, Paris, 1804, vol. i. p. 339.

" *Effects.* According to the observations of Romagnosi the magnet experiences a declination, and according to those of Mojon needles not previously magnetised acquire by this means a sort of magnetic polarity." *

In a paper read before the Royal Academy of Munich, in May 1805, Ritter, a Bavarian philosopher, advanced some curious speculations, which, although always quoted, as suggestive of electro-magnetism, are really as wide of the mark as the experiments of Romagnosi, Schweigger, and Mojon. We find them thus described in the *Philosophical Magazine*, for 1806 :†—

"The pile with which M. Ritter commonly performs his experiments consists of 100 pairs of plates of metal, two inches in diameter; the pieces of zinc have

* Mr. Sabine appears to have studied Izarn, yet he writes thus, at p. 23 of his *History and Progress of the Electric Telegraph*, 2nd edit., London, 1869 :—" After explaining the way to prepare the apparatus, *which consists simply in putting a freely suspended magnet needle parallel and close to a straight metallic conductor* through which a galvanic current is circulating, he describes the effects in the following words," &c. The words that we have italicised are altogether misleading.

† Vol. xxiii. p. 51. "An ingenious and extraordinary man, from whom much might have been expected, had nature permitted the continuance of his scrutiny into her secret operations. A premature death deprived the world of one whose constitutional singularity of opinion, ardency of research, and originality of invention, rendered him at once systematic in eccentricity, inexhaustible in discovery, and ingenious even in error."—Donovan's *Essay on the Origin, Progress, and Present State of Galvanism*, Dublin, 1816, p. 107.

Johann Wilhelm Ritter was born December 16, 1776, and died at Munich, January 23, 1810.

a rim to prevent the liquid pressed out from flowing away, and the apparatus is insulated by several plates of glass.

" As he resides at present near Jena I have not had an opportunity of seeing experiments with his great battery of 2000 pieces, or with his battery of 50 pieces, each thirty-six inches square, the action of which continues very perceptible for a fortnight. Neither have I seen his experiments with the new battery of his invention, consisting of a single metal, and which he calls *the charging pile*.*

"I have, however, seen him galvanise a louis d'or. He places it between two pieces of pasteboard thoroughly wetted, and keeps it six or eight minutes in the circuit of the pile. Thus it becomes charged, though not immediately in contact with the conducting wires. If applied to the recently bared crural nerves of a frog the usual contractions ensue. I put a louis d'or thus galvanised into my pocket, and Ritter told me, some minutes after, that I might discover it from the rest by trying them in succession upon the frog. I made the trial, and actually distinguished, among several others, one in which only the exciting quality was evident.

"The charge is retained in proportion to the time that the coin has been in the circuit of the pile. Thus,

---

* The charging pile, or, as we now call it, the secondary battery, was first described by Gautherot in 1801. See Izarn's *Manuel du Galvanisme*, Paris, 1804, p. 250; also *Phil. Mag.*, for 1806, vol. xxiv. p. 185.

of three different coins, which Ritter charged in my
presence, none lost its charge under five minutes.

"A metal thus retaining the galvanic charge, though
touched by the hand and other metals, shows that this
communication of galvanic virtue has more affinity
with magnetism than with electricity, and assigns to
the galvanic fluid an intermediate rank between the
two.

"Ritter can, in the way I have just described, charge
at once any number of pieces. It is only necessary
that the two extreme pieces of the number communicate with the pile through the intervention of wet
pasteboards. It is with metallic discs charged in this
manner, and placed upon one another, with pieces of
wet pasteboard alternately interposed, that he con-
structs his charging pile, which ought, in remembrance
of its inventor, to be called the *Ritterian pile.* The
construction of this pile shows that each metal galvanised in this way acquires polarity, as the needle does
when touched with a magnet.*

    *   *   *   *   *   *

"After showing me his experiments on the different
contractibility of various muscles, Ritter made me

* We may here dispose of a paragraph which has hitherto puzzled
a good many writers, who have supposed it to refer to some kind of
magneto-electric machine. It occurs in *The Monthly Magazine,* for
April 1802, p. 268, and reads as follows :—
    "Galvanism is at present a subject of occupation of all the German
philosophers and chemists. At Vienna an important discovery has
been announced—an artificial magnet—employed instead of Volta's

observe that the piece of gold galvanised by communication with the pile exerts at once the action of two metals, or of one voltaic couple, and that the face, which in the voltaic circuit was next the negative pole, became positive; and the face towards the positive pole, negative.

"Having discovered a way to galvanise metals, as iron is rendered magnetic, and having found that the galvanised metals always exhibit two poles as the magnetised needle does, Ritter suspended a galvanised gold needle on a pivot, and perceived that it had a certain dip and variation, or deflection, and that the angle of deviation was always the same in all his experiments. It differed, however, from that of the magnetic needle, and it was the positive pole that always dipped." *

Ritter also observed that a needle composed of silver and zinc arranged itself in the magnetic meridian, and was slightly attracted and repelled by the poles of a magnet; and, again, that a metallic wire through which a current had been passed took up of itself a N.E. and S.W. direction.

----

pile, decomposes water equally well as that pile, or the electrical machine; whence it has been concluded that the electric, galvanic, and magnetic fluids are the same." Clearly the artificial magnet here mentioned can be none other than Ritter's secondary pile. One thing is certain, it cannot be a magneto-electric machine, for magneto-electricity was not known in 1802.

* C. Bernoulli, in *Van Mons' Journal*, vol. vi. See further on this subject in *Phil. Mag.*, vol. xxv. pp. 368–9.

As the result of all these observations the Bavarian philosopher concluded that "electrical combinations, when not exhibiting their electric tension, were in a magnetic state ; and that there existed a kind of electro-magnetic meridian depending on the electricity of the earth, and at right angles to the magnetic poles." * These speculations are, as we see, sufficiently obscure, and, like those that we have hitherto described, failed to throw any light on the relation so anxiously sought after.

Nor can we give Oersted credit at this period for any more distinct apprehensions.   In a work which

---

* *Phil. Mag*, vol. lviii. p. 43.   It is curious to note that the English philosophers entirely neglected this study, being content to follow the brilliant lead of Sir Humphry Davy in another branch of the science.   Indeed, it seems to have been the general opinion in this country, as late as the year 1818, that there was nothing more to be discovered.   Bostock, in his *Account of the History and Present State of Galvanism*, published in London in that year, says :—

"Although it may be somewhat hazardous to form predictions respecting the progress of science, I may remark that the impulse, which was given in the first instance by Galvani's original experiments, was revived by Volta's discovery of the pile, and was carried to the highest pitch by Sir H. Davy's application of it to chemical decomposition, seems to have, in a great measure, subsided.   It may be conjectured that we have carried the power of the instrument to the utmost extent of which it admits ; and it does not appear that we are at present in the way of making any important additions to our knowledge of its effects, or of obtaining any new light upon the theory of its action" (p. 102).

Napoleon did not hold these views.   In the First Consul's letter to the Minister, Chaptal, founding two prizes to encourage new researches in galvanism, he said :—"Galvanism, in my opinion, will lead to great discoveries."

he published in German, in 1807, on the identity of chemical and electrical forces, he observes :*—

"When a plate composed of several thin layers is electrified, and the layers afterwards separated, each is found to possess an electric polarity, just as each fragment of a magnet possesses a magnetic polarity.

"There is, however, one fact which would appear to be opposed to the theory of the identity of magnetism and electricity. It is that electrified bodies act upon magnetic bodies, as if they [? the magnetic bodies] were endowed with no force in particular. It would be very interesting to science to explain away this difficulty ; but the present state of physics will not enable us to do so. It is, meanwhile, only a difficulty, and not a fact absolutely opposed to theory ; for we see in frictional electricity and in that of contact [galvanism] analogous phenomena. Thus, we can alter the tension of the electric pile by bringing near it an excited glass rod, and yet not affect in any way the chemical action. A long column of water, or a wetted thread of flax or wool, will also suffer a change in its electricity without experiencing any chemical changes.

"It would appear, then, that the forces can be superposed without interfering with each other when they operate under forms of different activities.

"The form of galvanic activity holds a middle place between those of magnetism and [static] electricity.

* Chap. viii. pp. 235-6 of the French edition, *Recherches sur l'Identité des Forces Chimiques et Électriques*, Paris, 1813.

The force is in that form more latent than as electricity, and less so than as magnetism. It is, therefore, probable that the electric force, when superposed, will exercise a less influence on magnetism than on galvanism. In the galvanic pile, it is the electric state [tension] which it acquires that is affected by the approach of an excited glass rod; more, it is not that interior distribution of forces constituting magnetism that we can change by electricity, but it is the electric state which belongs to the magnet as to bodies in general.

"We do not pretend to decide anything in this matter; we only wish to clear up, as far as possible, a very obscure subject, and, in a question of such importance, we shall be very well satisfied if we have made it apparent that the principal objection to the identity of the forces which produce electricity and magnetism is rather a difficulty of reconciling facts than of the facts themselves."

And again, on p. 238, he says:—"Steel when heated loses its magnetism, showing that it becomes a better conductor by the elevation of temperature, like electrical bodies. Magnetism, too, like electricity, exists in all bodies in nature, as Bruckmann and Coulomb have shown. From this it seems that the magnetic force is as general as the electric; and it remains to be seen whether electricity in its most latent state [*i. e.*, as galvanism] will not affect the magnetic needle *as such*.

" This experiment will not be made without difficulty, for the electrical actions will blend and render the observations very complicated. In comparing the attractions on magnetic and non-magnetic bodies, some *data* will probably be obtained."

In trying experiments with a view to the illustration of these hazy notions Oersted is said to have succeeded in obtaining indications of the action of the conducting wires of the pile, during the passage of electricity, on the needle ; but the phenomena were, at first view, not a little perplexing ; and it was not till after repeated investigation that, in the winter of 1819–20, the real nature of the action was satisfactorily made out.*

Even then Oersted seems not to have clearly understood the full significance of his own experiment. Unlike Davy, who, when he first saw the fiery drops of potassium flow under the action of his battery, recorded his triumph in a few glowing words in his laboratory journal,† Oersted took no immediate steps,

* " Professor Forchhammer, the pupil and friend of Oersted, states that, in 1818 and 1819, it was well known in Copenhagen that he was engaged in a special study of the connection of magnetism and electricity. Yet we must ascribe it to a happy impulse—the result, no doubt, of much anxious thought—that, at a private lecture to a few advanced students in the winter of 1819–20, he made the observation that a wire uniting the ends of a voltaic battery in a state of activity affected a magnet in its vicinity."—*Ency. Brit*, 8th ed., Dissertation vi. p. 973.

† On 16th October, 1807, while investigating the compound nature of the alkalies. On seeing the globules of potassium burst through the crust of the potash, and take fire as they entered the atmosphere, he could not contain his joy, but danced about the room in wild delight,

T

either to complete, or to publish, his discovery. "Although," he says, "the effect was unquestionable, it appeared to me, nevertheless, so confused that I deferred a minute examination of it to a period at which I hoped for more leisure." * And when he had made this minute examination and published the results, he could not explain the phenomena by a better hypothesis than that negative electricity acts only on the northern pole, and positive only on the southern pole of the needle.†

This most important discovery may be thus briefly defined :—Supposing the electric current to pass from north to south through a wire, placed horizontally in the magnetic meridian, then a compass needle suspended above it will have its north end turned towards the west; if below the wire, to the east; if on the east side of it, the north end will be raised; and if on the west side, depressed. These results Oersted first published in a Latin tract, dated the 21st July, 1820, a copy of which (with translation in English), will be found in the *Journal of the Society of Telegraph Engineers*, vol. v. pp. 459–69.

---

and some time elapsed before he could sufficiently compose himself to continue his experiments.—Bakewell's *Manual of Electricity*, London, 1857, p. 34.

* Tyndall's Lectures on Voltaic Electricity at the Royal Institution, 1876.

† See concluding paragraph of his paper in the *Journal of the Soc. of Tel. Engs*, vol. v. p. 468.

CHAPTER X.

ELECTRO-MAGNETISM AND MAGNETO-ELECTRICITY —HISTORY IN RELATION TO TELEGRAPHY (*continued*).

THE effect of Oersted's pamphlet was most wonderful. The enthusiasm, says Lardner,* which had been lighted up by the great discovery of Volta twenty years before, and which time had moderated, was relumined, and the experimental resources of every cabinet and laboratory were brought to bear on the pursuit of the consequences of this new relation between sciences so long suspected of closer ties. The inquiry was taken up, more particularly, by Ampère and Arago, in France; by Davy, Faraday, Cumming, and Sturgeon, in England; and by Seebeck, Schweigger, De la Rive, Henry, and numerous other philosophers in all parts of Europe and America.

Among these, Ampère has assumed the first and highest place. No sooner was the fact discovered by Oersted made known, than that philosopher commenced the beautiful series of researches which has surrounded his name with so much lustre, and

* *Electricity, Magnetism, and Meteorology*, vol. i. p. 205.

T 2

brought electro-dynamics within the pale of mathematical physics. On the 18th of September, 1820,
within less than two months of the publication of
Oersted's experiments, he communicated his first
memoir on electro-magnetism to the Academy of
Sciences.

In this paper was explained the law which determined the position of the magnetic needle in relation
to the electric current. In order to illustrate this, he
proposed that a man should imagine the current to
be transmitted through his body, the positive pole
being applied to his feet, and the negative pole to
his head, so that the current shall pass upwards from
the feet to the head. This being premised, a magnetic
needle, freely supported on its centre of gravity, and
placed before him, will throw itself at right angles to
him ; the north pole pointing towards his left, and the
south pole towards his right.

If the person through whose body the current thus
passes turn round, so as to present his face in different
directions, a magnetic needle, still placed before him,
will have its direction determined by the same condition ; the north pole pointing always to the left, and
the south to the right.

In the same memoir were described several instruments intended to be constructed ; especially spiral, or
helical, wires, through which it was proposed to transmit the electric currents, and which, it was expected,
would thereby acquire the properties of magnets, and

retain these properties so long as the current might be transmitted through them. The author also explained his theory of magnets, ascribing their attractive and directive powers to currents of electricity circulating constantly round their molecules, in planes at right angles to the line joining their poles; the position of the poles, on the one side or the other of these planes,* depending on the direction of the revolving current.

While Ampère was proceeding with these researches, Arago directed his inquiries to the state of the wire through which the current was transmitted, so as to determine whether every part of its surface was endowed with the same magnetic properties. With this view, he placed iron filings around the wire, and found that they adhered to it so long as the current flowed, and fell away immediately the connection with the battery was broken. He also found that on placing small steel needles across the wire through which a current from a voltaic pile, or a discharge from a Leyden jar, was sent, they were attracted, and, on removal, were found to be permanently magnetised. Acting upon Ampère's theory of magnetism, he placed in a glass tube an ordinary sewing needle, and wound round the tube a copper wire. On sending a current through this wire the needle was magnetised, its polarity depending on the

* *Annales de Chimie et de Physique*, Paris, 1820, vol. xv. pp. 59 and 170.

direction of the current. If the helix were right-handed, the north pole was found at the end at which the current entered; and if left-handed, the same end was a south pole. In the same way he was able to impart a temporary magnetism to soft iron wires.*

Another important discovery, which followed fast on the heels of Oersted's experiments, was that of Schweigger, of Halle, announced on the 16th September, 1820. Observing that the deflection produced by the outward current of a battery flowing over the needle was the same as that of the return current under the needle, he made the wire proceed from and to his battery above and beneath the needle, and obtained, as he expected, twice the effect; by giving the wire another turn round the needle the effect was again doubled; a third turn produced six times the original deviation; a fourth, eight times, and so on. This effect may be thus formulated : — If a magnetised needle, free to move, be surrounded by a number of convolutions of insulated wire, the power of the current to deflect it will increase in proportion to the number of con-

---

* *Annales de Chimie et de Physique*, vol. xv. p. 93. Soon after, and before any knowledge of Arago's experiments had reached England, Davy also succeeded in magnetising needles by the voltaic current, as well as by ordinary frictional electricity, and showed the effect of the conducting wire on iron filings. See his letter to Wollaston, dated November 12, 1820, in the *Phil. Trans.*, for 1821. About the same time Seebeck communicated a paper to the Berlin Academy on the same subject.

volutions.* In this way the effect of a very feeble current may be so multiplied as to produce as great a deviation of the magnetic needle as would otherwise be produced by a very strong current.

On this principle are constructed instruments for indicating and measuring currents of electricity, called *electro-magnetic multipliers*, or, more commonly, *galvanometers* †—the former being the name originally given to the arrangement by Schweigger. His first contrivance was a very humble affair, consisting of a small compass-box, round which were coiled several turns of copper wire in a direction parallel to the meridian line of the card ‡ Yet this was the prototype of the beautiful instruments of Du Bois-Reymond and Sir William Thomson, in the former

---

* The practical reader is, of course, aware that this definition is not strictly true,—for three reasons : 1st, as the convolutions increase, the strength of the current decreases, by reason of the increased resistance in the circuit ; 2nd, each convolution has less and less effect, as it is farther and farther removed from the needle ; and, 3rd, the current exerts less and less force on the needle, as it is deflected farther and farther from the plane of the current.

† In the early part of the century this name was applied to measuring instruments based on the chemical and calorific properties of the current ; but these are now denominated *voltameters*, and the name *galvanometer* is reserved exclusively for the class of apparatus described in the text.

‡ Schweigger's *Journal fur Chemie und Physik*, vol. xxxi pp. 1-17. A galvanometer of different form, called a *galvano-magnetic condensator*, with vertical coils and unmagnetised needle, was shortly after, but independently, devised by the celebrated Poggendorff, then a student at Berlin. As the *published* description of his apparatus preceded that of Schweigger's, he is sometimes regarded as the first inventor (Gilbert's *Annalen der Physik*, vol. lxvii. pp. 422-29).

of which as many as 30,000 convolutions are sometimes employed.

There was, however, still wanting another discovery
to bring the galvanometer to its present perfection,
and this want was soon supplied. In deflecting a
magnetic needle the current acts against the directive force of terrestrial magnetism ; hence it is clear
that if this force could be neutralised the deflection
would be greater ; in other words, a galvanometer, in
which the needle is freed from the controlling action
of the earth's magnetism, would be more sensitive
than the same galvanometer when its needle was
not so freed. Ampère suspended a single needle
so that the earth's magnetism acted perpendicularly
to it, and had, therefore, no directive force upon it ;
and he found that it set accurately at right angles to
the current.

This led him to the invention of the double, or
astatic, needles, which he thus describes in his memoir
of 1821 :—

"When a magnetic needle is withdrawn from the
directive action of the earth, it sets itself, by the
action of a voltaic conductor, in a direction which
makes a right angle with the direction of the conductor, and has its south pole to the left of the
current against which it is placed ; so that if M.
Oersted, in the experiments which he published in
1820, only obtained deviations of the needle which
were less than a right angle, on placing it above or

below a conducting wire parallel to its direction, it was solely because the needle which he subjected to the action of the current was not withdrawn from that of the earth, and took consequently an intermediate position between the directions which the two forces tended to give it. There are several means of withdrawing a magnetic needle from the earth's action. A very simple one consists in attaching to a stout brass wire, which has its upper part curved and fitted with a steel point of suspension, two magnetic needles of equal strength, in such a manner that their poles are in opposite directions, so that the directive force of the earth upon one is destroyed by the action in the opposite direction which it exercises on the other. The needles are so arranged that the lower one is just below the conducting wires, and the upper one close above them. On sending a current through the convolutions the needles turn, until they take a direction at right angles with the conducting wire." *

Having oscillated a magnetised needle, freely suspended in a circular copper cage, the bottom and sides

---

* *Annales de Chimie et de Physique*, vol. xviii. p. 320. In Professor Cumming's paper *On the connection of Galvanism and Magnetism*, read before the Cambridge Philosophical Society on April 2, 1821, he described a near approach to the astatic needle. In order to neutralise the terrestrial magnetism he placed a small magnetised needle under the galvanometer needle.—*Trans. Cam. Phil. Soc*, vol. i. p 279. The credit of Ampère's discovery is usually attributed to Nobili. As in Noad's *Manual of Electricity*, London, 1859, p 327; also Roget's *Electro-Magnetism*, in Library of Useful Knowledge, London, 1832, p. 42.

of which were very near the needle, Arago, in 1824, noticed that the oscillations rapidly diminished in extent, and very quickly ceased, as if the medium in which they were being produced had become more and more resistant. The proximity of the copper, while thus checking the *amplitude* of the oscillations, was observed to have no effect on their *duration*, they being accomplished in exactly the same time as in free air. By making the needle oscillate at different distances above discs of different materials, Arago found that distance considerably diminished the effect ; and that metals acted with more energy than wood, glass, &c.*

Arago now conceived the idea of trying whether the disc which possessed this remarkable property would not draw the needle with it if itself rotated. The experiment was tried and resulted in the discovery of a new class of phenomena to which its author gave the name of *magnetism by rotation.* If we fix to a rotation apparatus, such as a table made for experiments on centrifugal force, a copper disc, about twelve inches diameter, and one-tenth inch thick, and just above it suspend, by a silk fibre, a magnetic needle, in such a manner that its point of suspension is exactly above the centre of the disc (care being taken to interpose a

---

* *Annales de Chim. et de Physique,* vol. xxvii. p. 363. Seebeck, of Berlin, on repeating these experiments two years later, obtained analogous results. See *Pogg. Ann* , vol. vii. We shall see further on, pp. 321, and 336-7, the use that has been made of this fact in the telegraphs of Gauss and Weber, and Steinheil.

screen of glass or paper, so that the agitation of the air resulting from the motion impressed upon the disc may have no effect upon the needle), and then put the disc in rotation, the needle is seen to deviate in the direction of this rotation, and to make with the magnetic meridian a greater or less angle according to the velocity with which the disc is revolved. If this movement be very rapid, the needle is deflected more and more, until finally it rotates with the disc.

The effect diminishes very rapidly with the distance of the needle from the disc ; and is still further lessened by cutting slits in the latter in the direction of rays— a fact which, as our practical readers know, is of the highest importance in the construction of electro-magnets.

Whilst Arago was analysing the force that he had discovered, Babbage and Herschel, Barlow, Harris,* and others, undertook an investigation of the causes that may vary its intensity. Messrs. Babbage and Herschel repeated Arago's experiment by inverting it. They found that discs of copper, or other substances, when freely suspended over a rotating horse-shoe magnet, turned in the same direction as the magnet, with a movement at first slow, but which gradually increased in rapidity. The interposition of plates of glass and of non-magnetic metallic bodies in no degree

---

* For researches of the three first-named philosophers, see *Philosophical Transactions*, for 1825 ; for those of Sir W. Snow Harris, see same for 1831.

affected the results ; but it was not the same with plates of *iron*. The action was then greatly reduced, or even entirely annihilated.

These two philosophers confirmed the accuracy of Arago's observations on the influence of solutions of continuity, either partial or total, in the discs subjected to experiment. Thus, a light disc of copper, suspended at a given distance above a magnet, executed its (6) revolutions in 55". When cut in eight places, in the direction of radii near the centre, it required 121" to execute the same number ; but, on the parts cut out being again soldered in with tin, the original effect was almost attained, the disc performing its revolutions in 57". The same effects were obtained with other metals.

Sir W. Snow Harris, who made a great number of experiments on this subject, not only found great differences between bodies with regard to their power of drawing the needle after them when rotating, but also with regard to the property they possess of intercepting this action. He observed that iron, and magnetic substances generally, are not the only ones that are thus able to arrest the effect of magnetism by rotation. Plates of non-magnetic substances, such as copper, silver, zinc, will do the same, provided only they be sufficiently thick, as from three to five inches.

From a study of all these experiments Christie deduced the law that the force with which different substances draw along the magnetic needle in their rotatory movement is proportional to their conducting power for electricity. But a full explanation of these phenomena could not be given until after Faraday's discoveries in 1831, when it was seen that they were, one and all, the result of the electric currents induced in the disc by its rotation in the field of the magnet. In the case of those discs in which slits, or rays, were cut, the free circulation of these currents was prevented, and, consequently, there was no effect on the needle.*

In November 1825, a great advance was made on Arago's experiment of magnetising soft iron, by the invention of the electro-magnet—an instrument which, in one form or another, has become the basis of nearly every system of electric telegraphy. We owe this most important contrivance to Sturgeon, a well-known electrician of Woolwich, who had worked in his earlier days at the cobbler's last,† as Franklin had done at the printing stick, and Faraday at bookbinding. Fig. 10 shows the earliest form of the instrument—a piece of stout iron wire, bent into the form of a horse-shoe,

---

* See Faraday's *Experimental Researches*, 1831 ; also Henry's classical paper on *Electro-dynamic Induction*, in *Trans. Amer. Phil. Society*, for 1839, vol. vi. p. 318.

† He was apprenticed to a shoemaker, and disliking the employment, at the age of nineteen entered the Westmoreland Militia, and two years later enlisted in the Royal Artillery While in this corps he devoted his leisure to scientific studies, and made himself familiar with all the great facts of electricity and magnetism, which were then opening on the world. His subsequent career has created for him an undying name in the annals of electricity.

coated with an insulating varnish, and then bound round loosely with bare copper wire, the turns (of which there were sixteen) being, of course, separated from each other. This electro-magnet, when excited by a

FIG. 10.

single voltaic pair of large (130 square inches) surface, was capable of supporting a weight of nine pounds, a wonderful performance in those days.*

Some of the further steps in the perfection of the electro-magnet as used in telegraphy were made by Professor Henry in America, between the years 1828 and 1831, and it will be interesting to retrace them here, if only to see how little learned professors, fifty years ago, understood the conditions underlying the conversion of voltaic into magnetic force, and consequently how much groping in the dark, and stumbling to conclusions, where now Ohm's celebrated law makes everything so clear.

Henry was led to his first improvements in electro-magnets by a study of Schweigger's galvanometer,

---

* *Transactions Society of Arts*, 1825, vol. xliii. pp. 38–52.

which resulted in the idea that a much nearer approximation to the requirements of Ampère's theory could be attained by insulating the conducting wire itself, instead of the rod to be magnetised, and by covering the whole surface of the iron with a series of coils in close contact.

In June 1828, he exhibited at the Albany Institute of New York, of which he was then professor, his electro-magnet, constructed on this principle. It consisted of a piece of soft iron, bent in the form of a horse-shoe, and closely wound with silk-covered copper wire, one-thirtieth of an inch in diameter. In this way he was able to employ a much larger number of convolutions, while each turn was more nearly at right angles with the magnetic axis of the bar. The lifting power of this magnet was, conformably to Henry's anticipations, much greater, *cæteris paribus*, than that of Sturgeon.

In March 1829, he exhibited, at the same place, a somewhat larger magnet of the same character. A round piece of iron, about one quarter inch diameter, was bent into the usual horse-shoe form, and tightly wound with thirty-five feet of silk-covered wire, in about four hundred turns, with silk ribbon between. A pair of small battery plates, which could be dipped into a tumbler of dilute acid, were soldered, one to each end of the wire, and the whole mounted on a stand. With this small battery the magnet could be much more powerfully excited than another of the same sized core, wound according to the method of Sturgeon and excited by a battery of twenty-eight plates of copper and zinc, each plate eight inches square.*

"In the arrangement," says Henry, "of Arago and Sturgeon, the several turns of wire were not precisely at right angles to the axis of the rod, as they should be to produce the effect required by the theory, but slightly oblique, and, therefore, each tended to develop a separate magnetism not coincident with the axis of the bar. But in winding the wire over itself, the obliquity of the several turns compensated each other, and the resultant action was at the required right angles. The arrangement, then, introduced by myself was superior to those of Arago and Sturgeon, first, in the greater multiplicity of turns of wire, and second, in the better application of these turns to the development of magnetism.†

"The maximum effect, however, with this arrangement and a single battery was not yet obtained. After a certain length of wire had been coiled upon the iron, the power diminished with a further increase of the number of turns. This was due to the increased resistance which the longer wire offered to the con-

---

* *Smithsonian Report*, 1878, p. 282.

† "When this conception," said Henry, "came into my brain, I was so pleased with it that I could not help rising to my feet and giving it my hearty approbation." It was his first discovery. See Professor Mayer's Eulogy of Henry, before the American Association for the Advancement of Science, 1880.

duction of electricity. Two methods of improvement, therefore, suggested themselves. The first consisted, not in increasing the length of the coil, but in using a number of separate coils on the same piece of iron. By this arrangement the resistance to the conduction of the electricity was diminished, and a greater quantity made to circulate around the iron from the same battery. The second method of producing a similar result consisted in increasing the number of elements of the battery, or, in other words, the projectile force of the electricity, which enabled it to pass through an increased number of turns of wire, and thus to develop the maximum power of the iron." [*]

Employing a horse-shoe, formed from a cylindrical bar of iron, half an inch in diameter, and about ten inches long, and wound with thirty feet of fine copper wire, he found that, with a current from only $2\frac{1}{2}$ square inches of zinc, the magnet held 14 lbs.[†] Winding upon its arms a second wire of the same length (30 feet) whose ends were similarly joined to the same galvanic pair, the magnet lifted 28 lbs. On these results Henry remarks :—

" These experiments conclusively proved that a great development of magnetism could be effected by a very small galvanic pair, and also that the power of the coil was materially increased by multiplying the

[*] *Smithsonian Report*, for 1857, p. 102.

[†] It must not be forgotten that at the time when this experimental magnet was made, the strongest electro-magnet in Europe was that of Sturgeon mentioned on p. 285, and then considered a prodigy.

U

number of wires, without increasing the length of each. The multiplication of the wires increases the power in two ways : first, by conducting a greater quantity of galvanism, and secondly, by giving it a more proper direction ; for, since the action of a galvanic current is directly at right angles to the axis of a magnetic needle, by using several shorter wires we can wind one on each inch of the length of the bar to be magnetised, so that the magnetism of each inch will be developed by a separate wire. In this way the action of each particular coil becomes directed very nearly at right angles to the axis of the bar, and consequently the effect is the greatest possible. This principle is of much greater importance when large bars are used. The advantage of a greater conducting power from using several wires might, in a less degree, be obtained by substituting for them one large wire of equal sectional area ; but in this case the obliquity of the spiral would be much greater, and consequently the magnetic action less." *

In the following year, 1830, Henry pressed forward his researches to still higher results, assisted by his friend, Dr. Philip Ten-Eyck. "A bar of soft iron, 2 inches square, and 20 inches long, was bent into the form of a horse-shoe $9\frac{1}{2}$ inches high (the sharp edges of the bar were first a little rounded by the hammer); it weighed 21 lbs. A piece of iron from the same bar weighing 7 lbs., was filed perfectly flat on one surface

* Silliman's *American Journal of Science*, Jan. 1831, vol. xix p. 402.

for an armature, or lifter. The extremities of the legs of the horse-shoe were also truly ground to the surface of the armature. Around this horse-shoe 540 feet of copper bell-wire were wound in nine coils of 60 feet each; these coils were not continued around the whole length of the bar, but each strand of wire (according to the principle before mentioned) occupied about two inches, and was coiled several times backward and forward over itself. The several ends of the wires were left projecting, and all numbered, so that the first and the last end of each strand might be readily distinguished. In this manner we formed an experimental magnet on a large scale, with which several combinations of wire could be made by merely uniting the different projecting ends. Thus, if the second end of the first wire be soldered to the first end of the second wire, and so on through all the series, the whole will form a continued coil of one long wire. By a different arrangement the whole may be formed into a double coil of half the length, or into a triple coil of one-third the length, and so on. The horse-shoe was suspended in a strong rectangular frame of wood, 3 feet 9 inches high, and 20 inches wide." *

The accompanying figure, which we copy from the *Scientific American*, December 11, 1880, is an exact representation of this instrument, which is at present preserved in the College of New Jersey.

Two of the wires, one from each leg, being soldered

* Silliman's *Journal*, for 1831.

FIG. 11.*

* The coil at the right of the engraving represents the original silk-covered ribbon wire used by Henry in his celebrated experiments on induction. In the middle of the foreground is one of his pole-changers, which could also be used as a circuit breaker. He was accustomed to delight himself and his classes with this by making and breaking the current so quickly that a 28-lb. armature could not fall off, but was freed and attracted with a sharp snap.

together so as to form a single circuit of 120 feet, gave a lifting power of 60 lbs. The same two wires, when connected with the battery so as to form double circuits of 60 feet each, produced a lifting power of 200 lbs.; and four wires used in the same way supported as much as 500 lbs. Six wires united in three pairs, so as to form three circuits of 180 feet each, gave a lifting power of only 290 lbs.; while the same wires, when separately connected, as six parallel circuits, supported 570 lbs., or nearly double. When all the nine wires were joined up in parallel circuits with the battery, a lifting power of 650 lbs. was produced.*

In all these experiments a small single pair was used, consisting of two concentric copper cylinders, with a zinc one between, the active surface of which (on both sides) amounted to only two-fifths of a square foot. The exciting liquid consisted of half a pint of dilute sulphuric acid.

A maximum portative force of 750 lbs. was obtained from a zinc-copper pair of 144 inches of active surface, all nine coils being joined in multiple arc.†

---

* Henry was called to the chair of Natural Philosophy in the College of New Jersey, at Princeton, in 1832, and there he made two larger magnets for use in his investigations. One weighing 59½ lbs., and capable of sustaining 2063 lbs., is now in the cabinet of Yale College. The other, made in 1833, weighed 100 lbs., and could support 3500 lbs. It was many years before any magnet approaching this in power was constructed.

† Silliman's *Journal*, Jan. 1831. With a pair of plates, exposing exactly one square inch surface, the same arrangement of the coils could sustain a weight of 85 lbs.!

The only European physicist, who, up to this time, 1830, had obtained any results even approaching these, was Gerard Moll, professor of natural philosophy in the University of Utrecht, who having seen in London, in 1828, an electro-magnet of Sturgeon which could support 9 lbs., determined to try the effects of a larger galvanic apparatus. Having formed a horse-shoe, $12\frac{1}{2}$ inches high, and $2\frac{1}{4}$ inches diameter, he surrounded it with 26 feet of insulated copper wire, one-eighth of an inch thick, in a close coil of forty-four turns. The weight of the whole was about 26 lbs. With a current from a pair of 11 square feet of active (zinc) surface, this magnet sustained 154 lbs. This result was considered astonishing in Europe, yet Henry's horse-shoe, less in size and weight, supported nearly five times this load, with one-eleventh of Moll's battery power.[*]

After finding that the maximum attractive power was obtained by his artifice of multiple coils, Henry proceeded to experiment with electro-magnets formed of one long coil; and soon he was rewarded by a new discovery, namely, that, though multiple coils yielded the greatest attractive power close to the battery, one long continuous coil permitted a weaker attractive power to be exercised at a great distance, or through a great length of intervening wire.

Employing his earlier and smaller magnet of 1829,

[*] Brewster's *Edinburgh Journal of Science*, October 1830, p. 214.

formed of a quarter-inch rod, and wound with 8 feet of insulated copper wire; he tried the effects of different battery powers, of different lengths of external wire, and of different lengths of coil. Excited with a single pair of zinc and copper, having 56 square inches of active surface, the magnet alone in the circuit sustained $4\frac{1}{2}$ lbs. With 500 feet of copper wire, ·045 inch diameter, interposed between battery and magnet, the weight supported was only two ounces, or thirty-six times less than in the first case. With 1000 feet of wire interposed, the lifting power of the magnet was only half an ounce.

Using now a trough battery of twenty-five pairs, the magnet in direct connection (which, with a single pair, had supported $4\frac{1}{2}$ lbs.) lifted seven ounces, while with the thousand feet of interposed wire it sustained *eight* ounces.

"From this experiment," says Henry,[*] "it appears that the current from a galvanic *trough* is capable of producing greater magnetic effect on soft iron after traversing more than one-fifth of a mile of intervening wire than when it passes only through the wire surrounding the magnet. It is possible that the different states of the trough with respect to dryness may have exerted some influence on this remarkable result; but that the effect of a current from a trough, if not increased, is but slightly diminished in passing through

---

[*] Silliman's *Journal*, January 1831, p. 403. Here is an instance of the stumbling to conclusions of which we spoke on p. 286.

a long wire is certain. * * * From these experiments it is evident that, in forming the coil, we may either use one very long wire, or several short ones, as circumstances may require. In the first case, our galvanic combination must consist of a number of plates, so as to give 'projectile' force; in the second, it must be formed of a single pair."

Henry was thus the first to practically work out the different functions of two entirely different kinds of electro-magnet; the one, of numerous short coils, which he called the *quantity* magnet, and the other, of one very long coil, which he designated the *intensity* magnet. The former and more powerful, although little affected by a battery of many plates, was fully charged by a single pair; while the latter and feebler, which was but slightly affected by a single pair, was not only greatly excited by a battery of numerous elements, but was capable of receiving this excitation from a distant source.

In fact, Henry * had experimentally established the important principles at which Ohm had, a short time before, arrived from purely theoretical considerations, and which are now so universally applied under the name of *Ohm's Laws.* A corollary of these, *viz.*, that, by combining an *intensity* battery, of many small pairs, with an *intensity* magnet, of a long fine wire, a very long intervening conductor can be employed without sensible diminution of the effect—

* For more about Henry, see Appendix A.

is a fact which lies at the root of every system of electro-magnetic telegraphy.*

In the course of these pages we have had abundant evidence of the fact that motion could produce electricity, and electricity motion. Dessaignes showed us how difference of temperature, or heat, could produce electricity; † and Peltier gave us the strict converse of this in the conversion of electricity into heat, including both its relations—hot and cold; again, we have seen how the nervous force in certain fishes could

---

* In 1827, Georg Simon Ohm, professor of physics at Munich, published his celebrated formulæ; but for many years they failed to attract attention, and were no doubt unknown to Henry in 1830, as they were to Wheatstone in 1837. Numerous researches have, since Henry's time, been made with the view of determining in a rigorous manner the conditions necessary for obtaining the greatest electro-magnetic force For these, see Ganot's *Physics*, London, 1881, p. 783; Noad's *Text Book of Electricity*, London, 1879, p. 285; Du Moncel's *Elements of Construction for Electro-Magnets*, London, 1883, *passim;* and the back volumes of *The Electrician*, for papers by Schwendler, Heaviside, &c.

† In 1815, or six years before Seebeck, who is always credited with the observation. (See Bostock's *History of Galvanism*, London, 1818, p. 101.) Many observations bearing on *thermo-electricity* had been made even long before Dessaignes. Passing by that of Theophrastus, 321 B.C., that tourmaline could be electrified by friction, as irrelevant, since he does not appear to have had any idea that the effect might be due to heat produced by the friction, we find that, in the year 1707, the thermo-electric properties of tourmaline were unmistakably pointed out by a German author, "J. G. S," in his *Curious Speculations during Sleepless Nights*. In 1759, Æpinus called attention to the same phenomena, and pointed out that electricity of opposite kinds was developed at opposite ends of the crystal. In 1760, Canton observed the same properties in the topaz; and between 1789 and 1791, Haüy showed the thermo-electric properties of various other substances, as mesotype, prehnite, Iceland spar, and boracite.— Priestley's *History of Electricity*, 1767, pp. 314–26.

produce electricity, the converse of which was long and vainly sought after by Galvani and his disciples.

When, therefore, Oersted discovered the property of electricity to deflect a magnetic needle, and Arago its corollary—the magnetising power of the current, the conviction became strong that magnetism must be able in some way to produce electricity.

The credit of completely establishing this connection fell to the lot of our distinguished countryman, Michael Faraday. In his brilliant series of *Experimental Researches* commenced in 1831, he says :— " Certain effects of the induction of electrical currents have already been recognised and described ; as those of magnetisation, Ampère's experiments of bringing a copper disc near to a flat spiral, his repetition with electro-magnets of Arago's extraordinary experiments, and perhaps a few others. Still, it appeared unlikely that these could be all the effects which induction by currents could produce. * * *

" These considerations, with their consequence, the hope of obtaining electricity from ordinary magnetism, have stimulated me at various times to investigate experimentally the inductive effect of electric currents."

Faraday thus describes his first successful experiment :—" 203 feet of copper wire in one length were coiled round a large block of wood ; other 203 feet of similar wire were interposed as a spiral between the turns of the first coil, and metallic contact everywhere prevented by twine. One of these helices was connected with a galvanometer, and the other with a battery of 100 pairs. * * * When the contact was made, there was a sudden and very slight effect at the galvanometer, and there was also a similar slight effect when the contact with the battery was broken.   But whilst the current continued to flow through the one helix, no galvanometrical appearances, nor any effect like induction upon the other helix, could be perceived."

The same effects were produced in another way. Several feet of copper wire were stretched in wide zigzag forms, representing the letter W, on the surface of a broad board ; a second wire was stretched in precisely similar forms on a second board, so that when brought near the first, the wires should everywhere touch, except that a sheet of thick paper was interposed.   One of these wires was connected with a galvanometer and the other with a voltaic battery. The first wire was then moved towards the second, and as it approached the needle was deflected.   Being then removed, the needle was deflected in the opposite direction.   As the wires approximated, the induced current was in the *contrary* direction to the inducing current; and as they receded, the induced current was in the *same* direction as the inducing current.

Faraday next took a ring of soft iron, round the two halves of which he disposed two copper-wire coils. In passing a current through one coil, and thus magnetising the ring, a current was induced in the other coil, but, as in the former cases, only for an instant.

When the primary current ceased, and the magnet was unmade, an opposite current shot through the secondary coil. The primary coil was now suppressed, and the piece of soft iron embraced by the secondary coil was magnetised by a couple of powerful bar magnets, with which contact was alternately made and broken. Upon making contact the needle of the galvanometer was deflected ; continuing the contact, the needle became indifferent and resumed its first position, and on breaking contact it was again deflected in the opposite direction, and then became once more indifferent. When the magnetic contacts were reversed the deflections of the needle were also reversed.

In order to prove that the induced current was not occasioned by any peculiar effect taking place during the formation of the magnet, Faraday made another experiment in which soft iron was rejected, and nothing but a permanent steel magnet employed. The ends of the empty helix being connected as before with the galvanometer, either pole of the magnet was thrust into the axis, and immediately the needle was momentarily deflected. On rapidly withdrawing the magnet, a second and instantaneous deflection ensued, and in the opposite direction.

The strength of these induced currents depended on many circumstances ; as on the length and diameter of the wires of the coils, the energy of the inducing current, or the strength of the magnet, &c.

Hitherto, in order to produce the phenomenon of

induction by electric currents we have spoken of two conductors—one for the inducing, and another for the induced current; but experiment has shown that the same result can be obtained with only one conductor, and in this case the phenomenon is termed the induction of a current upon itself. In the sparking of relays and commutators of dynamo machines, &c., we have familiar examples of this action. Its discovery we owe to Professor Henry as far back as 1832, for he was the first to observe that, when the poles of a battery are united by means of a copper wire and mercury cups, a brilliant spark is obtained at the moment the circuit is broken by raising one end of the wire out of its cup of mercury. To obtain this effect it was found that the wire must not be less than twelve or fourteen yards long, and further, that if coiled into a helix the effect would be greatly increased.*

Faraday made a particular study of this phenomenon, and showed the existence of the *extra current* not only on the breaking, but also on the making of the circuit. To the former he gave the name of *extra current direct*, to the latter *extra current inverse*. The latter, of course, cannot be directly perceived, since it flows in the same circuit as the current of the battery itself, and cannot be developed until this current is established, and, consequently, not until the circuit is closed. Its presence, however, is shown in an indirect way by the well-known phenomenon of retardation in magnetisations by means of the electric current.

* Silliman's *Journal*, vol. xxii.

CHAPTER XI.

TELEGRAPHS BASED ON ELECTRO-MAGNETISM AND MAGNETO-ELECTRICITY.

> " The invention all admired ; and each how he
> To be the inventor missed ;—so easy seemed
> Once found, which yet unfound most would have thought
> Impossible."—Milton's *Paradise Lost*, book vi.

1820.—*Ampère's Telegraph.*

VERY soon after Oersted's discovery of the deflecting power of the current, La Place, the distinguished French mathematician, suggested its employment for telegraphic purposes ; and, on the 2nd October in the same year (1820), Ampère, in a paper read before the Paris Academy of Sciences, sketched out roughly a telegraph in which the signals were to be indicated by the deflection of small magnets placed under the wires. His idea was a purely theoretical one, and was thrown out simply *par parenthèse* in the course of his memoir.

He says :—" According to the success of the experiment to which La Place drew my attention, one could, by means of as many pairs of conducting wires and magnetic needles as there are letters, and by placing each letter on a separate needle, establish, by the aid of a pile placed at a distance, and which could be made to communicate by its two extremities with those of each pair of conductors, a sort of telegraph, which would be capable of indicating all the details that one would wish to transmit through any number of obstacles to a distant observer. By connecting with the pile a key-board whose keys would carry the same letters and establish the connection (with the various wires) by their depression, this means of correspondence could be established with great facility, and would only occupy the time necessary for touching at one end, and reading at the other, each letter." *

It will be seen from this passage, which we have literally translated from the original, that Ampère makes no mention of surrounding the needles with *coils of wire*, as is so frequently stated by writers on the telegraph. Indeed he could not then have even heard of the galvanometer ; for, although Schweigger's paper on the subject was read at Halle on the 16th September, 1820, it was not published until the November following.

1830.—*Ritchie's Telegraph.*

In order to increase the effect of the current on the needles, and to enable this effect to be delivered through a great length of intervening wire, Professor Fechner, of Leipsic, suggested, in 1829, enclosing the needles in the multiplier coils of Schweigger. He says, in his *Lehrbuch des Galvanismus :*—"There is

* *Annales de Chimie et de Physique*, vol. xv. p. 73.

no doubt that if the insulated wires of twenty-four multipliers, corresponding to the several letters of the alphabet, and situated at Leipsic, were conducted underground to Dresden, and there connected to a battery, we could thus obtain a means, probably not very expensive comparatively speaking, of transmitting intelligence from the one place to the other, by means of signals properly arranged beforehand. I confess it is a very seductive idea, to imagine that by some future development of such a system, a communication between the central point and the distant parts of a country can be established, which shall consume no time, like communication between the central point of our organism and its members by means of the nerves, by what appears to be a very analogous arrangement" (p. 269).*

Acting on this suggestion, Professor Ritchie, of the Royal Institution, London, improved upon Ampère's plan ; and, on the 12th of February, 1830, exhibited a model of a telegraph in which were twenty-six metallic circuits and twenty-six magnetic needles, each surrounded by a coil of wire.

The exhibit is thus referred to in the *Philosophical Magazine*, for 1830 (vol. vii. p. 212) :—"Feb. 12.— This evening Ritchie briefly developed the first principles of electro-magnetism, with a view of setting forth, in a distinct and practical manner, M. Ampère's proposal of carrying on telegraphic communication

* See note on p. 239.

by means of this extraordinary power. Of course, the principle consists in laying down wires, which at their extremities shall have coats [coils] of wire and magnetic needles so arranged, that, when voltaic connections are made at one end of the system, magnetic needles shall move at the other. This was done by a small telegraph constructed for the purpose, where, however, the communication was made only through a small distance, the principle being all that could be shown in a lecture-room." *

In his paper *On a Torsion Galvanometer*, Ritchie refers to the subject in these words :—" We need scarcely despair of seeing the electro-magnetic telegraph established for regular communication from one town to another, at a great distance. With a small battery, consisting of two plates of an inch square, we can deflect finely–suspended needles at the distance of several hundred feet, and consequently a battery of moderate power would act on needles at the distance of a mile, and a battery of *ten* times the power would deflect needles with the same force, at the distance of a *hundred* miles, and one of *twenty* times the force, at the distance of *four hundred* miles, provided the law we have established for distances of seventy or eighty feet hold equally with all distances whatever." †

* See also *Quarterly Journal of the Royal Institution*, for March 1830, vol. xxix. p. 185.

† *Journal of the Royal Institution*, October 1830, pp. 37–8.

X

In speaking thus guardedly, the learned professor had evidently in view Barlow's experiments of 1824, which seemed to prove the utter impracticability of all such projects. Barlow then wrote :—" In a very early stage of electro-magnetic experiments it has been suggested [by Ampère] that an instantaneous telegraph might be constructed by means of conducting wires and compasses. The details of this contrivance are so obvious, and the principles on which it is founded so well understood, that there was only one question which could render the result doubtful, and this was, is there any diminution of effect by lengthening the conducting wire? It had been said that the electric fluid from a common electrical battery had been transmitted through a wire four miles in length without any sensible diminution of effect, and to every appearance instantaneously, and if this should be found to be the case with the galvanic circuit then no question could be entertained of the practicability and utility of the suggestion above adverted to.

" I was, therefore, induced to make the trial, but I found such a sensible diminution with only 200 feet of wire as at once to convince me of the impracticability of the scheme."

---

* *Edinburgh Phil. Journal*, for 1825, vol. xii. p. 105.   It may save some future inquirer a good deal of trouble if we here refer to a supposed early suggestion of an electric telegraph, which we find thus recorded in *Notes and Queries*, for October 30, 1858, p. 359 :—

"In *Notes to Assist the Memory*, 2nd edit., 1827 (the first edition of

Dr. Jacob Green, of Jefferson College, Philadelphia, re-echoed this opinion in 1827; and there can be no doubt that the opinions of such men, so clearly enunciated, and supported by such apparently irrefutable experiments, had the effect of retarding for a while the introduction of electric telegraphs.

1825–37.—*Schilling's Telegraph.*

The invention that we are now about to describe is a very interesting one, not only because it was by far the most practicable of the proposals that had hitherto been made, but because it was the prototype of our well-known needle instruments, and was the immediate cause of the introduction of electric telegraphs into England.

---

which was published in 1819), the following note is added to the article on telegraphs :—' The electric fluid *has been* conducted by a wire four miles in length, apparently instantaneously, and without any diminution of effect. If this should be found to be the case with the galvanic circuit, *an instantaneous telegraph* might be constructed by means of wires and compasses.' "

Now if this passage occurred in the 1819 edition it would be prior to Oersted's discovery ! Our curiosity was aroused in the highest degree, and we instituted a fatiguing search for the book in the British Museum. We found at last a small 12mo volume, of which the following is the full title : *Notes to Assist the Memory in Various Sciences*, London, John Murray, 1825. On the face is written in pencil " By Walter Hamilton, M.R.A.S." The note above quoted appears on p. 110. This is the only book of the name in the British Museum, and there is no mention anywhere of a previous edition. The writer in *Notes and Queries* must therefore be wrong in his figures, and this will appear all the more certain on comparing the words of the note with those of Barlow, which we give in the text.

X 2

According to Dr. Hamel,* Baron Pawel Lwowitch Schilling (of Canstadt), then an *attaché* of the Russian Embassy at Munich, saw for the first time, on the 13th August, 1810, a telegraph (Sómmerring's) in action, and so impressed was he with the beauty and utility of the contrivance that, from that day, electricity and its applications became one of his most favoured studies. In the following five, or six, years his duties frequently took him to Munich, and at these times he was a constant visitor at Sommerring's house, whither he delighted to bring his friends from all parts of Europe to witness the performances of the telegraph. Indeed, during much of this time he may be said to have lived in an electrical atmosphere in the society of Sömmerring, Schweigger, and other kindred spirits.

Schilling's first application of electricity was to warlike ends. We learn from Hamel† that the war impending between France and Russia, in 1812, made him anxious to devise a conducting wire which could be laid, not only through moist earth, but through long stretches of water; and which should serve for telegraphic correspondence between fortified places and the field, as well as for exploding powder mines.

So diligently did he work at this task that before the autumn of the same year he had "contrived a

* *Historical Account of the Introduction of the Galvanic and Electro-Magnetic Telegraph into England,* Cooke's reprint, London, 1859, p. 13.

† Hamel, Cooke's reprint, pp. 20–2.

subaqueous galvanic conducting cord" (a copper wire insulated with a solution of india-rubber and varnish), and an arrangement of charcoal points, by means of which he was able to explode powder mines across the Neva, near St. Petersburg. At Paris, during the occupation of the allied troops, in 1814, he also frequently ignited gunpowder across the Seine with this *electric exploder*, to the great astonishment of the *gamins*.

In the next ten years (1815-25), Schilling divided the spare moments of a busy diplomatic and military career between lithography, an art then recently developed at Munich and which he was anxious to introduce into Russia, and electricity. In these circumstances, then, the surprise is, not that he brought out an electric telegraph of his own construction, but that he did not do so at an earlier date than the one usually assigned. Dr. Hamel vaguely fixes this date at about 1825, for he says that the Emperor Alexander (who died on 1st December, 1825) "had been pleased to notice the invention in its earlier stage."*

Schilling's apparatus was based on the property of a voltaic current to deflect a magnetic needle. It is sometimes described as a single-needle telegraph, and sometimes as one with five or six needles. What seems probable is that he tried many arrangements, that he first constructed a telegraph with one needle and was thence led on to combine several into one

---

* Hamel, Cooke's reprint, p. 41.

system, so as to be able to transmit a number of signals at once.

The signal-indicating part of the single-needle telegraph consisted of an ordinary Schweigger galvanometer. The needle was suspended horizontally by a silken thread, to which was attached, parallel to the

FIG. 12.

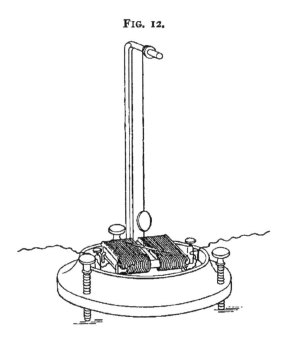

needle, a little disc of paper, painted black on one side, and white on the other. By the deflection of the needle to the right, or to the left, according to the direction in which the current moved in the coil, either face of the disc could be shown at pleasure, and two primary signals could be thus obtained, whose repetitions variously combined would represent the twenty-six letters of the alphabet, the ten ciphers, and four conventional signs.

The following is Schilling's alphabet as given by Vail, at p. 156 of his *American Electro-Magnetic Telegraph :*—

A = b w		N = w b
B = b b b		O = b w b
C = b w w		P = w w b b
D = b b w		Q = w w w b
E = b		R = w b b
F = b b b b		S = w w
G = w w w w		T = w
H = b w w w		U = w w b
I = b b		V = w w w
J = b b w w		W = b w b w
K = b b b w		X = w b w b
L = w b b b		Y = w b b w
M = w b w		Z = w b w w

In order to prevent the prolonged, or violent, swinging of the needle after each deflection, Schilling fixed to the lower extremity of the axle a thin platinum plate, or scoop, which, dipping into mercury placed beneath, deadened the motions, changing what might

* The bi-signal alphabet is popularly supposed to have come into existence with the Morse telegraph, but, in reality, its invention is almost as old as the hills. It was constantly employed in all kinds of semaphoric, luminous, and acoustic signalling from the days of the Greeks and Romans down to our own time. Lord Bacon gives an example in the 6th book of his *Advancement and Proficience of Learning,* published in 1605 ; and a still better one will be found in *Cryptographia Frederici* (p. 234), published in 1685.

otherwise be prolonged oscillations into *dead-beat* movements—a method since adopted in a modified form in some of Sir William Thomson's mirror galvanometers.

As a means of attracting attention, Schilling added a contrivance, the idea of which he clearly borrowed from Sommerring's alarum. It differed from the instrument that we have been describing only in that

FIG. 13.

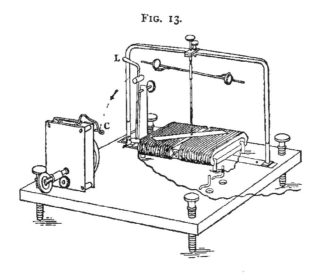

the rod whence the needle was suspended was made of rigid metal (wire), and carried a horizontal arm, which, when the needle was deflected, struck a finely balanced lever, and caused a leaden ball resting upon it to fall upon another lever, and so release the detent of an ordinary clockwork alarum.

At first the currents were transmitted by touching

the ends of the line wires (outgoing and returning) direct to the poles of the battery in one way or another, according to the direction in which the currents were required to flow through the coil. But soon this primitive arrangement was superseded by a simple commutator, consisting of (1) a wooden board having four small holes arranged in a square and filled with mercury, into which dipped, severally, the terminal wires of the battery, and the ends of the line wire; and (2) another similar board provided with a handle on one side, and two metallic strips on the other, the ends of which were turned at right angles to the face of the board. They thus formed bridges which were adjusted to dip into the holes of the first board and so establish connection in one sense or another between the poles of the battery and the ends of the line. As a guide to the operator the top of the second board was painted black and white. This commutator is shown in two pieces in Fig. 14.

These instruments were placed on view by the Russian Government at the Paris Electrical Exhibition of 1881, together with a model of Schilling's six-needle telegraph. We translate the following account of this instrument from *La Lumière Électrique*, for March 17, 1883 :—

"This apparatus," says the official (Russian) description of it, "consists of six multiplier-coils, each enclosing a magnetic needle, suspended by a silken thread from a copper support. A little above each needle is placed the paper disc, painted black on one side, and white on the other, as in the one-needle telegraph.

FIG. 14.

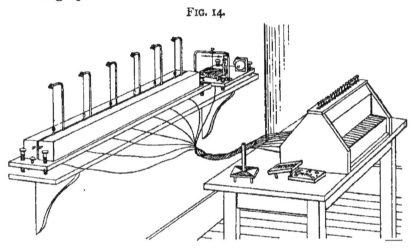

" The sending arrangement consists of a key-board like that of a pianoforte, having sixteen keys, in pairs of one black and one white. Each key, on being depressed, closes the circuit of a galvanic battery, its poles being connected to the lower contacts of the black and white keys respectively. Thus, for example, the negative pole may be connected to all the black keys, and the positive pole to all the white ones. The first six pairs of keys are joined to the six line wires (of copper), which are connected at the distant station to the six multiplier-coils ; the seventh pair serves to work the alarum through its own line wire ; and the eighth and last pair is joined to the return wire." *

The official account from which we are quoting is very obscurely written, so that it is impossible to gather in what way Schilling proposed to work this telegraph. It would seem that he wished to show from one to six signals at a time, sometimes black, sometimes white, and sometimes both combined ; but he could not effect the latter had he employed only one battery, as the account we are following would have us believe. He must, therefore, have used two separate batteries, and granting this, it is easy to see the immense number of permutations and combinations of which his apparatus was susceptible.

In 1830, Schilling set out for a voyage in China, and took with him a small model of his (? single-needle) telegraph, with whose performances he astonished the natives wherever he went. He returned to Europe in March 1832, and again occupied himself with telegraphic experiments, and hence, possibly, the date, 1832, which many writers assign to his inventions.†

---

* A short length of the original wires was shown in connection with the apparatus at the Paris Exhibition. There were eight copper wires, each separately insulated by a coating of resin, and all afterwards made up into a cable and bound with hemp also soaked in resin.

† Besides this error in date, it is often stated that Schilling employed *vertical* needles, that there were thirty-six of them, enclosed in as many multiplier-coils ! and that the line wires (? thirty-six also) were of *platinum ! !* insulated with silk. All these mistakes are contained in one short paragraph, which originally appeared in the *Journal des*

In May 1835, he started for a tour in southern and western Europe, taking with him a working model of his one-needle telegraph. At Vienna, he engaged in a series of experiments upon it in conjunction with Baron Jacquin and Professor A. von Ettingshausen. Amongst others they tried the comparative merits of leading the wires over the roofs of the houses, and burying them in the earth. The result was, as may be supposed, in favour of the former plan, for, owing to the defective insulation afforded by a thin coating of india-rubber and varnish, the earth, in the latter case, conducted the current from one wire to the other which lay parallel to it, and at a little distance.*

In September 1835, he attended the meeting of German naturalists at Bonn, and there, on the 23rd instant, exhibited his apparatus before the Section of Natural Philosophy and Chemistry, over which Professor Muncke, of Heidelberg, presided. Muncke was so pleased with its performance that he had a model made for exhibition at his own lectures at Heidelberg; and other members of the Congress took away with them to their respective homes such wonderful

---

Travaux de l'Académie de l'Industrie Française, for March 1839, p. 43, and which has since been copied unquestioningly into nearly every history of the telegraph that we have seen.

* These experiments are always described as if they had reference to an entirely new system of telegraph, the invention of Jacquin and Ettingshausen (see Dr. Hamel's *Historical Account,* &c., p. 60, Cooke's reprint) Andreas von Ettingshausen, a physicist of European fame, died at Vienna, May 25, 1878, aged eighty-two years.

accounts of its action that Schilling's telegraph was henceforth an object of great curiosity, and became a stock subject for popular lectures, and for articles in all the scientific papers of the period.

Dr. Hamel tells us* that, on his return home from Germany in 1836, Schilling received two letters urging him to bring his inventions to England, but he declined the suggestion, saying that he preferred to try to introduce them first in his own country. He was soon after honoured by a visit from the Emperor Nicholas, who witnessed with the greatest interest the performances of the telegraph "through a great length of wire," and ended by expressing the desire of having it established between St. Petersburg and Peterhoff. "Of all the high dignitaries," says Jacobi, "who surrounded him, His Majesty was the only one who foresaw the future of what was then looked on only as a toy."†

As was the custom in such cases, a commission of inquiry was appointed, which consisted of Lieut.- General Shubert, Adjutant-General Count Klein-

---

* In his lecture on *The Telegraph and Baron Paul Schilling*, before the Imperial Academy of Sciences, St Petersburg.

† Du Moncel's *Traité Théorique et Pratique de Télégraphie Électrique*, Paris, 1864, p. 217. Yet Russia was one of the last countries to adopt, generally, the electric telegraph. Why? "Because the Emperor Nicholas saw in it only an instrument of subversion, and by a *ukase* it was, during his reign, absolutely prohibited to give the public any information relative to electric telegraph apparatus, a prohibition which extended even to the translation of the notices respecting it, which, at this time, were appearing in the European journals."—Colonel Komaroff's *La Presse Scientifique des Deux Mondes*, as quoted in the *Annales Télégraphiques*, for November-December 1861, p. 670.

Michel, and Flugel-Adjutants Heyden and Treskine, with Prince Alexander Menshikoff as president. Schilling, in due course, submitted his plans, and gave the commissioners a choice of two modes of effecting the communication—(1) either the wires should be covered with silk and varnished, then bound together, tarred, and deposited along the bottom of the Gulf of Finland, or (2) "foreseeing the difficulties of such a plan, they were to be suspended on posts erected along the Peterhoff Road."*

An experimental telegraph was set up at the Admiralty, the line consisting partly of overground, and partly of "cable," which was submerged in the canal. The ends of the line with the appropriate instruments were placed at a great distance apart, one being at the window of Prince Menshikoff's study, in the N.W. corner of the building, and the other in a room near the great entrance of "the building office."

The results appear to have been eminently successful, for, in due course, Prince Menshikoff presented to the Emperor a most favourable report, upon the strength of which an Imperial Decree was issued (in May 1837), ordaining the establishment of a telegraph to connect Cronstadt with the capital by

---

* Du Moncel, p. 217. Jacobi, whom Du Moncel is quoting, says :— "The latter proposition was received by the Commission with shouts of derision, one of the members saying in my presence, ' Your proposition is foolish, your wires in the air are truly ridiculous.'" We wonder what he would say, could he take a walk through some of our London streets, where alas! this *folly* has attained to *ridiculously* gigantic proportions.

means of a "cable," laid along the bottom of the Gulf of Finland. But Schilling died on the 6th of August, 1837, and he and his country missed the glory of establishing not only the first really practicable telegraph, but also the first submarine line.*

1833–8.—*Gauss and Weber's Telegraph.*

In this year Messrs. Gauss and Weber constructed an apparatus at Gottingen, which, although at first intended for purely scientific purposes, soon came to be employed as a means of ordinary correspondence as well. The telegraph, as we must call it, contrived by these well-known physicists is remarkable for three reasons—1st, as being the first in which magneto-electricity was used; 2nd, for the ingenious, yet simple, method of increasing the deviations of the signalling needle—a plan which was long afterwards adopted by Sir William Thomson in his beautiful mirror galvanometers, † and last, though not least, for having had an *actual* existence for several years, during which it rendered most excellent service.

The line, consisting of two copper wires, main and

---

* Of course a wire insulated with (1) a thin coating of india-rubber and varnish, as in the Neva and Seine experiments of 1812–14; or (2) with resin, and hemp saturated with resin, as in the "cable" shown at the Paris Exhibition of 1881; or (3) with silk varnished and tarred, as just mentioned, would not last long, but necessity is the mother of invention, and practice makes perfect. The cable laid from Dover to Calais in 1850 was only a little less crude.

† As early as 1826 Poggendorff applied a mirror to the magnetic needle for accurately determining minute variations in its horizontal declination.—Pogg. *Ann. der Phys. und Chem.*, vol. vii. pp. 121–30.

return, was carried upon posts over the houses, and extended from the Physical Cabinet to the Observatory; whence, in 1834, it was continued to the Magnetic Observatory of Professor Weber—a distance

FIG. 15.

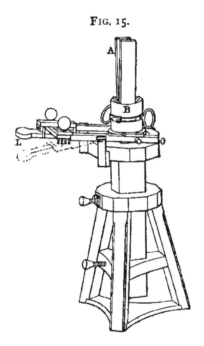

altogether of one mile and a quarter English. An ordinary voltaic pair was employed for generating the current until 1835, when it was replaced by a magneto-electric machine made by Steinheil of Munich.*

* In our account of Gauss and Weber's telegraph, we follow Sabine's *History and Progress of the Electric Telegraph*, London, 1869, 2nd edit., pp. 33–8 ; and *La Lumière Électrique*, for March 17, 1883, p. 334. See also Pogg. *Ann.*, xxxii. 568 ; and Dingler's *Journal*, lv. 394.

This instrument, called the *Inductor*, consisted of a compound two-bar magnet A, Fig. 15, weighing 75 lbs., fixed vertically on a stool ; a wooden bobbin B, supplied with a handle L, and wound with 3500 turns (and later with 7000 turns) of insulated copper wire (No. 14, silvered), rested on the stool and encircled the magnet, as shown in the figure. On lifting the bobbin by depressing the handle, a momentary current would be induced in the coil in one direction, and on lowering it again to its position of rest another momentary current would be induced in the opposite direction. The ends of the coil B, were connected through the commutator, Fig. 17, to the line wires, and the distant ends of these were similarly joined to the ends of the coil of the receiver.

The receiver, shown in Fig. 16, consisted of a large copper frame B, B,* upon which was wound 3000 feet of insulated copper wire, like that of the inductor. A permanent magnet A, 18 inches long, and $3''' \times 5'''$ transverse section, and weighing one hundred pounds,

---

* The copper frame, which Gauss called *the damper*, was necessary in order to prevent the great number of oscillations which the magnet would have made across the meridian had no such check been introduced. The checking action of masses of metal, and indeed of any other solid or liquid substance, in the vicinity of an oscillating magnet was discovered by Arago, in 1824. Sir William Snow Harris found that the oscillations of a freely suspended magnetic needle were reduced, from 420 without a damper, to 14 with a damper. In the present case the great mass of the magnet and the minuteness of the deviation must have aided materially in bringing it quickly to a state of rest. See pp. 282–3 *ante*.

Y

was suspended, in the interior of the coil, by a number of untwisted silk fibres, from a hook above it. To enable the observer to read off with care the minute

FIG. 16.

deviations of the magnet, a small mirror M, was affixed to the supporting shaft, and in this was seen, through a telescope, at ten or twelve feet distance, the reflection of a scale placed above it. Notwithstanding the weight of the magnet, its movements were thus made beautifully energetic and distinct, a very small force, such as that supplied from a single cell, causing a deviation of over a thousand divisions of the scale.

The commutator, by means of which the electric currents were directed through the line wires in one sense or another, was similar to that of Schilling, being simply an arrangement for bringing two points alternately in communication with two others. Let *a*, and *c*, Fig. 17, be two points in connection with the two poles of a battery, or other electromotive system, and *b*, and *d*, the ends of any other circuit ; if the metal bars *e*, and *f*, be pressed upon the ends *a*, *b*, and *c*, *d*, respectively, the current will pass in the direction B $+ a e b$ R $d f c -$ B. But if the bars *e*, and *f*, be removed from these positions and placed at right angles, that is to say, *e*, between *b*, and *c*, and *f*, between *a*, and *d*, as shown by the dotted lines, the current will go through B $+ a d$ R (in the opposite direction) *b c* $-$ B.

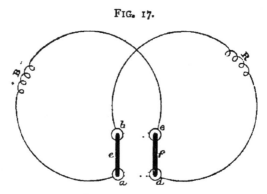

FIG. 17.

The *modus operandi* was as follows :—On lifting up the coil B, Fig. 15, by depressing the handle L, to the position shown by the dotted lines, a current was induced in the wire. This current passed by the commutator, placed as in Fig. 17, from *a*, to *b*, through one of the line wires and the multiplier R, of the

Y 2

receiving station, deflecting the magnet for an instant in one direction, and returned by the other wire over *d*, and *c*, of the commutator. When it was wished to deflect the needle of the receiving instrument in the opposite direction, this was attained by simply lowering the coil B, again to its original place, and the observer at the receiving station read off one deflection to the right for instance, and one to the left. But, in constructing a code of signals, it was necessary that two or more deflections to the right or left should frequently follow each other. This was done by means of the commutator. Thus, on lifting the coil, if we suppose a deflection of the magnet was produced to the right, by reversing the commutator and then lowering the coil again, another deflection in the same direction would be observed. To produce a third deflection in the same direction it would be necessary, evidently, to reverse the commutator again before raising up the inductor. After this fashion Gauss and Weber were enabled, by combining the deflections to the right and to the left, to form the following alphabet and numerals, with a maximum of four elementary signals : —

$r = a$	$rlr = f, v$	$rrlr = s$	$lrlr = 3$
$l = e$	$lrr = g$	$rlrr = t$	$llrr = 4$
$rr = i$	$lll = h$	$lrrr = w$	$lllr = 5$
$rl = o$	$llr = l$	$rrll = z$	$llrl = 6$
$lr = u$	$lrl = m$	$rlrl = o$	$lrll = 7$
$ll = b$	$rll = n$	$rllr = 1$	$rlll = 8$
$rrr = c, k$	$rrrr = p$	$lrrl = 2$	$llll = 9$
$rrl = d$	$rrrl = r$		

r, represents the swing of the north pole of the magnet towards the right, and *l*, the swing of the same pole towards the left of the magnetic meridian. Various lengths of the pauses between the signals indicated the conclusion of words and sentences.

In 1835, an alarum was added, which, according to some accounts, consisted in giving to the magnet A, Fig. 16, a more than ordinary deviation, and so making it strike a bell ; while, according to others, it it was very similar to that of Schilling, the magnet, when largely deflected, upsetting a delicately-poised lever in train with the detent of an ordinary clockwork alarum.

Gauss and Weber's apparatus was in daily use for telegraphic and astronomical purposes down to the year 1838.

CHAPTER XII.

TELEGRAPHS BASED ON ELECTRO-MAGNETISM AND MAGNETO-ELECTRICITY (*continued*).

1836.—*Steinheil's Telegraph.*

THE apparatus last described was, as we have said, established for other than telegraphic purposes, and it was for this reason, that Gauss, unable himself to afford the time, invited Steinheil, of Munich, to pursue the subject, and endow with a practical form an invention which he believed capable of great results.

The perfection to which this ingenious inventor brought Gauss and Weber's telegraph has rendered it as much, or more his than theirs. His own estimate, however, of the changes effected in the telegraph erected at Munich towards the end of 1836, is very modest and is worth quoting:—" To Gauss and Weber," he says, " is due the merit of having actually constructed the first simplified galvano-magnetic telegraph. It was Gauss who first employed magneto-electricity, and who demonstrated that the appropriate combination of a limited number of signs is all that is required for the transmission of intelligence.* Weber's discovery that a copper wire 7460 feet long, which he had led across the houses and steeples of Göttingen, required no special insulation, was one of great importance, and at once established the practicability of a galvanic telegraph in a most convenient form.

" All, therefore, that was required was (1) an appropriate method of inducing, or exciting, the galvanic current, with the power of changing its direction without the need of any special contrivances ; and (2) a mode of rendering the signals audible [or legible]. The latter was a task that apparently presented no very particular difficulty, inasmuch as in the very scheme itself a mechanical motion—namely, the deflection of a magnetic bar—was given. All that we had to do, therefore, was to contrive that this motion should be made available for striking bells, or for marking indelible dots.

" This falls within the province of mechanics, and there are, therefore, more ways than one of solving the problem. Hence the alterations that I have made in the telegraph of Gauss and Weber, and by which it has assumed its present form, may be said to be founded on my perception and improvement of its imperfections. I by no means, however, look on the arrangement I have selected as complete ; but as it

* Steinheil was apparently unaware of all that Baron Schilling had done in this direction.

answers the purpose I had in view, it may be well to abide by it till some simpler arrangement is contrived." *

We condense the following account of Steinheil's apparatus from an English translation of the author's own classical memoir, which appeared in Sturgeon's *Annals of Electricity*, for March and April, 1839.†

---

* To Steinheil's lasting honour be it said, that when some ten years later "a simpler arrangement" in the shape of Morse's telegraph *was* brought to his attention he was the first to appreciate it, and to urge upon the Bavarian Government its adoption, to the abandonment of a portion of his own beautiful system.

Apropos of this, we find an amusing story in Reid's *Telegraph in America*, pp. 85–6 :—"The (Morse) relay could not be patented in Germany, and, therefore, could not with safety be exposed. In 1848, two young Americans had gone there with Morse machinery, and built a line from Hamburg to Cuxhaxen, a distance of 90 miles, for the transmission of marine news. The line worked charmingly, the registers clicked out loud and strong at either end, but the relays were carefully concealed in locked boxes. The German electricians scratched their heads and wondered. Finally Steinheil was sent forward to reconnoitre ; he looked carefully around, and his keen eyes soon detected the locked boxes. He asked to see their contents, but the view was courteously declined. So he returned and reported that the Yankees kept their secret locked, but that the action was magnificent. And when at a later date he *did* know all, he showed the grand stuff of which he was made. He gave Morse his hand, confessed himself beaten, and the two were friends for ever after."

† Vol. iii. pp. 439–52, and 509–20. See also *Comptes Rendus*, September 1838 ; and Shaffner's *Telegraph Manual*, New York, 1859, pp. 157–78. Shaffner says that "the first published notice of this important invention will be found in the third volume of *The Magazine of Popular Science*, in a letter from Munich, under date December 23, 1836." This letter appears on pp. 108–10 ; is chiefly concerned with electro-magnetic experiments ; and in the last two paragraphs briefly mentions Steinheil's telegraph.

" The telegraph is composed of three principal parts :—

1. A metallic connection between the stations.
2. The apparatus for exciting the galvanic current.
3. The indicator, or receiving apparatus.

1. *Connecting Wire.*

" This so-called connecting wire may be looked on as the wire completing the circuit of a voltaic battery extended to a very great length. What applies to the one holds good of the other. With equal thicknesses of the same metal, the resistance offered to the passage of the galvanic current is proportional to the length of the wire ; and with equal lengths of the same metal, the resistance diminishes inversely with the section. The conducting power of metals is very different. Thus, according to Fechner, copper conducts six times better than iron, and four times better than brass, while the conducting power of lead is even lower ; so that the only metals which can well vie with each other in their technical use are copper and iron.   But although iron is about six times as cheap as copper, it will be requisite to give the iron wire six times the weight of a copper one to gain the same conducting power with equal lengths.   We thus see that as far as the expense is concerned it comes to the same thing, whichever of these metals is chosen.   The preference will be given to copper, as this metal is less liable to oxydation from exposure to the atmosphere.

"This latter difficulty may, however, be surmounted by simple means, *viz.*, by galvanising the iron. It would even appear that the simple transmission of the galvanic current, when the telegraph is in use, is sufficient to preserve the iron from rust; such at least is observed to be the case with the iron portion of the wire used for the telegraph here, and which has already been exposed in all weathers for nearly a twelvemonth.

"If the galvanic current is to traverse the entire metallic circuit without any diminution of strength, the wire during its whole course must not be allowed to come into contact with [*i. e.*, short-circuit] itself; neither should it be in frequent contact with semi-conductors, for, since the power called into action always completes its circuit by the shortest course, the remote parts of the wire would be thus deprived of a portion of the current.

"Numerous trials to insulate wires and to lay them below the surface of the ground have led me to the conviction that such attempts can never answer at great distances, inasmuch as our most perfect insulators are at best but very bad conductors. And since in a wire of very great length, its surface in contact with the so-called insulator is uncommonly large when compared with its section, there necessarily must arise a gradual diminution of the force, inasmuch as the outgoing and returning wire, although but slightly, yet *do* communicate in intermediate points.*

"It would be wrong to think that this difficulty could be got over by placing the two wires very far apart. The distance between them is, as we shall see in the sequel, almost a matter of indifference. As, then, we shall never succeed in laying down conductors that are sufficiently insulated beneath the surface of the ground, there is but one other course open to us, *viz.*, leading them through the air. Upon this plan, it is true, the conductor must be supported from time to time, is liable to be injured by the evil disposed, and is apt to suffer from violent storms, or from ice which forms upon it. As we, however, have no other method that we can avail ourselves of, we must endeavour by suitable arrangements to get the better of these difficulties in the best way we can.

" The conducting chain of the telegraph erected here consists of three parts—one leads from the Royal Academy to the Royal Observatory, at Bogenhausen, and back. The total length of its wire is 32,506 feet, and the weight amounts to 260 lbs. [*sic*]. Both wires (there and back) are stretched across the steeples of the town at a distance apart of 4 feet 1 inch. The distance from support to support ranges from 640 to 1279 feet : this is undoubtedly far too great for a single-strand wire, inasmuch as the ice that forms

---

* It should be remembered that when Steinheil wrote, gutta-percha and india-rubber were both unknown as insulators.

upon it materially increases its weight, and considerably augments its diameter, so that it becomes liable to be torn asunder by high winds.*   Over those places where there are no high buildings the wire is supported upon poles forty or fifty feet high, which are let five feet into the ground, and at the top of which it is fastened by twisting on cross wooden bars.   At the points of support the wire rests on pieces of felt.

" The conducting wire thus mounted is by no means completely insulated.   When, for example, the circuit is broken at Bogenhausen, an induction-shock given in Munich ought to produce no galvanic excitation whatever in the parts of the chain then disconnected, yet Gauss's galvanometer gives indication of a weak current.   Measurements, indeed, go to show that this current goes on increasing, as the point at which the interruption of the stream is made recedes from the inductor.   The total amount of this current [leakage] is not constant, being, generally, greatest in damp weather.

" At moderate distances of a few miles this loss of power is of almost no importance, more especially as the construction of the inductor places currents of almost any strength we choose at our command. When the distance, however, amounts to upwards of

---

* All these evils could be got over by making the connection by at least a triple strand (and not by a single wire), supporting it at intervals of 300 feet, and giving it a tension not exceeding one-third of what it would bear without giving way.   This, however, in the experimental telegraph erected here, was not practicable.

200 miles [*sic*], the greatest part of the effect would be dissipated. In such cases, therefore, much greater precaution must be taken with regard to the points of support of the metallic circuit.*

" A second portion of the conducting chain leads from the Royal Academy to my house and observatory in the Lerchenstrasse. This is of iron wire, its length, there and back, is 5745 feet, and it is stretched over steeples and other high buildings, as has already been described.

"Lastly, a third portion of the chain, running through the interior of the buildings connected with the Royal Academy, leads to the mechanical workshop attached to the cabinet of Natural Philosophy. It is composed of a fine copper wire 598 feet long, let into the joinings of the floor, and, in part, imbedded in the walls.

2. *Apparatus for Generating the Galvanic Current.*

" Hydro-galvanism, or the galvanic current generated by the action of the voltaic pile, is by no means fitted for traversing *very long* connecting wires, because the resistance in the pile, even when many hundred pairs of plates are employed, would be always inconsiderable compared with the resistance offered by the wire itself. The principal disadvantages, however,

* When thunder-storms occur, atmospheric electricity collects on this semi-insulated chain as upon a conductor, but the passage of the galvanic current is not at all affected thereby.

attendant on the use of the pile, or trough apparatus, are (1) the fluctuations of its current, and (2) its speedy loss of power.

"All these difficulties are got over by having recourse to Faraday's important discovery of magneto-electric induction, that is to say, by moving magnets in the neighbourhood of conducting coils. The better way is, not to move the magnets as Pixii does, but rather to give motion to coils of wire in close proximity to a fixed magnet. This arrangement, known as Clarke's, is the one which with some modifications we have adopted

"The magnet is built up of seventeen horse-shoe bars of hardened steel, Fig. 21. With its iron armature its weight is about 74 lbs., and it is capable of supporting about 370 lbs. Between the arms, or poles, is fastened a piece of metal which supports the axis on which the coils revolve. These coils, of which there are two, have in all 15,000 turns of silk-covered wire, a metre of which weighs 15½ grains. The two ends of the wire are passed up through the interior of the axis, and terminate in two hook-shaped pieces which just dip into semicircular cups of mercury, separated from each other by a wooden partition. From these cups there proceed short wires to which the line wires are connected. The mercury, owing to its capillarity, stands at a higher level in the cups than the partitions, so that the terminal hooks pass over the latter without touching them whenever the coils are revolved. The hooks are thus brought into the cups alternately at every half turn of the coils, and as a consequence the induced current preserves its sign as long as the coils are turned in one direction, and changes it on the motion being reversed. The current, as we shall see when treating of the indicator, should only be permitted to act during as short a time as possible, while during that time it should have the greatest intensity we can give it. To effect this the mercury cups are arranged as shown in the dark portion of Fig. 18. The terminal hooks travel in the white annular space, and make contact only at the moments when passing over the points *a*, and *b*.

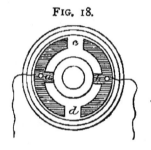

FIG. 18.

" In order to cut off the inductor when not in action, its axis is made to carry a cross-piece of metal, at right angles to the terminal hooks of the coils, which, when the inductor is at rest, dips into the mercury cups. Whence it follows that the current, on being transmitted from any other station, passes directly from one cup to the other without traversing the wire of the inductor coils. In order to put the coils in motion without trouble a fly-bar terminating in two metal balls is attached horizontally to the vertical axis of the coils (see Fig. 21).

" At every half turn a spark occurs as the hooks of

the coils leave the mercury. As this is for many reasons objectionable, we have latterly designed a commutator of a far simpler construction. The ends of the coil are in this case fastened to two strips of copper let into the periphery of a wooden ring, directly opposite each other. This ring is placed upon the axis of the coils and made fast to it by clamps, and the two ends of the line wire are so disposed as to press like springs against the copper strips as the coils are revolved. With this arrangement the ends of the coils are in metallic connection with the line only during a small portion of each revolution, while during the rest of the time the metal cross-piece, with which also the wooden ring is provided, brings the two ends of the line wire into direct connection. This form of commutator, in which mercury is entirely dispensed with, is, on account of its greater simplicity and durability, preferable to the arrangement just described, and is employed in the apparatus of the stations at Bogenhausen and in the Lerchenstrasse.

3. *The Indicator.*

" Figs. 19 and 20 represent vertical and horizontal sections of the indicator, containing two magnets, movable on axes *m, m,* and which from their construction are applicable either to strike bells, or to note down signals. Round a frame formed of sheet

brass are wound six hundred turns of the same insulated copper wire as is used in the inductor.   The
magnetic bars are, as Fig. 20 shows, so placed that
the north pole of the one is presented to the south

FIG. 19.

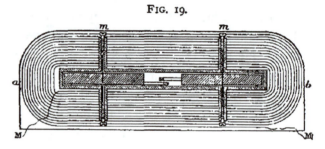

pole of the other.   To these ends are screwed two
slight brass arms, supporting little cups which are
provided with extremely fine perforated beaks *c, c*

FIG. 20.

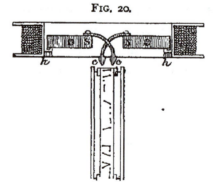

When printing ink is put into these cups it insinuates
itself into the beaks owing to capillary attraction,
and, without running out, forms at their orifices a

projection of a semi-globular shape. The slightest contact, therefore, suffices for noting down a black dot.

"Two plates, or pins, *h, h,* prevent the magnets from being deflected in a direction opposite to that in which they are to print, as the deflection by the current would otherwise cause them to swing, and, perhaps, record false dots while thus oscillating. As a further check to these oscillations, and in order to bring back the bars quickly to their normal positions after each deflection, recourse is had to smaller movable magnets (Fig. 21) whose distance and position with regard to the others are to be varied until the desired effect is produced. Owing to the disposition of the magnet bars and the controlling action of the pins *h, h,* a current sent through the coil deflects only one magnet at a time, the other being simply pressed tightly against its pin ; and on the current being reversed, the reverse takes place, the last-mentioned magnet being deflected, while the first is held back.

"Much nicety is required in obtaining the magnets of exactly the right size. They must not, for example, be too large, because their inertia would be too great ; nor too small, because then their mechanical force would not be sufficient for printing or sounding the signals.

"For the recording of the signals, a flat surface of paper must be kept moving with a uniform velocity in front of the little beaks, Fig. 21. The best way of doing this is to employ very long strips of the so-called endless paper which is to be wound round a cylinder of wood and then cut upon the lathe into bands of suitable width. One of these strips of paper must be made to unwind itself from a cylinder, pass close in

FIG. 21.

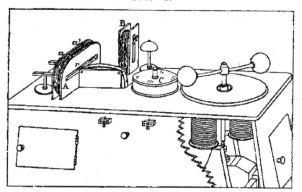

front of the beaks, run along a certain distance in a horizontal position, so that the dots noted down may be read off, and lastly wind itself up again on to a second cylinder. This second cylinder is put in motion by clockwork, the regularity of whose action is insured by a centrifugal fly-wheel.

"If this apparatus be employed for producing two sounds easily distinguishable to the ear by striking on bells, it will be right to select clock-bells, or bells of glass, both of which easily emit sounds, and whose notes differ about a sixth. This interval is by no means a matter of indifference. The sixth is more easily distinguished than any other interval; fifths and octaves would be frequently confounded by those not versed in such matters. The bells are to be supported on little pillars, and their position with respect to the bars is to be determined by experiment. The knobs let into the bars for striking the bells must give the blow at the place which most easily emits a sound. They are not, however, to be too close to the bells, as in that case a repetition of the signal can easily ensue. A few trials will soon get over this difficulty.

" It is evident that the same magnetic bars cannot be at once employed for striking bells and for writing, the little power they exert being already exhausted by either of these operations. But to combine them both, all we have to do is, to introduce a second indicator coil into the chain ; this can, however, only be done at the cost of an increased resistance, and, in order that this increase may be as little as possible, it would in future be better that the coils of the indicator should be made of very thick copper wire, or of strips of copper plate. Fig. 21 shows two coils in circuit, the one marked B, being used as an alarum [which no doubt the attendant could short-circuit after replying to its call].

"At the central station in the Physical Cabinet a commutator, C, Fig. 21, is placed which enables us by simple transpositions to effect the following changes in the wires and apparatus :—

"(1.) The currents emanating at the central station

traverse the receiving instruments of both the Bogenhausen and Lerchenstrasse stations at the same time.

"(2.) The currents traverse the Lerchenstrasse line
and instrument only, and the Bogenhausen line is
connected through the receiving instrument of the
central station, so that while one attendant at the
latter station is *sending to* Lerchenstrasse, another
attendant may be *receiving from* Bogenhausen.

"(3.) Is the reverse of the last-named arrangement.

"(4.) Bogenhausen and Lerchenstrasse are joined
direct, and the apparatus at the Physical Cabinet
cut out of circuit altogether.

"We have said before that at every half turn of the
fly-bar from (say) *right to left*, one of the magnets of
the indicator is deflected. Now we have so connected
the apparatus that every time this movement takes
place the high-toned bell should be struck, if the
receiver be arranged as an acoustic instrument; or the
corresponding beak shall print a dot on the paper
strip, if the receiver be arranged as a recording instrument. On turning the fly-bar from *left to right*, the
low-toned bell sounds, or the corresponding beak
prints a dot, not upon the same line as the first, but
on a lower one. High tones, therefore, correspond
with the dots on the upper line, and low tones with
the dots on the lower line, as in a musical score.

"As long as the intervals between the sounds or the
signs remain equal, the said sounds or signs are to be
read together as one signal, a longer interval indicates

the completion of a letter or signal. We are thus enabled by appropriately selected groups to represent all the letters of the alphabet, or stenographic characters, and thereby to repeat and render permanent at all parts of the chain where an apparatus like that above described is inserted any information that we choose. The alphabet that I have chosen represents the letters that occur the oftenest in German by the simplest signs. By the similarity of shape between these signs and that of the Roman letters, they become impressed upon the memory without difficulty. The distribution of the letters and numbers into groups consisting of not more than four dots is shown below.

FIG. 22.

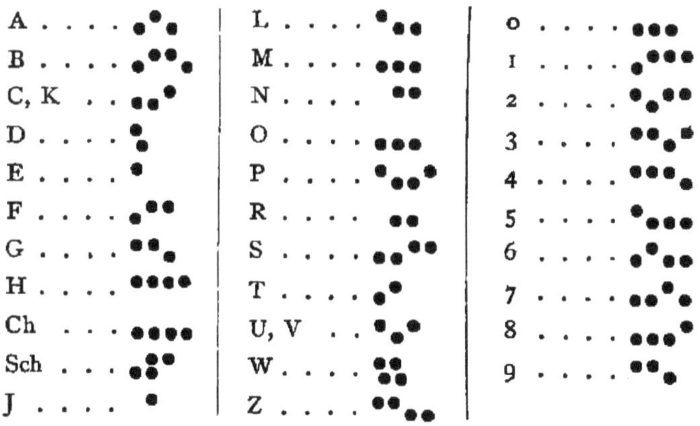

"Messages were printed with this apparatus at the rate of ninety-two words in a quarter of an hour, or over six words per minute."

Discovery of the Earth Circuit.

In order not to interrupt the continuity of our description of Steinheil's beautiful apparatus, we have reserved for a special paragraph our notice of this most important discovery.

As we have seen in our second, third, fourth, and fifth chapters, the earth circuit was used, with few exceptions, in all experiments with static electricity. Its function, however, was either unsuspected or misunderstood.* Of all the telegraphic proposals based on static electricity, those of Bozolus, 1767, and of the anonymous Frenchman, 1782, are the only ones in which complete metallic circuits were proposed. Reusser, 1794, used one common return wire; while all the others employed the earth, Volta, Cavallo, and Salvá making distinct mention of their doing so.

The power of the earth to complete the circuit for dynamic electricity has also been known for a very long time. Thus, on the 27th of February, 1803, Aldini sent a current from a battery of eighty silver and zinc plates from the West Mole of Calais harbour to Fort Rouge through a wire supported on the masts of boats, and made it return through 200 feet of intervening water.†

Basse, of Hamel, made similar experiments, and

* As in Watson's experiments, described at pp. 111–13 of Priestley's *History of Electricity*, 1767.

† Aldini's *Account of late Improvements in Galvanism*, London, 1803, p. 218.

about the same time, on the frozen water of the ditch, or moat, surrounding that town. He suspended 500 feet of wire, on fir posts, at a height of six feet above the surface of the ice, then making two holes in the ice and dipping into them the ends of the wire, in the circuit of which were included a galvanic battery and a suitable electroscope, he found that the current circulated freely. Similar experiments were made in the Weser; then with two wells, 21 feet deep, and 200 feet apart ; and, lastly, across a meadow 3000 feet wide. Whenever the ground was dry it was only necessary to wet it in order to feel a shock sent through an insulated wire from the distant battery. Erman, of Berlin, in 1803, and Sömmerring, of Munich, in 1811, performed like experiments, the one in the water of the Havel, and the other along the river Isar.*

All these are very early and very striking instances of the use of the earth circuit for dynamic electricity ; but the most surprising and apposite instance of all has yet to be mentioned, in which the use of the earth

* Gilbert's *Ann. der Physik*, vol. xiv. pp. 26 and 385 ; and Hamel's *Historical Account*, &c , p. 17 of Cooke's reprint. Fechner, of Leipsic, after referring to Basse's and Erman's experiments in his *Lehrbuch des Galvanismus* (p. 268), goes on to explain the conductibility of the earth in accordance with Ohm's laws. As he immediately after alludes to the proposals for electric telegraphs, he has sometimes been credited with the knowledge of the fact that the earth could be used to complete the circuit in such cases. This, however, is not the fact, as we learn from a letter which Fechner addressed to Professor Zetzsche on the 19th February, 1872 (Zetzsche's *Geschichte der Elektrischen Telegraphie*, p. 19).

is suggested precisely as we employ it to-day. In a letter signed " Corpusculum," and dated December 8, 1837, in the *Mechanics' Magazine*,* we read :—

" It seems many persons have formed designs for telegraphs. I, too, formed mine, and prepared a specification of it five years ago, and that included the plan of making one wire only serve for the returning wire for all the rest, as in Alexander's telegraph ; *but even that might, I think, be dispensed with where a good discharging train, as gas, or water, pipes, at each end of the telegraph could be obtained.*"

In July 1838, or seven months after the publication of " Corpusculum's " letter, Steinheil made his *accidental* discovery in a way which we find thus related by De la Rive :†—

" Gauss having suggested the idea that the two rails of a railway might be employed as conductors for the electric telegraph, Steinheil, in 1838, tried the experiment on the railroad from Nuremburg to Furth, but was unable to obtain an insulation of the rails sufficiently perfect for the current to reach from one station to the other. The great conductibility, with which he remarked that the earth was endowed, caused him to presume that it would be possible to employ it instead of the return wire. The trials that he made in order to prove the accuracy of this conclusion were followed

---

* For 1837, p. 219. The full text of this interesting letter will be found at p. 477, *infra*.

† *Treatise on Electricity*, London, 1853–58, vol. iii. p. 351.

by complete success ; and he then introduced into electric telegraphy one of its greatest improvements."

In Steinheil's own account of this discovery, he begins by pointing out that Ampère required for his telegraphic proposal more than sixty line wires ; that Sómmerring reduced the number to thirty or so ; Cooke and Wheatstone to five ; and Schilling, Gauss, and Morse to " one single wire running to the distant station and back."

He then goes on to say :—" One might imagine that this part of the arrangement could not be further simplified ; such, however, is by no means the case. I have found that even the half of this length of wire may be dispensed with, and that, with certain precautions, its place is supplied by the ground itself. We know in theory that the conducting powers of the ground and of water are very small compared with that of the metals, especially copper. It seems, however, to have been previously overlooked that we have it within our reach to make a perfectly good conductor out of water, or any other of the so-called semi-conductors.

" All that is required is that the surface that its section presents should be as much greater than that of the metal as its conducting power is less. In that case the resistance offered by the semi-conductor will equal that of the perfect conductor ; and as we can make conductors of the ground of any size we please, simply by adapting to the ends of the wires plates presenting a sufficient surface of contact, it is evident that we can diminish the resistance offered by the ground, or water, to any extent we like. We can indeed so reduce this resistance as to make it quite insensible when compared to that offered by a metallic wire, so that not only is half the wire circuit spared, but even the resistance that such a circuit would present is diminished by one half.

"The inquiry into the laws of dispersion according to which the ground, whose mass is unlimited, is acted upon by the passage of the galvanic current, appeared to be a subject replete with interest. The galvanic excitation cannot be confined to the portions of earth situated between the two ends of the wire; on the contrary, it cannot but extend itself indefinitely, and it, therefore, only depends on the law that obtains in this excitation of the ground, and the distance of the exciting terminations of the wire, *whether it is necessary or not to have any metallic communication at all for carrying on telegraphic intercourse.*

"An apparatus can, it is true, be constructed in which the inductor, having no other metallic connection with the multiplier than the excitation transmitted through the ground, shall produce galvanic currents in that multiplier sufficient to cause a visible deflection of the bar. This is a hitherto unobserved fact, and may be classed amongst the most extraordinary phenomena that science has revealed to us. It only holds good, however, for small distances;

and it must be left to the future to decide whether we shall ever succeed in telegraphing at great distances without any metallic communication at all. My experiments prove that such a thing is possible up to distances of 50 feet. For greater distances we can only conceive it feasible by augmenting the power of the galvanic induction, or by appropriate multipliers constructed for the purpose, or, in conclusion, by increasing the surface of contact presented by the ends of the multipliers. At all events the phenomenon merits our best attention, and its influence will not perhaps be altogether overlooked in the theoretic views we may form with regard to galvanism itself." *

* Sturgeon's *Annals of Electricity*, vol. iii. pp. 450–2. Dr. O'Shaughnessy (afterwards Sir William O'S. Brooke), the organiser of the East Indian telegraphs, claims to have independently discovered the earth circuit, and points for evidence to his paper in the *Journal of the Asiatic Society of Bengal*, for September 1839, pp. 714–31. See his *Electric Telegraph in British India*, London, 1853, p. 21.

CHAPTER XIII.

EDWARD DAVY AND THE ELECTRIC TELEGRAPH, 1836–1839.

" It seldom happens that the author of a great discovery, after failing to attract attention to his application of science, lives to see his own invention universally adopted. Mr. R——— appears to be the least pushing of original inventors, and it is just that in his later years he should have the satisfaction of knowing that he is appreciated by his countrymen."—*Saturday Review*, November 17, 1866.

FEW of our readers have heard of the name of Edward Davy in connection with the history of the telegraph, and for the sufficient reason that, beyond a few very short and very imperfect accounts * of a needle telegraph which he exhibited in London in 1837–38, and extracts, more or less copious, from the specification of his electro-chemical recording telegraph, patented in July 1838, nothing has been published regarding him and his early labours. Yet it is certain that, in those days, he had a clearer grasp of the requirements and capabilities of an electric telegraph than, probably, Cooke and Wheatstone themselves ; and had he been taken up by capitalists, and his ideas licked into shape by actual practice, as they and theirs were, he would have successfully competed with them for a share of the profits and

---

* *Mechanics' Magazine*, for January 20, and February 3, and 17, 1838 ; on which are based the very meagre and, of course, incorrect descriptions in all books on the telegraph. Also the *Penny Mechanic*, for February 10, 1838.

honours, which have so largely accrued to them as the practical introducers of the electric telegraph.

This, at all events, is the conclusion that we have come to, after the perusal of a number of most interesting MS. documents, which have been obligingly placed in our hands by Dr. Henry Davy, of Exeter, Edward Davy's nephew; and it is a conclusion which, we believe, our readers will cordially indorse when they have read the extracts from them, which we are now about to give.* We feel a peculiar satisfaction at being thus the means of re-introducing to his countrymen one who deserves a most honourable recognition. Mr. Davy, we are glad to say, is alive and well, and, though now 78 years of age, is still following his profession, as a surgeon, in a far-off colony, whither, for reasons which need not concern us, he emigrated in 1839.

The idea of an electric telegraph first occurred to him about 1836, when he sketched out a plan to be worked by static, or frictional, electricity. We give it almost in the author's own words :—

" *Outline of a New Plan of Telegraphic Communication, by which Intelligence may be Conveyed, with Precision, to Unlimited Distances, in an Instant of Time, Independent of Fog or Darkness.*

" The agent is electricity, which is well known to pass through a conducting medium with the rapidity

* At our suggestion Dr. Davy has presented all these valuable MSS. to the Society of Telegraph-Engineers and Electricians for deposition in the library, where they may now be consulted.

of lightning. The only difficulty is in the mode of applying it to the end proposed. Our method is as follows :—

" Let us suppose a number of copper wires, each covered with silk, and varnished, to be laid underground, side by side, from London to Liverpool. For greater protection, they may be enclosed in an iron pipe. If there be a small brass ball at each end of each wire, an electric spark applied to the ball at the London end of any of them might be drawn at the same instant from the corresponding ball at the Liverpool end.

" If there be twenty-four such wires, there will be one for each letter in the alphabet ; but six wires would be more than sufficient in practice, owing to the numerous changes that might be made upon them by combination. We will, however, suppose that there are twenty-four, for the sake of illustration, and that the intelligence is to be sent from London to Liverpool.

" The questions, then, are—

" 1st. How the wires are to receive the signals in London.

" 2nd. How they are to deliver them in Liverpool.

" A single letter may be indicated at a time, each letter being taken down by the attendant as it arrives, so as to form words and sentences; but it will be easy to see that, from the infinite changes upon a number of letters, a great number of ordinary communications [whole sentences] may be conveyed by a single previously concerted signal [consisting of one, or more letters in a group].

"Let *a, b,* Fig. 23, represent one of the wires—*a,* the London end, and *b,* the Liverpool end. At *a,* is fixed a small metallic cup of mercury ; *c, d, e,* is a bar of metal moving on a hinge at *d,* so that when the end *c,* is elevated, *e,* will dip into the mercury ; *f,* is a chain, or wire, communicating with the prime conductor of an electrical machine, or with a powerful electro-

FIG. 23.

(Drawn from original Manuscript.)

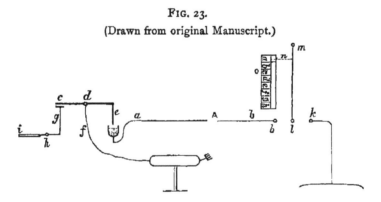

phorus. The hinge, or pivot, *d,* may be continuous metal, and common to all the bars belonging to all the wires. *g, h, i,* is made of glass, or partly of sealing wax, and turns on a hinge, or pivot, *h,* something like the key of a pianoforte. The pressure of the finger at *i,* will then raise *g,* and *c,* and depress *e,* which, by dipping into the mercury, will communicate the electric spark from *f,* to the wire *a, b.* This wire may stand for the letter A; and each of the others will be connected to a similar key apparatus, the same source of electricity sufficing for all.

"At the Liverpool end is a small brass ball *b*; and *k*, is another, communicating by means of a metallic conductor with the earth ; *l*, is a light [pith] ball suspended from *m*, by a rigid rod. When the electricity arrives at *b*, *l*, is attracted, and immediately after repelled to *k*, where it discharges itself, afterwards resuming its normal position midway between *b*, and *k*. *o*, contains the letters ranged in a row, and each letter is connected to the rod of the corresponding electrometer by a stiff hair, as shown in the dotted line, *n*. It is evident, then, that, at every movement of the rod towards *k*, the letter will be drawn from its place of concealment, and exposed to view in the open space above."

This, as Davy himself distinctly says, was not the plan that he would recommend in practice, and was described merely as an aid to a clearer perception of the principles involved. Accordingly, we find him, very soon afterwards, drawing up a proposal for a telegraph, based on the electro-magnetic properties of the voltaic current.

This was to consist of as many line wires as there were letters of the alphabet. Twenty, he says, would have sufficed, or a still fewer number would do, by having recourse to the various combinations of which they would be, obviously, susceptible. Besides the letter wires, there was to be one for the alarum, and another for the return circuit, which was to be common to all.   As to the *form* of the wires, Davy says :—"Since the electricity is believed to move on the surface, and not in the substance, of a conductor, I conceive that where there is a long distance to travel, instead of wires, it will be preferable to use broad ribbons, such as would be obtained by passing a thick copper wire between the rollers of a flatting mill."*

Each ribbon, for its perfect insulation and protection, was to be varnished with shellac, then covered with silk, or woollen, and laid in a slight frame of wood, well dried and varnished ; or each ribbon might be bound from end to end with listing, saturated with melted pitch, or surrounded with caoutchouc, or with cloth saturated with this substance.   All the ribbons, in whatever way insulated, were to be laid together underground in air-tight and water-tight pipes of iron, or earthenware.

The best source of electricity Davy considered would be, either one of Daniell's constant-current cells, then just discovered, or, perhaps, a magneto-electric machine as constructed by Newman or Clarke. The electricity so obtained was to be set in motion in the wires by a set of keys, resembling those of a pianoforte, and connected to their respective wires in the way that we have just described.   On pressing a key, its wire would dip into a cup of mercury con-

---

* It is but fair to mention that this and one or two others are the only cases of false reasoning to be found in all Davy's MSS.

nected with the source of, say, positive electricity, and thus a communication could be readily established, which would be instantly broken when the finger ceased to press on the key. The return wire was to be permanently connected to the source of negative electricity.

At the receiving end the signals might be indicated in various ways, in the adoption of one or other of which much, he says, would depend on the actual amount of electricity that would be available. Davy gave the preference to the following method :—Each line wire was to terminate in another of smaller size, formed into a rectangular coil of from five to two hundred turns, according to circumstances. All the coils were to be ranged in a row in the magnetic meridian, and in each was to be suspended a delicate magnetic needle. The whole was then to be covered by a board, the straight edge of which just concealed the needles in their positions of rest, but, on any of them being deflected, allowed one end to project, and so expose the letter marked upon it. The feeblest current, says Davy, would usually suffice to cause this deflection ; but if, *from the great distance that the electricity had to travel, it became too feeble, its effect on the needles could be increased by multiplying the convolutions of the coils.*

" The quantity of electricity," he goes on to say, " requisite to deflect a magnetic needle is so inconsiderable, that if the current of a moderately-sized pair of plates were sent into one end of a wire, and only one-hundredth part of it came out at the other end, it would still be sufficient. It is for this reason that I prefer the method just described, of indicating the signals, to others which occur to me, and which, as they may answer under certain circumstances, I will briefly describe :—

" 1. A coil of wire from each conductor may be wound round a vertical glass tube, a light needle, or slip, of iron inserted in this tube will be lifted up while the electricity is passing through the coil ; a letter fixed to the iron by a bristle will then appear above the lip of the tube and be the signal.* The only objection to this plan, in other respects the neatest and simplest, is, that the force of the current after passing a long distance may not be sufficient to raise the iron.

" 2. On the same principle, a piece of soft iron may be surrounded by a helix of copper wire, so as to form a temporary magnet, which will attract and relinquish a small piece of iron carrying the signal letter, at every make and break of the current.

" 3. Instead of steel needles, coils of copper wire may be deflected in the neighbourhood of fixed magnets ; thus, *a b*, Fig. 24, are mercury cups, into which dip the line wires *c, d*, and the ends of the coil *e*, which are provided with steel points, and rest on agate surfaces, so that the coil can revolve with perfect freedom ; a

---

* Here we have the germ of the axial magnet used in Royal House's telegraph.

slender spring, *f*, keeps the coil in its normal position of rest. During the passage of electricity in the coil it [the coil] will be subject to the influence of *g*, and be deflected." *

FIG. 24. (Drawn from original manuscript.)

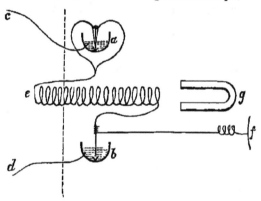

The alarum was to consist of a coil and needle, similar to those used for the letters, only that the needle was to carry a little fulminating silver card, which, on the passage of a current through the coil, was brought into the flame of a small lamp, and exploded.

Davy concludes the document from which we have been quoting with a few words on the general question of conservancy, in which he says that the best situation for the lines would be along railways, where a

* Here we have the germ of the Brown and Allan relay. Davy says this plan is suggested on the possibility of steel magnets being influenced by the wrong wires, which might happen when all are close together in one parallel row.

number of men are constantly watching, and would prevent damage.

At short intervals, as every half mile, more or less, he would have a contrivance (not described), for ascertaining in what precise spot a fault existed, in case of any derangement of the wires. For this purpose the continuity of the copper was to be interrupted, and the two ends made to communicate by means of a cup of mercury. The places where these interruptions occurred were to be under lock and key in the possession of the surveyor.

According to the "Statement" which we find amongst Davy's MSS., giving the order in which his discoveries were made, he had not long finished the preceding paper when he saw how the number of wires might be reduced one half, by employing reverse currents to produce right and left deflections of each needle, each of which deflections could represent a letter. Other and very important improvements followed in quick succession, until, at the commencement of 1837, his ideas had not only assumed a really practical form, but his apparatus was so far complete that he was able to submit it to the test of actual experiment. For this purpose he obtained permission of the Commissioners of Woods and Forests to lay down a mile of copper wire, around the inner circle of the Regent's Park, through which, with the help of his friend, Mr. Grave, he performed many successful experiments.

Soon after this, in March 1837, Davy appears to have been alarmed by the rumours which got abroad of Professor Wheatstone being engaged on an electric telegraph, and, in order to secure for himself a priority, he hastened to lodge a *caveat*, and, at the same time, deposited with Mr. Aikin, the Secretary of the Society of Arts, a sealed description of his invention, in its then state.

Davy now added the relay, or, as he called it, the "Electrical Renewer," which was the only thing wanted to make his apparatus complete and practicable, and the idea of which, it appears, occurred to him after a conversation on the subject of a telegraph, with a Mr. Bush of the Great Western Railway.*

* Writing to the author, on June 11, 1883, Mr. Davy says on this subject :—" I procured access to the private part of the Regent's Park, and laid down a mile of copper wire on the ground, without any insulation. I, of course, found that the magnetic power was so much reduced by the length of the wire that there might be difficulty at great distances in working the contrivance I had in view for marking down signals. The power was, however, sufficient to deflect a not very delicate galvanometer needle.

" It then occurred to me that the smallest motion (to a hair's breadth) of the needle would suffice to bring into contact two metallic surfaces so as to establish a new circuit, dependent on a local battery ; and so on *ad infinitum*.

" In Cooke and Wheatstone's first patent there is a proposal to produce a powerful alarum, by striking a bell through the intervention of a local battery ; but *not a word about any renewer, or relay, as applicable to general electric telegraphy.*

" The relay was equally new to Mr. Morse, and was subsidiary, if not essential, to his admirable method of dots and dashes. On the occasion of my opposing his application for an English patent in 1838, the Solicitor-General told me that he (Morse) had then no idea of the relay.

" The principle of the relay rendered demonstrably practicable the

In May 1837, Cooke and Wheatstone applied for their first patent, and to this Davy entered an opposition, lodging with the Solicitor-General, of the time, a full description of his own apparatus. A copy of this important document* is now before us, and as, when taken in connection with what we have already written, it shows, in a very clear way, Davy's position in May 1837, we shall copy it *in extenso,* merely omitting a few preparatory remarks on the general principles of an electric telegraph, which, though necessary, because little understood, in Davy's time, need not now be repeated. We have also made a few verbal alterations here and there, which, while they in no way affect the sense, will, we hope, render the writer's meaning more clear. A comparison of this paper with Cooke and Wheatstone's first specification will be a curious study, and, all things considered, will certainly not be to Davy's disadvantage.

---

system of overland communication by electricity over unlimited distances, and, doubtless, had the effect of removing hesitation from the minds of those who might otherwise have thought the success of electric telegraphy problematical."

There can be no doubt that to Davy belongs the credit of the discovery of the ·relay system. The first Electric Telegraph Company bought up his patent, chiefly for the reason that it covered the use of the relay. See p. 366 *infra.*

* "This document is either a copy, or the identical one, left with the Solicitor-General. It may not originally have been prepared for that purpose, but, being ready and suitable, was so used. Perhaps it was returned to me and so found its place amongst my other papers."— Extract from Mr. Davy's letter of October 10, 1883, to the author.

"OUTLINE DESCRIPTION OF MY IMPROVED ELEC-
TRICAL TELEGRAPH.

" The parts of the telegraph may be divided into three, *viz.* :—

    1st. Signal and alarum arrangements.
    2nd. Originating mechanism.
    3rd. Conducting and continuing arrangements.

" 1. *Signals and Alarums.*

" These are effected by a number of galvanometer needles, and conducting wires, each of which terminates in a double coil, which acts upon two needles. The form of the coil is that of the figure 8 (only the loops are made rectangular), and in each loop a magnetic needle is suspended. The whole is so arranged that all the needles point north and south. By means of pins, or stops, each needle is prevented from having its signal pole deflected except in one direction. It is then evident that if the electric current be passing in one direction the upper needle only, and, if in the other direction, the lower needle only, will be deflected. This method greatly obviates the objectionable vibration of such needles when suspended in the ordinary way, which vibration is a great impediment to transmitting the signals with sufficient rapidity of succession.

" The needles are as follows, being altogether twelve in number :—

> 4 pairs of letter needles,
> 1 pair of colour needles,
> 1 pair of alarum needles.

"*Letter and Colour Needles.*—The deflectable extremity of each letter needle bears three letters, A, B, C, and so on, each differently coloured, say, A red, B black, C white. Which of the three letters is intended for the signal is decided by simultaneously observing the colour needles, one of which has its deflectable extremity red, and the other black. If neither of the colour needles move, the white letter is the intended signal. By this plan the number of needles, and, consequently, of conducting wires, is reduced to the minimum consistent with convenience. One of the colour needles acting alone may, of course, convey some arbitrary meaning, according to the exigencies of the establishment.

"*Alarum Needles.*—Of these only one is essential, so that the other is available for any other purpose required. The alarum needle is shaped so that a suitable piece of card, loaded with fulminating silver, may readily be slipped on to its extremity, which, when the needle turns, is carried round into the heat of a lamp constantly burning beside it. The card is to be renewed as often as it has been exploded, and a number are always at hand for the purpose. This alarum serves to call the attention of the person who is to watch the signals, and the same principle is evidently applicable to other purposes independent of the telegraph.

"There are other modes of producing different kinds of alarums, which, by the application of the principle on which the electric currents are continued [*i.e.*, relayed], can be accomplished without difficulty.

" 2. *Originating Mechanism.*

" This consists of—2 galvanic pairs of plates,

2 reversers,

4 pairs of communicating keys,

1 pair of colour keys,

1 pair of alarum keys.

"One pair of plates and one reverser belong to the keys of the colour needles exclusively, the other pair of plates and reverser being common to all the other keys. As the action of the two reversers is the same, they may both be included under one description.

"*The Reverser.*—The zinc and copper plates are, each, in communication with separate cups of mercury, which we may call, severally, the zinc and copper cups. There are two wires, one of which we may call the common communicator, and the other the return wire, each terminating in a double, or forked, extremity. To a wooden beam, capable of a certain degree of revolution on its axis, both of these forked pieces are fixed, so that if, in one position, the common communicator is in connection with the zinc cup, and the return wire with the copper cup, a partial revolution of the beam reverses the connections, making the wire from the common communicator now dip into the copper cup, and the return wire into the zinc cup. The reverser is actuated by every alternate communicating key.

" *The Communicating Keys.*—These somewhat resemble the keys of a pianoforte, and there is a pair for each wire.   Pressure upon the first of each pair causes the point of the line wire connected with it to dip into a cup of mercury continuous with the common communicator, which enables us to establish, or cut off, in the most convenient way, the connection between the galvanic pair at one end, and the signal needles at the other.   The second key of each pair, when pressed, turns the reverser before effecting the connection with the line wire, and, consequently, causes a reversed current to flow into the line.

" The colour and alarum keys are operated in the same way.

" 3. *Conducting and Continuing Arrangements.*

" To command the signals not more than eight conducting wires will be required (probably less).   All the letter and the alarum conductors, having, severally, formed their double coils, terminate in a cup of mercury from which a single conducting wire, called the return wire, reaches back to the originating station, where it is in connection with either the copper, or the zinc cup, according as the reverser may be set. The colour needles have a separate return wire,* as well as a separate reverser and pair of plates.

"The best mode of laying the conducting wires remains to be determined by experience on a large scale, and the localities through which they may have to be brought. Either they may be somewhat flattened between rollers, and bound together with interposed pieces of cloth, soaked in pitch or rosin, &c., the whole being enveloped with canvas tarred, or impregnated with melted caoutchouc and linseed oil, or the like ; or they may be secured in a tube (jointed laterally) of iron or earthenware. In the former case they would admit of *being suspended in the air, from post to post, protected by lightning conductors,* while in the latter they would be laid along, or underground ; or they may be separately coiled round with cotton, and bound together, each being of a different colour for guidance in case of repairs.

"As it may be more than doubted that an electric current in a circuit of great length, as between London and Dover, or London and Liverpool, would retain sufficient magnetic power to effect the signals

---

* Davy soon suppressed this, as he found that one return wire would suffice for *all.* He thus reduced the number of line wires to seven, *viz.,* four for letters, one for the alarum, one for the colour needles, and one return wire.

after travelling so far, it may be made to renew itself at given intervals, by the following self-acting contrivance :—

"*The Electrical Renewer.*—The principle of this contrivance is that, the total distance being divided into a number of shorter ones, there be a separate galvanic circuit for each, and that, at the termination of each length of wire, its current be made to produce a motion, which establishes a communication between a fresh source of electricity and the wire which extends through the next succeeding distance. For instance, the first portion of wire terminates in a rectangular figure of 8 coil, fixed horizontally, so as to act upon needles, so suspended as to be capable only of vertical motion. Each needle is rendered incapable of motion except in one direction, so that one, or the other, will be deflected according to the direction in which the current is passing. At the end of each needle is fixed a cross piece of copper wire, whose ends are turned downwards. One of these ends is constantly immersed in a cup of mercury, which is connected with one of the plates of a galvanic pair, while the other end dips into a second cup of mercury every time the needle is deflected by a current in the coil. This second cup is the commencement of the next circuit.

"To complete the circuit, a corresponding but reverse connection must be made with the return wire, so that while the needle of the signal wire establishes a communication with the zinc, that of the return wire establishes simultaneously a communication with the copper, and *vice versâ*.

"By this contrivance it is clear that there can be no physical limit to the distance to which electric currents may be carried; and, therefore, the expense of long distances will cease to be in an increased ratio to that of short ones.*

"*Additional Observations.*

"(1). The coils herein described may be either simple, or multiplied, as the case may require. It will probably be better that they should be multiplied, than that the needles should be too delicately suspended.

"(2). For stopping more effectually the vibration of the needles immediately on their relapse, on the cessation of the current, the following plan is proposed :—A portion of the wire is coiled round a small piece of soft iron, which is rendered magnetic during the passage of the electricity, its polarity varying with the direction of the current. This is so arranged that the end of the needle bears against it, and is held by it, when no current of electricity is passing, or when it is passing in one particular direction, and that, when passing in the other direction, the iron will bé so polarised as to repel, instead of attract, the point of the needle.†

* See note on p. 359 *supra.*

† There is a contrivance like this, and for exactly the same purpose, in Cooke and Wheatstone's first patent of 1837.

" (3). There is a variety of modes in which different kinds of alarums may be made, when once the principle of the Electrical Renewer is applied, as there is then no limit to the power which may be obtained, and it requires little reflection to suggest a multiplicity of methods of making this power produce sound. A piece of soft iron may be rendered alternately magnetic and non-magnetic, so as to withdraw, when required, the peg of an alarum clock, &c., or a needle may be made to carry round a red-hot wire, or match, so as to explode a cannon, &c.

" (4). *Portable Telegraphs.* — Such a contrivance might occasionally be useful in warfare. The conductors should then be made in short lengths, each conductor differently coloured for facility of distinction.

" (5). *Marine Telegraphs.*—Communications may be effected through, or under, the water by enclosing the conductors in ropes well coated, or soaked, in an insulating and protecting varnish, as melted caoutchouc, &c. The ropes could then be sunk to a certain depth by weights, and supported by small floats, or buoys. In connection with the rope we may have an air-tight and water-tight electrical renewing apparatus [*i. e.*, Relay] at each requisite interval.* On this subject further experiments are necessary.

" (6). Land telegraphs may, of course, be made to indicate similar signals and produce similar alarums at numerous places at the same time.

* As in Van Choate's patent No. 156 of January 19, 1865.

"(7). Estimated expense of particular lines of communication at 70*l.* per mile, which includes two sets of wires for communicating in each direction:*—

From	Miles.	£
London to Dover	71	5,000
,, Brighton	51	3,500
,, Bristol	119	8,500
,, Portsmouth	72	5,000
,, Birmingham	109	7,770
,, Liverpool	206	14,000
,, York	199	14,000
,, Newcastle	273	19,000
,, Edinburgh	367	26,000
,, Glasgow	400	28,000
,, Exeter	164	11,500
Liverpool to Manchester	30	2,200
		£144,000

"(8). Annual expense of each station, comprising salary of four or five clerks, attendants, rent, and workman to feed the batteries and keep the apparatus in order, 600*l.*

	£
Outlay from London through Birmingham to Liverpool 14,000*l.*, or say 20,000*l.*, of which the interest at 5 per cent. per annum	1,000
Expense of three stations, one in each town, at 600*l.* each	1,800
Contingencies	200
	£3,000

* It is not that Davy did not know how to make his apparatus reciprocating, so that one set of wires should suffice for to and fro correspondence, but because, as he explains in other places, he thought that when once the telegraph was established there would be more traffic than one set of wires could carry ; and he, therefore, recommends here, as elsewhere, the laying down at once of an *up* and *down* set of wires, for exactly the same reasons that we have *up* and *down* lines of railway.

2 B

which, including Sundays, is about 10*l.* per day ; and this, divided by 3, gives 3*l.* 6*s.* 8*d.* as the sum necessary to be received, on an average per day, at each station, in order to pay expenses and return an interest of 5 per cent. for the outlay.

" (9). *Capabilities.*—The telegraph constructed on the plan herein described is capable of transmitting about fifty letters in two and a half minutes, but, by an improvement devised subsequently to writing the early part of this description, they may be sent several times as rapidly.   In fact, the only limit now appears to be the quickness with which the eye can catch the letter, and the hand note it down."

During the summer and autumn of 1837, as stated in the last paragraph, Davy effected some important changes in the mode of making the signals.   In the plan just described the needles were made to turn horizontally, and the eye was obliged to attend to two movements, at the same time, in order to distinguish by the colour needle, which of the three letters was meant.   The oscillation of the needles in settling down to their positions of rest caused a waste of time, and was otherwise a bar to rapid signalling.   By the following plan both these disadvantages were obviated without introducing fresh ones:—

The needles were all suspended, somewhat like balance beams, so as to turn vertically, instead of horizontally, and they were made long, so that their ends might describe large arcs, although the movement at the centre might be small. The letter ends were weighted so as, under ordinary circumstances, to dip under cover, but, on the passage of a current, they were raised, so as to bring the letters opposite an illuminated sight groove.

What used to be the colour needles were now provided with small screens, which could be raised, or depressed, in front of the letter needles, so as to conceal, or expose, them at pleasure. Thus, if the

A

letter needle bearing B were shown, and neither of the

C

distinguishing needles moved, their screens would lay so as to cover A, and C, B, only being visible. If now one screen, say the top one, were deflected it would cover B, and expose A, and, similarly, if the other screen were moved it would expose C, and cover B.

In this arrangement the eye had to watch only one signal, and, as all the letters could be arranged more compactly, the field of view was greatly reduced, and the letters could be easily caught and noted down, without the necessity of even turning the eyes.

The next alteration was in the same direction of simplification and perfection of the signalling apparatus, and was a decided improvement even upon the last. It consisted in making the letters immovable, and covering them by three screens in such a way as to be able to expose any desired letter at will.

Fig. 25 shows the arrangement. 1, 2, 3, 4, were pairs of screens, which, when at rest, covered all the letters, ranged in rows of three behind them, and one, or other, of which, in each pair, could be moved aside, according to the direction of the current in the line wire to which it belonged. The screens 5, 5, and 6, 6, answering to the colour needles of the old plan, but now called *triplicators*, were so arranged as, in their normal position, to cover the top and bottom rows of letters, and, on the passage of a current in their coils, to move inward and cover the centre row.

FIG. 25.

(Drawn from original manuscript.)

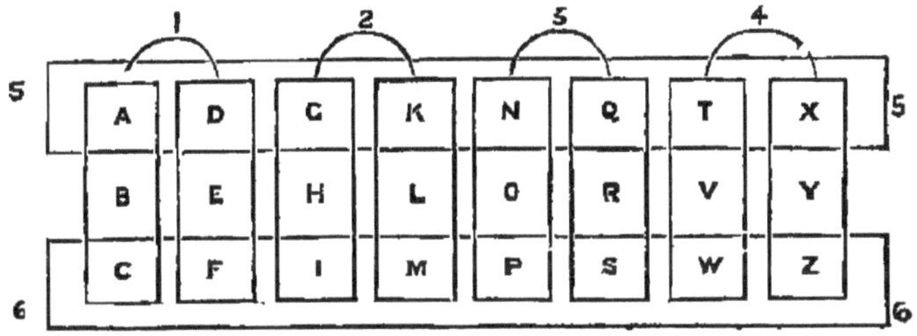

If, now, any one of the letter screens, say that on the extreme left of the figure, were moved aside, only one letter, B, would appear ; because A, and all the letters in the top row, would be covered by the tripli-cator 5, 5, and C, and its fellows in the bottom row, by the triplicator 6, 6 ; but if one of the triplicators, as 5, 5, had been simultaneously moved, then only the letter A, would appear. Thus, a single movement of a letter screen will expose any letter in the centre row, and a combined movement of a letter screen and a triplicator will exhibit any letter in the top, or bottom, row according to the triplicator employed.

But, as each wire was to be provided with a separate battery, and as the currents could be sent in one direction, or another, in one, or more, wires at the same time, it is clear that one to four letters could be shown at once, *provided* they were all in the same row, and were not covered by the same pairs of screens. Thus, the word B O Y could be signalled at once, by sending, say, a positive current into the first wire (on the left), exposing B, a positive current into the third wire, exposing O, and a negative current into the fourth wire, exposing Y. Again, the word A N T could be shown at once, by sending a current into the triplicator wire, so as to cause the screen 5, 5, to move down, then a positive current sent into the first, third, and fourth letter wires would uncover the letters A, N, T, respectively. In this way, besides being able to show all the letters of the alphabet singly, about 200 different groups of letters could be displayed at one operation, which, by having certain meanings attached to them, would greatly expedite a correspondence.

The needles by which the screens were actuated were, as before, suspended in the manner of ordinary balance beams with horizontal axes; but these axes were now prolonged, and carried tall upright rods, at the free ends of which the screens were fastened.

Thus, the slightest movement of the axes produced a considerable deviation of the screens, while their extension permitted of the needles (with, of course, their coils) being placed at sufficient distances apart to prevent their mutual disturbance.

In the working model which Davy had constructed for exhibition all the letter-indicating mechanism was enclosed in a mahogany case, which could serve also as a desk for writing down the signals as they appeared. In the front of the case there was an aperture about sixteen inches long, and three or four inches wide, and this, at ordinary times, was so dark that difference of surfaces of the screens could not be detected, which led to the deception that only one screen was used—a deception which the author purposely planned and encouraged, in order that the *modus operandi* of his instrument might not be divined, which would prevent him taking out a patent for it afterwards, as he contemplated doing.* Behind the screens was a plate of glass, covered with black card-board, out of which spaces, representing the letters of the alphabet, were cut, and behind the glass was a white cardboard, on which the light of a lamp was thrown. The result was that, whenever a screen was turned aside, a beautifully white letter appeared to the spectator at the aperture.

Attention, in the first instance, was called by three strokes on a little electric bell, the termination of a

---

* This little *ruse* explains the fogginess of all accounts of Davy's telegraph hitherto published. See p. 349 and foot-note.

word was indicated by a single stroke, and the end of the communication by two strokes.

A working model, embodying all the author's improvements to date, was shown about November-December 1837, at the Belgrave Institution, London, and from the great interest which it there excited, Davy resolved upon a more public exhibition ; accordingly, he rented a room for one year in Exeter Hall, and there installed his telegraph from December 29, 1837, to November 10, 1838.*

A writer in the *Mechanics' Magazine*, for February 17, 1838, thus describes the exhibit. It will be observed that, for the reason which we have given above, his language in places lacks clearness, but this is of little consequence, for we, who are now in the secret, can easily follow him :—

" Davy's Electrical Telegraph.

" Sir,—The favourable notice of your correspondent, 'Moderator,'† on the subject of Mr. Davy's electrical telegraph, induced me to visit Exeter Hall, for the purpose of carefully inspecting the invention ; and I am enabled to bear testimony to the general accuracy of your correspondent's remarks, and also of the great

---

* Davy's MSS., No. 5. See also *Mechanics' Magazine*, for January 20, 1838, and "The Electric Telegraph Company *versus* Nott and others," Nott's and Grane's affidavits. The room occupied by Davy was that known as No. 5, for which he paid rent at the rate of 35*l.* per annum.

† Correctly, "Moderatus," in *Mechanics' Magazine*, for February 3, 1838, p. 296.

pleasure I experienced in the investigation of the apparatus. Under these circumstances I beg to offer a few additional remarks, in some measure corrective of those made by 'Moderator.'

"As a preliminary observation, I would suggest to the inventor the necessity of removing to some other part of the building, or, if that cannot be accomplished, of quitting the place altogether, and locating himself in some situation where his light may not, literally, be 'hid under a bushel.' He appears to be surrounded by rooms under repair or alteration, and his delicate apparatus is, consequently, smothered with dust; the room is also small, dark, and altogether of most unpromising appearance.

"In front of the oblong trough, or box, a lamp, described by your correspondent, is placed, and that side of the box next the lamp is of ground glass, through which the light is transmitted for the purpose of illuminating the letters. The oblong box is open at the top, but a plate of glass is interposed between the letters and the spectator, through which the latter reads off the letters as they are successively exposed to his view. At the opposite side of the room a small key-board is placed (similar to that of a pianoforte, but smaller) furnished with twelve keys; eight of these have, each, three letters of the alphabet on their upper surfaces, marked thus, A. D., and so

B. E.

C. F.

on. By depressing these keys in various ways the signals, or letters, are produced at the opposite desk as previously described. How this is effected is not described by the inventor, as he intimated that the construction of certain parts of the apparatus must remain *secret*. By the side of the key-board there is placed a small galvanic battery from which proceeds the wire, 25 yards in length, passing round the walls of the room. Along this wire the shock is passed, and operates upon that part of the apparatus which discloses the letters, or signals.

" The shock is distributed as follows :—The under side of each signal key is furnished with a small projecting piece of wire, which, on depressing the key, is made to enter a small vessel filled with mercury, placed under the outer end of the row of keys. A shock is instantly communicated along the wire, and a letter, or signal, is as instantly disclosed in the oblong box. By attentively looking at the effect produced, it appeared as if a dark slide were withdrawn, thereby disclosing the illuminated letter. A slight vibration of the (apparent) slide occasionally obscuring the letter indicated a great delicacy of action in this part of the contrivance, and, although not distinctly pointed out by the inventor, is to be accounted for in the following manner :—When the two ends of the wire of the galvanic apparatus are brought together over a compass needle the position of the needle is immediately turned at right angles

to its former one; and again, if the needle is placed
with the north point southward, and the ends of the
wire are again brought over it, the needle is again
forced round to a position at right angles to its original one [*sic*]. Thus it would appear that the slide,
or cover, over the letters is poised similarly to the
common needle, and that, by the depression of the
key, a shock is given in such a way as to cause a
motion from right to left, and *vice versâ*, disclosing
those letters immediately under the needle so
operated upon.

"A gentleman present hazarded a doubt as to the
shock being energetic enough for a considerable
distance. The inventor replied that he was in possession of means that would enable him to convey
intelligence to any distance that may be required.
Whether this was to be effected by *coils* of wire at
intervals was not stated; such, however, appears to
me a reasonable supposition. The difficulty of *tubing*
for the protection of the wire was discussed. I took
the liberty of suggesting the employment of a proper-
sized *tobacco-pipe tubing*, which was received with
satisfaction. It was also stated by a gentleman
present that he was in possession of a smaller battery
than that at Exeter Hall, and had obtained from it
a power equal to *forging* iron plate; it will, he said,
be shortly produced.—Yours respectively [*sic*],

"CHRIS. DAVY.

" 3, Furnival's Inn, Feb. 5, 1838."

CHAPTER XIV.

EDWARD DAVY AND THE ELECTRIC TELEGRAPH, 1836–1839 (*continued*).

RETURNING to our examination of the Davy MSS., we find a memorandum of another modification of the screen arrangement, which would require only two line wires (and one return wire), and yet would yield twelve elementary signals. This contrivance, which is fully explained, need not, however, detain us further than to indicate the highly ingenious plan adopted for producing some of the necessary changes of the screens. It consisted in the employment, at certain times, of batteries of *different strengths* (they being of necessity of opposite signs), so as to determine, at those times, a current of positive, or negative, sign in the return wire, and thereby actuate screens which, if the currents had been of *equal strength*, would, of course, be inoperative. This neat and effective arrangement was utilised in another of Davy's instruments, of which we must now say a few words.

The recording telegraph is a very beautiful piece of mechanism—the first of a long line of chemical telegraphs,—and we cannot help thinking that, had it had a fair start in 1838, and been fined down by practice, as it could have been, and as Cooke and Wheatstone's first inventions were, it would have given to English telegraphy a somewhat different character from that impressed upon it by the rival plans, and chemical telegraphs might now be the rule instead of the exception.

Like all his other inventions in telegraphy, Davy perfected this apparatus before December 1837, or, as he says in his "Statement," before the enrolment of Cooke and Wheatstone's first specification, of the nature of which he was, at the time, in perfect ignorance, "except in so far as it could be gathered from paragraphs in the newspapers, which conveyed really no information."

He wished to take out a patent at once for this instrument, but, owing to legal formalities, and the opposition of Cooke and Wheatstone, the specification was not sealed until July 4, 1838. The opposition was based on the plea that some parts of Davy's mechanism were infringements of their patent of June 12, 1837, but, on a reference to Professor Faraday, who gave it as his opinion that the two inventions were distinct, the Solicitor-General quashed the opposition, and allowed the application to pass.*

* Davy's MSS., No. 10, contain some warm passages on this most unfair charge. Writing to the author, on June 11, 1883, Mr. Davy says further:—"On applying for my patent Messrs. Cooke and Wheatstone opposed it, before Sir J. Rolfe, the then Solicitor-General. That gentleman, however, told me that he would at once pass my application if I confined myself to the renewer, as on some other matters he had his doubts. I did not feel disposed to relinquish

The following passages, which we have extracted, by kind permission of Mr. Latimer Clark, from the MSS. correspondence of Messrs. Cooke and Wheatstone, are explanatory of this point :—

"20, Conduit Street, Jan. 20, 1838.

"My dear Sir,—

*　　*　　*　　*　　*　　*

"Davy has advertised an exhibition of an electric telegraph at Exeter Hall, which is to be opened on Monday next. I am told that he employs six wires, by means of which he obtains upwards of two hundred simple and compound signals, and that he rings a bell. I scarcely think that he can effect either of these things without infringing our patent ; if he has done so, I think some step should be taken. As the point of resemblance in Davy's instrument is, no doubt, a 'return wire,' I do not think that an injunction could be procured to restrain him, without proceeding also against the exhibitors of Mr. Alexander's.

"The latter case is very clear. Previous to our patent no person had ever proposed otherwise than to employ a complete circuit (*i. e.*, two wires) for each

---

any of my claims, so it was arranged to refer the whole matter for advice to Mr. Faraday, with whom both parties were to communicate. It appears that Messrs. Cooke and Wheatstone were under the impression that I wanted to patent only what had been exhibited at Exeter Hall I spent two or three hours with Mr. Faraday, and left my papers (rough specification) with him, when he said that he would take a week to consider, and report to the Solicitor-General. He accordingly reported that my inventions were quite original, and entitled to a patent. Probably, some notes of this transaction may be found in the records of the Solicitor-General's office."

magnetic needle. The most important original feature in my instrument was that the same wire should be capable of forming different circuits according as it was conjoined with other wires.

"After the patent was sealed, a notice of some of my experiments appeared in the *Scotsman;* and some weeks subsequently there appeared in the same paper an account of what Mr. Alexander intended to do, and, after a long interval, a description of a model which he had produced.

"There is no doubt that in our Scotch patent we must limit ourselves to the application of the permutating principle ; but as our English patent was sealed before the slightest publicity was given to Mr. Alexander's intentions, I think no lawyer can doubt our priority [in England].

"If Mr. Davy has taken the return wire because he has seen it in Alexander's instrument, and therefore thinks that we do not claim it, the point will be an easy one to settle ; but it will be more difficult if he had an idea of it before the hearing by the Solicitor-General. Think over the matter, and let me know your opinion before any proceedings are commenced.

*   *   *   *   *   *

"I remain, my dear Sir,

"Yours very truly,

"C. WHEATSTONE.

"W. F. Cooke, Esq.,

"Compton Street, Brunswick Square."

In a letter dated March 10, 1838, Wheatstone
writes :—

"Let me know the title of Davy's patent, and also
when it is likely the opposition will be heard, as I wish
to make some preparations in time. I have heard
that a physician, residing in your neighbourhood, is
the party who encourages Davy, and furnishes him
with cash."

On March 24, 1838, he wrote :—

"My dear Sir,—The Solicitor-General was with me
twice yesterday at the College. Davy was extremely
anxious to obtain a decision on the plea* of going out
of town immediately. The Solicitor-General, however, has not yet given an answer, and on applying at
his office this afternoon I was informed he had left
word that he should not decide the question for several
days. I shall endeavour to see him again to-morrow,
as I know his difficulty, and have another argument
to offer him.

  *   *   *   *   *   *

     "Yours very truly,

       "C. WHEATSTONE.
"W. F. Cooke, Esq."

The following account of the construction and
modus operandi of the apparatus we condense from
Davy's specification, to which we refer our readers for

---

* We have seen Mr. Davy's private letters of this date, and know
how true the plea was.

fuller details.* It contains all the essentials of a complete telegraphic system, and can be procured at the Patent Office for a small sum.

"The drawing, Fig. 26, represents the apparatus employed at the place of making a communication, say London, and that at the place where the communication is received, say Birmingham, or Liverpool. The wires A, B, C, are those which are laid down between those places †

" The principle on which this apparatus works is this, that there be two, or more, wires which communicate with another wire, and, for distinction sake, we will call the former the signal wires, and the latter the common communicator, for it should be understood that no metallic circuit can be formed between the signal wires of themselves, but only by the aid of the common communicator ; and further, whatever be the number of wires employed so having a connection with a common communicating-wire, that there be a suitable electric apparatus, such as a voltaic battery, to each signal wire. The drawing shows the apparatus to consist of two signal wires, A, and B, and a common communicating-wire C ; and the

* See also Vail's *American Electro-Magnetic Telegraph*, Philadelphia, 1845, pp. 187–99, or Shaffner's *Telegraph Manual*, New York, 1859, pp. 255–68.

† Sabine, on p. 50 of his *History and Progress of the Electric Telegraph*, 2nd edit., London, 1869, states erroneously that at least four wires were required. That excellent French journal, *La Lumière Électrique* (April 7, 1883), has recently committed the same mistake, and shows four wires in its illustration, Fig. 33.

drawing further shows the apparatus to have three separate batteries; the object in using the third is to obtain a greater extent of signals than can be obtained by the employment of only two. In this case, the common communicating-wire may have needles and suitable apparatus, and may thus become a means of communicating signals as well as the signal wires.

"D, E, are the pair of finger-keys, which cause electric currents to pass through a circuit, partly made up of the signal wire A, and the common communicating-wire C, and it will be found that the parts are so arranged that depressing the key D, will bring the signal wire A, in metallic communication with the negative pole of the battery No. 1, and at the same time cause the common communicating-wire C, to be in metallic communication with the positive pole of the same battery; consequently, the currents will pass positively through the wire C, and negatively through the wire A. The course of the currents may be reversed by depressing the key E, instead of D.

"The keys H, and I, act on the wires B, and C, and form metallic circuits through which currents from the battery No. 2 may be transmitted in like manner to what has just been described in respect to the wires A, and C, and the battery No. 1, by the keys D, E.

"The wires A, B, C, just before being connected together at the distant station, are each formed into

2 C

two coils, or convolutions, similar to what are employed
for galvanometers, in order that the electric current
may operate with sufficient power on the needles
placed within them as to deflect them in a direction
corresponding to that in which the current is passing.
M, N, show these coils, or convolutions.

FIG. 26.

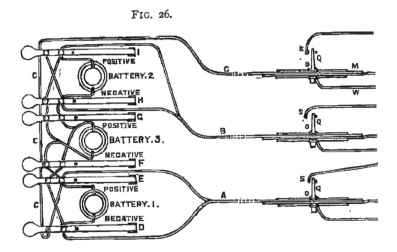

"The needles O, P, are little magnetised plates of
steel, moving on points similar to a magnetic needle ;
and, when at rest, are to be in a line with the coils in
which they move. To the upper part of each needle
is affixed the upright contacting-piece Q, which at its
lower end dips into a little cup of mercury. S, T, are
wires against which the upper ends of the contacting-
pieces, at times, come in contact in order to form
local metallic circuits for the purpose of producing

marks on chemically-prepared fabrics, as hereinafter explained.

"V, is a compound battery from the positive pole of which a wire W, communicates with each and all of the cups containing mercury. Consequently, when any one of the contacting-pieces is caused to touch

FIG. 26.

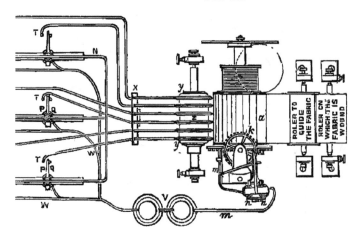

its wire S, or T, as the case may be, there will be a metallic contact with the positive pole of the battery and the said wire S, or T. Thus, supposing a positive current to be passing through the wire A, the con-tacting-piece Q, of the needle O, of the wire A, would be deflected towards, and would come in con-tact with, the wire S, and would form a metallic contact between it and the positive pole of the battery V ; and, on the other hand, if a negative current pass through the wire A, the contacting-piece Q, of the needle P, would be brought in contact with the wire T. Thus it will be seen that each line wire has a capability of giving two separate indications, and these may be increased, by compounding, to eight, according to the order in which they are communicated. The number may be still further increased to twelve by applying needles and coils to the wire C, and employing a third battery, marked No. 3, and two extra keys F, G.

"The object of using a third battery is to give greater quantity of electricity to certain currents. Thus, supposing a positive current to be passing through the wire A, and a negative current through the wire B, there would be two currents passing to the wire C, in opposite directions; consequently, the needles on that wire would not be acted on in such manner as to produce a certain and definite indication, *unless* one of the currents so passing be made more powerful than the other. And this may readily be effected by the keys F, G, which can bring the wires A, C, and B, C, in connection with the battery No. 3, in addition to their own.

"The pairs of wires S, T, pass through a block of wood X, which acts as a support. Their ends are forked, and embrace (touch) the metallic rings (of platina) *y, y,* affixed to the wooden cylinder Z. These rings press closely against the metallic cylinder *a,* which turns in suitable bearings carried by the framing, as shown in the drawing. This cylinder has a constant tendency to revolve in one direction, communicated to it by a spring or weighted cord as in Fig. 27, but is only permitted to turn a certain distance each time that a signal has been made through the wires.

"It should be stated that there is a metallic contact between the negative pole of the battery V, and the metallic cylinder *a*, by means of the wire *m*, which is coiled round a bent bar, or horse-shoe, of soft iron, in order to produce an electro-magnet *n, n*, and from thence the wire *m*, passes to, and is held in contact with, the end of the cylinder *a*; consequently, whenever any one or more of the contacting-pieces of the needles come in contact with their wires S, or T, a metallic circuit or circuits will be formed ; and it will be evident that if properly prepared fabrics, such as calico impregnated with hydriodate of potass and muriate of lime,* be placed between the metallic

---

* "Although I have recommended the use of calico prepared in the manner above stated as the fabric to be used for receiving the marks, I do not confine myself thereto, as other fabrics may be used and the chemical materials employed may be varied, so long as they will be similarly marked by the passage of electric currents. The fabric so employed may be printed with subdivisions, as is shown in the drawing, or it may be used plain, because the marks made, whether a single one or more than one at a time, will be in rows across the fabric; and each row, whatever be the number of marks, will be a signal, and this mode of receiving marks in rows across and lengthwise of the fabric constitutes an important feature in my invention ; for although I prefer that the marks should be produced by the chemical action of the electric currents acting on fabrics properly prepared, yet it will be evident that other means of producing a series of marks in rows, crossways and lengthwise of the fabric, may be resorted to, such as pencils or ink

rings *y*, *y*, and the cylinder *a*, whichever of these rings are for the time being in circuit will, by pressing against the cylinder *a*, pass the current through the prepared fabric, and produce marks thereon. It will only be necessary to assign to each mark so produced a definite cypher referable to a proper key-book, as is well understood in telegraphic communications.

" At the same time that the signal is being thus recorded the armature D, Fig. 27, of the electro-magnet M, is attracted, the forked piece J, in which it terminates, goes up from the pallet *a*, of the fly-vane G, and so allows the cylinder K, to revolve (carrying with it the prepared fabric) until the pallet *a*, coming in contact with the forked piece E, stops the mechanism. When the hand of the person making the signal is removed from the key or keys, the contacting-pieces resume their vertical positions, thus opening the local circuits. As a consequence, marks cease to be made on the prepared fabric, and at the same time the armature is drawn back by the spring S ; the fly-vane G, is thus again liberated, and the cylinder K,

---

connected to, or carried by, proper holders acted on by electro-magnets, one pencil, or other marking instrument, to each wire S, T ; by which means every time a metallic circuit was produced by the aid of any of the wires S, T, they would cause their electro-magnets to bring the marking instrument in contact with paper, or other suitable fabric, and give marks thereto across such fabric ; and as the fabric was moved forward, the next row of marks would be made at a distance from the preceding row, and separated therefrom, the same instrument producing its mark at all times in the same longitudinal row."— Pp. 11, 12 of Davy's specification.

revolves through another space, until once more stopped by the pallet *a*, catching in the arm J. The first movement is for the making of the signals, the second marks the intervals between them.* The fabric,

FIG. 27.

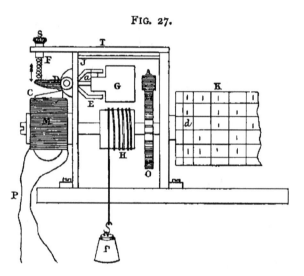

as it is carried forward by the cylinder K, is conducted away over a guide-roller, and drawn forward by a weight, or in any other suitable manner."

Following the original drafts of this apparatus, which are preserved amongst the Davy MSS., we find

---

* As the description of this portion of the instrument is somewhat involved in Davy's specification, we have in the text used our own words, which, with the illustration (borrowed from Schellen), will we hope make the action clear. Vail's *American Electro-Magnetic Telegraph*, pp. 187–99, gives a full and well illustrated account, to which we would refer our readers anxious for further information.

a description of another telegraphic project, which, as our readers will observe, is based on the same electrical principles as the diplex and quadruplex systems of the present day. It is a mode of obtaining one, two, or more signals through a single pair of wires by means of currents of one, two, or more degrees of strength, acting on needles so weighted as only to respond to the currents destined to move them.

For the signal-indicating part of this plan, Davy proposed to employ a new form of galvanometer, which he called an "electro-magnetometer," and which he had designed, in the first place, "for measuring with greater precision than heretofore the exact quantity of electricity passing in any given circuit." This instrument is figured and described in his patent of July 4, 1838, pp. 16, 17.

Foreseeing that to operate a telegraph of this kind it would be necessary to regulate, and keep regulated, the currents with great accuracy, Davy devised, for this purpose, a "self-regulating galvanic battery." "The principle of the contrivance," to quote his own words, "is that as soon as the magnetic energy of its electric current rises above, or falls below, a certain required standard, one of the metals of the galvanic pair is, by the agency of this magnetic energy, either raised out of, or further depressed into, the acid, or exciting liquid, in the cell; so as to become, thereby, less, or more, exposed to the action of the liquid ; the extent of its exposure regulating the quantity of electricity generated.

"To effect this intention, there are two coils in the conducting wire proceeding from the battery, in which are two of my electro-magnetometer needles. Of these needles the dipping end of one is a certain degree heavier than that of the other, and the intention is that the electric current should remain within the limits of the two, so as just to act upon the lighter, but not on the heavier.

FIG. 28.   (Drawn from original manuscript.)

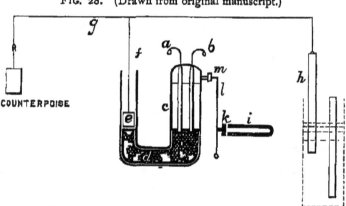

"Now, if the current be too powerful the heavier needle, by dipping, will cause a communication between a fresh, or distinct, source of electricity and the two wires *a*, and *b*, Fig. 28, attached to platina plates in dilute sulphuric acid contained in the air-tight tube, *c*. Then, by the decomposition of the water, gases will be evolved, and depress the liquid at *c*, and also the mercury below it, *d*, so as to elevate the piston, *e*, and its rod, *f*, whereby the lever, *g*, is also elevated, and lifts the metallic plate, *h*, belonging to the galvanic battery, so as to diminish the energy of the said battery to the required degree.

" If, on the other hand, the electric current be too feeble, then the lighter needle will fall and open a communication with another distinct source of electricity through the coil of wire which surrounds the electro-magnet, *i*, whereby it is rendered temporarily magnetic and attracts the armature, *k*, which, through the lever, *l*, removes a caoutchouc (or other suitable) stopper from the minute aperture at *m*, so as to allow the gas in the tube, *c*, to escape until the metallic plate, *h*, has again sunk sufficiently into the liquid in the battery cell to generate the required, or standard, quantity of electricity, such standard being allowed to vary between these minute differences only.

" Having thus obtained the element of a uniform battery, the quantity of electricity to be transmitted through the circuit may be regulated, either by the number of such batteries uniting their currents, or else the current from one battery may be divided ; the mode of so dividing it is the remaining consideration.

" The electric current may be made to travel through pieces of platina, or other wire, of different diameters, and of given lengths, so that the thicker the wire, the greater will be the quantity of electricity to pass. The exact dimensions of these to be regulated by actual experiment,* and, in order to prevent their ignition and combustion, they may be arranged under water, or, should water be objectionable, under some non-conducting, and non-electro-decomposable liquid, such as sulphuret of carbon, or naphtha, or whatever other may be found advisable."

From amongst Davy's miscellaneous memoranda we select two or three, with which we must close this portion of our work. In the first our readers will, we doubt not, be amazed, as we were ourselves, to find how near the writer was to discovering the telephone in 1837–8.

" 20. The plan proposed (101) of propagating com-munications *by the conjoint agency of sound and electri-city—the original sound producing vibrations, which cause sympathetic vibrations in a unison sounding apparatus at a distance, this last vibration causing a renewing wire to dip† and magnetise soft iron so as to repeat the sound, and so on, in unlimited succession.*"

The sheet from which we copy these remarkable words is headed "Exclusive Claims," and seems to have served as an *aide mémoire* to the drawing up of

---

* Here we have the germ of the rheostat, or set of resistance coils, as used at the present day.

† *i. e.*, causing a relay to close a local circuit containing an electro-magnet. Davy always speaks of the relay as the "renewer," or the "renewing wire." By dip he means to dip into mercury, or, as we say nowadays, to close the circuit.

his patent specification. If our surmise be correct, it would fix the date of the paper as not later than the beginning of February 1838, for we shall see, later on, that he was, in that month, submitting his inventions to Mr. Carpmael, a well-known patent agent of that period. Unfortunately we can find no further mention of the "plan proposed," and can only suppose that Davy designed some kind of telephonic relay.

In the following memorandum the writer could only have in view a form of cell, which is now so well and so deservedly esteemed under the name of its recent inventor, M. Leclanché :*—

" *A New Galvanic Battery,*

"A particular mode of using oxide of manganese as the electro-negative element of the battery, or in connection with the electro-negative plate.

"Certain other improvements in the battery, which will be described, if there be any opposition on this head,

"*An Improved Magneto-Electric Machine,*

"To be described if there be any opposition on this head."

---

* "The new galvanic battery *was* on the principle of Leclanché's ; but, attention having been directed to other matters, it was never perfected by me."—Extract from Mr. Davy's letter of October 10, 1883, to the author.

The following extracts are from a paper headed—

" Elemental Forces and Alarums.

" There are two objects for which alarums may be required as essential appendages to the Electrical Telegraph.

> 1st. To give notice that communications are about to be sent, and call the attention of the person who is to receive them. For this purpose alarums of great loudness will not, generally, be required, unless the party be asleep, or not in the same room, and even in these cases a moderate loudness will suffice.
>
> 2nd. To give notice of accidents on a railway, or in other cases where the alarum may require to be heard by persons who may be at a distance at the time.

" 1st. With the first object an alarum is easily made. One of my horizontal dipping needles (surrounded by its coil) may have an upright rod, as a radius from its axis, with a little hammer on a spring to strike a small bell by the deflection, or dip, of the needle. Thus two needles in the same wire may strike two distinct bells and produce a kind of chime; either one, or both, or variations of which, may be advantageously used according to the intention of the alarum.

*   *   *   *   *   *

" 2nd. Whenever an almost irresistible, or, at least, very great power is required, either to produce alarums, or for any other purpose, I claim the following mode of effecting the object, which is also applicable in other cases where temporary magnetisation may prove insufficient.

"A piece of platina wire, *a*, Fig. 29, connected in circuit with the conducting wire, *b*, and *c*, is securely enclosed in an airtight, and strong vessel, *d, d*, in contact, or proximity, with a quantity of sulphuret of carbon, or other suitable volatile liquid, Then the current of electricity from *b*, to *c*, will ignite, or heat, the platina wire *a*, so as to convert a portion of the volatile liquid into vapour, which will then expand with a degree of force, proportioned to the heat of the platina wire, and its continuance in a heated state. This will force the mercury, which is below the volatile liquid, through the tube *e, e*, so as to elevate the piston at *f*, and *g*. The force thus obtained may be applied to any required purposes.

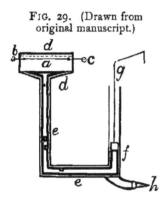

FIG. 29. (Drawn from original manuscript.)

"When the current of electricity ceases to pass, the sulphuret of carbon, or other volatile liquid, will re-condense, and the piston gradually resume its former position without the necessity for an attendant to liberate the vapour. Of course a safety valve may be attached, if necessary, either at *h*, or at *d*, or the self-regulating battery would be useful in combination with this contrivance.

"A continuous sound may be produced by applying either of the above-mentioned forces to open a valve so as to admit air, or gas, from a vessel containing such air, or gas, under compression through a whistle, horn, or other wind instrument. Air, or gas, under compression for this purpose may be provided by the action of dilute sulphuric acid on old iron, on the principle of the hydrogen instantaneous light apparatus, where, as soon as a certain quantity of gas is generated, the liquid is forced into another part of the vessel so as no longer to act on the metal; or air may be pumped in from time to time."

As we have in one or two places, in the course of these pages, referred to Davy's "Statement," we think it advisable to reproduce this important document, as, while confirming our chronology, it will also serve as an excellent *résumé* of the writer's leading discoveries :—

"*Statement.*

"The idea of an electrical telegraph first occurred to me about the year 1836, at which time I was not aware but that it was perfectly original. In the commencement of 1837, having tried some experiments with a mile of copper wire in the Regent's Park, aided by my friend, Mr. Grave, I entered a *caveat*, in March, and, about the same time, I deposited with Mr. Aikin, Secretary of the Society of Arts, a sealed description of my invention, in its then state.

"My earliest idea of applying the deflection of the needle for telegraphic purposes, was similar to that since claimed as a new invention by Alexander, with a common return wire. The next improvement was the obtaining the two actions upon each needle by the reverse currents. Then, the fixing two instead of one needle in each circuit, and subsequently, the system of permutation described, with the use of the colour needles, and the employment of more than one battery. It was at this stage (in March 1837) that I first heard of Professor Wheatstone being engaged on the same subject, which led me to enter the *caveat.* Shortly after this, the idea of the renewing needles [relay] occurred to me. This was after a conversation on the subject with Mr. Bush of the Great Western Railway.

"In May 1837, Messrs. Cooke and Wheatstone applied for a patent, to which I entered opposition, having provided myself with a written description of my inventions, and prepared to attest it by the evidence of several confidential friends. This evidence was partly direct, and partly corroborative. I had Dr. Grant, Mr. Thornthwaite, and Mr. Hebert, besides the workman who helped me to make the models, and the Solicitor-General on some specific, but all-sufficient, points. The paper was carefully inspected by my friends, who were also present at the hearing on the opposition.

" The Solicitor-General at the time gave an opinion that the two inventions were different, and allowed the patent to pass, although time has since shown that they contained some of the clearest identities.

" My remedies for the injustice thus sustained are, that I may move a writ of *scire facias* to set aside and annul Messrs Cooke and Co.'s patent, on the ground that the Crown was misled in granting it, or else, or after failing that, to act upon the [my] invention so that they may bring an action for infringement, which I have ample grounds for defending, and the failure of which will *virtually* render their patent void. Litigation of this kind, which will be highly injurious to one party, and but partially beneficial to the other, is what it is in every way desirable to avoid, if the matter can be otherwise adjusted.

" From the time of this decision (May 1837) up to the time of the enrolment of their specification in December, I was in perfect ignorance of the nature of their invention, except in so far as it could be gathered from paragraphs in the newspapers, which conveyed really no information. In the meantime I introduced into my plans first, the use of screens, then the means of determining the signals to specific places exclusively,* and finally, that which I believe is cal-

---

* Re-invented in 1853 by Wartmann. See De la Rive's *Treatise on Electricity*, vol. iii. p. 783.

culated to supersede all others, the recording telegraph by electro-chemical decomposition." *

Through the kindness of Mr Richard Herring, whose name will be familiar to our readers as the inventor of a beautiful recording telegraph, which ought to be better known, we have lately been in communication with Mr. Thornthwaite, one of the gentlemen just mentioned, then Davy's assistant, and now the chairman of the Gresham Life Assurance Society. At our request he has jotted down his reminiscences of this period, which, as corroborative of Davy's "Statement," may fittingly be given here :—

" To J. J. Fahie, Esq.

" London, December 14, 1883.

"My dear Sir,—I find on examination of some old papers that I was a pupil of Professor Daniell in 1834, and that, through the introduction of a mutual friend, I entered the service of Mr. Edward Davy about the end of the year 1835, as pupil and laboratory assistant. Very shortly after entering on my duties Mr. Davy informed me confidentially that he was engaged in some important investigations, the nature of which he could only communicate under a bond of secrecy and an understanding not to make use of the information

* To this may now be added (1) a block system for railways, (2) the telephonic relay, and (3) the oxide of manganese (Leclanché) cell, besides numberless suggestions of a more or less practical nature, many of which are noticed in these pages.

to his detriment, or to my own advantage. On my giving him the required undertaking he stated that his investigations and ideas had reference to the transmission of signals through great distances by electricity, and the employment of electricity as a motive power, both of which he expressed his opinion were of vast future moment.

" A short time after this conversation he took into his employ a workman of the name of Nickols to make a telegraph instrument to work by the galvanic current causing a deflection of horizontally suspended magnetised steel bars while circulating through coils of insulated copper wire. Each magnetised bar was to carry a light screen of thin paper to uncover and indicate a letter when thus deflected. This instrument, after many modifications of form, was afterwards publicly exhibited in action in the small room in Exeter Hall.

" My engagements in the laboratory prevented my giving much personal assistance in the experiments in Regent's Park, but I understood they were generally successful as demonstrating the possibility of sending for some considerable distance very distinct signals, amongst others firing a pistol by the agency of a galvanic current transmitted through a thin uncoated copper wire laid on the grass. These experiments were brought to an abrupt termination by our finding one morning that the cowherd had made the curious discovery of some copper wire lying on the grass, and had amused himself by coiling up and removing the same.

" I have no doubt the idea of using the fulminating silver card as an alarum* was suggested by a circumstance which occurred about this time.   Mr. Davy was sent for one morning by Mr. Minshell,† the magistrate of Bow Street Police Court, and, on his return, he placed on the counter a shallow wooden box, about six inches by three, telling me that it had come into the possession of one of the police officers in connection with some explosive letters lately put into the post, and that, when he arrived at Bow Street Court House, he found the box, containing a brownish powder, being handed about the Court, and its contents being tested even by the smell.   On his pronouncing the powder to be fulminate of silver, of sufficient quantity and power to blow the Court to pieces, and liable to explode with the smallest particle of grit and friction, the box was suddenly treated with the utmost respect, and various suggestions were made as to its disposal—the magistrate proposing that it should be taken by an officer and thrown over one of the bridges into the Thames.   No one, however, appeared willing to undertake the job.   In this state of perplexity, and on the appeal of Mr. Minshell, Mr. Davy took the box and contents under his charge.   Having told me these particulars, he said :—' Will you carefully separate the powder into small parcels of about a dram each, and

* See p. 362, *ante.*   † See p. 523, *infra.*

wrap each parcel in two or three papers, and place them separately in different parts of the house for safety.' I need hardly say that I felt an infinite amount of satisfaction when the last parcel was safely disposed of.

"You are quite at liberty to make what use you think fit of this letter, or any part thereof, that may further your efforts, to honour the name of my old friend and master, Mr. Edward Davy.

"I am, yours very truly,

"W. H. THORNTHWAITE."

As showing Davy's wonderful perception of the uses which the telegraph would subserve, as well in the internal economy of railways, as in the political economy of the nation, and of the world at large, we give below the concluding portion of a lecture, which bears evidence of having been written about the middle of 1838 : *—

" The point which now remains for consideration is, of what use will this electrical telegraph be ? What are its applications, how will society at large benefit by it, and what inducements does it hold out to private adventurers to take it up as a means of investing capital ?

" Now, at the outset of nearly all new propositions

---

* Referred to in his letter of 16th June, which see *infra*. " This was given at an institution near Oxford Street, name forgotten."— Extract from Mr. Davy's letter of October 10, 1883, to the author.

of this nature, there are two kinds of objections which we have to contend with. The first arises from the circumstance of the invention being a novelty, and different from all that people have previously been accustomed to. We get laughed at; the matter is treated as a dream. 'Really, sir,' says one, 'you cannot be serious in proposing to stop the escape of a thief, or swindler, by so small an electric spark, acting on a needle; if you had talked of sending a thunderbolt, or flash of lightning, after him, I might have thought there was some feasibility in it.' Another tells us that the experiments are very well across a room, but would not succeed on a large scale. Then, as soon as the practicability of the thing is undeniably established, the same people turn upon us with the question, 'What is the use of it?'* There must be some present who will recollect that the first introduction of gas was beset with the same objections. So also were the railroads, and to a certain extent they continue to be up to this time. So also was the steam engine, printing; in fact, almost everything new is discountenanced, or coldly received, by the public at large in the first instance. However, the time has, I believe, already arrived, when the practicability of this

---

* "As an instance of how new ideas are sometimes misjudged, even by very intelligent men, I may mention that, in conversation with me in 1837, Dr. Birkheck, of Mechanics' Institute celebrity, expressed the opinion that the electric telegraph, if successful, would be 'an unmixed evil' to society—would only be used by stock-jobbers and speculators—and that the present Post Office was all that public utility required."— Extract from Mr. Davy's letter of June 11, 1883, to the author.

electric telegraph is no longer doubted, either by scientific men, or by the major part of the public, who have given any attention to the facts upon which the invention rests.

"I have, therefore, to confine my remaining observations to the uses and application of it. And first, I have a few words to say upon what must be considered as a minor application, namely, the purposes it will answer upon a railway, for giving notices of trains, of accident, and stoppages. The numerous accidents which have occurred on railways seem to call for some remedy of the kind; and when future improvements shall have augmented the speed of railway travelling to a velocity which cannot at present be deemed safe, then every aid which science can afford must be called in to promote this object. Now, there is a contrivance, secured by patent,* by which, at every station along the railway line, it may be seen, by mere inspection of a dial, what is the exact situation of the engines running, either towards, or from, that station, and at what speed they are travelling.†

* In the drawing up of the specification specific mention of this invention was, most unaccountably, omitted. This enabled Wheatstone in 1840 to patent a similar step-by-step instrument, with dial face, &c.

† At every railway station there will be a dial, like the face of a clock, on which, by means of a hand, or pointer, it may be seen where any particular train, running, towards, or from, that station, may be at any particular instant. Every time the engine passes a milestone, the pointer on the dial moves forward to the next figure, a sound, or alarum, accompanying each successive movement.—Davy MSS., No. 11.

Not only this, but if two engines are approaching each other, by any casualty, on the same rails, then, at a distance of a mile or two, a timely notice can be given in each engine, by a sound, or alarum, from which the engineer would be apprised to slacken the speed ; or, if the engineer be asleep, or intoxicated, the same action might turn off the steam, independent of his attention, and thus prevent an accident.*

" I cannot, however, avoid looking at the system of electrical communication between distant places, in a more enlarged way, as a system which will, one of these days, become an especial element in social inter-course.   As the railways are already doing, it will tend still further to bring remote places, in effect, near together.   If the one may be said to diminish distance, the other may be said to annihilate it altogether, being instantaneous.   The finger of the London correspon-dent is on the finger key ; and, anon, in less time than he can remove it, the signal is already on the paper in Edinburgh ; and almost as fast as he can touch one key after another in succession, these signals are formed into words and intelligible sentences.   These may either have private interpretations attached to them, easily arranged between individuals, or they may be translated according to rule by a clerk of the establishment, supposing such an establishment to be instituted and thrown open to the public like the Post

---

* The most perfect block system of the present day does not do anything like this.

Office, on the principle, that any one might send a communication on paying some moderate fee, to be charged according to length. All the practical details of such an establishment are easily chalked out.

" Now, how far would there be sufficient employment, or business, to remunerate the projectors, and how far would the public at large be benefited ? Premising that it is a very shallow supposition to consider it as facilitating monopolies, inasmuch as it would be open to all, the first question is, what would be the cost, or original outlay, on a very complete system ? I believe about 100*l.* per mile. That would be 10,000*l.* from London to Birmingham, and about 10,000*l.* more, making 20,000*l.*, to bring these towns into communication with Liverpool and Manchester.

" Now, if there be 2000 miles of railway altogether open, or likely to be open ere long, then the capital requisite to carry such an enterprise generally throughout the kingdom would be 200,000*l.*, or about one-fifteenth of what has been expended on the London and Birmingham Railway alone. Let us first confine ourselves to the line of communication between the four great towns, London, Liverpool, Manchester, and Birmingham, at an outlay of about 20,000*l.* When once laid down, the repairs would be very inconsiderable, and very rare. The annual expenses, beyond the interest of the money, would be almost confined to the clerks and superintendents of the establishment, making a total, which, for argument's sake, we will call

2000*l.*, or 3000*l.* a year.   Whence will be the revenues to cover this expense, and leave a profit?

"In the first place, there is a certain amount of staple employment, which would be daily and regular. We should inevitably have to communicate the prices on exchanges, the market prices of commodities, rise and fall in stocks and shares.   There would be the earliest information of commercial stoppages, arrival of ships with cargoes, and their departures.   Then there would be Lloyd's shipping list, as a matter of course, Government despatches, and certain portions of banking correspondence and announcements.   Lastly, among the best regular customers would be the news-papers.   Public curiosity upon events of importance would ensure that the press would generally get the earliest possible information for their readers, and competition alone would oblige it.   There are certain events which would be communicated by telegraph to all the principal towns in the kingdom for publication in the newspapers, as regularly as the publishing day or hour came round.   There would be Parliamentary divisions, results of elections, public meetings, criminal news, results of trials of general interest, and the earliest foreign news of all kinds.   So much for the regular employment.

" But I conceive that the occasional employment of individuals, for private family correspondence, or for purposes of business, would make up in the aggregate even a far greater amount.   Here it is quite impossible to see how multifarious may be the occasions on which such a means of rapid communication would be of vital moment. Let any individual reflect whether in the course of his life, whether in the course of the past year, there has not been more than one occasion when he would eagerly have availed himself of it, if it had been in existence? Generally speaking, we know that the post is fast enough, and often letters are sent by private hands, when they are many days delayed, and it is of no consequence. But such occasions there are, and though, for argument's sake, I suppose them rare, yet in reality they are not so. If in the population of London, upon an average, only one private person in eight employed the telegraph only once in six months, and received an answer by the same means, at no higher charge than the present postage, say 1*s.*, we should have at once a revenue of 40,000*l.* a year, which I take to be infinitely within the mark.

"Now, what are the occasions on which private individuals would prefer the telegraph to the post? Let us say to announce a birth, or marriage, in a family connection, a death, or sudden illness. No one would be satisfied to convey intelligence of such an event to anxious relatives by any other than the most rapid communication, and if the medium was in existence people would be expected to use it. If one death in ten which take place in London were communicated by telegraph, and that to only one person at a distance, the amount of income from this single source alone would exceed 1000*l.* a year.   Announcements of dangerous illnesses, and daily communications thereon, which would often be transmitted, would considerably exceed even those of the deaths.   But this is not all; all sorts of family events, besides births, deaths, and marriages, and all business transactions, as urgent communications between commercial travellers and their principals, errors and oversights to correct before too late, &c., all these would be of no very unfrequent occurrence in every family, or business firm, and taken on the whole, among the great population of this active nation, they would supply the telegraph with as much employment as it could well get through.

" But now some one will say, supposing it all very true that these things can be done, supposing that it will pay very well to speculators, of what advantage will it be to society at large ?   Railroad travelling is quick enough in all conscience ; people used to say that stage coach travelling was quick enough ; and some years before that, they were no doubt very well satisfied with the waggons.   Now here is a means of communication compared with which the railroad travelling is as a snail's pace.   The electrical telegraph can be considered as only one means of facilitating intercourse between distant places ; and it is adapted for occasions where all other means would fail.   It will in some respects give to persons living at remote distances the same advantages as if they lived in the same street.   Should the system ever be adopted generally throughout Europe, what a vast field does it not open to us. Whatever is going on in Turkey, or in Russia, may be known in London the same hour; and, though it may seem a bold speculation, I can see no improbability that this will be realised wherever the line of country admits of it. In fact, the greater the distance the more valuable in proportion will be the information communicated.

"Goods ordered from a distant country will, of course, arrive in just half the time they otherwise would, because the outward voyage, or journey, for carrying out the order by letter is dispensed with. On general principles, whatever tends to promote intercourse between distant countries, or distant parts of the same country, will inevitably promote civilisation and increase the comforts of life.

"I must now conclude by stating that the electrical telegraph is already in progress of being established through a considerable line of this country, and there is every encouragement for supposing that it will, without delay, be brought into operation on a still more extended scale. I trust, therefore, that the company present will live long enough to see that, while we have not presumed to use the thunderbolts of Jupiter for destructive ends, we have acquired a command over the same electrical principle, for purposes infinitely more beneficial."

CHAPTER XV.

EDWARD DAVY AND THE ELECTRIC TELEGRAPH—
1836–1839 (*continued*).

HAVING now given a full and impartial account of Davy's many and wonderful discoveries in electric telegraphy, it will be interesting to follow him in the steps which he took to get his inventions adopted. For this purpose we must turn to another class of his MSS., *viz.*, his private letters to members of his family, and chiefly, to his father, Mr. Thomas Davy, surgeon, of Ottery St. Mary. In the extracts which we shall give from these the reader, who knows anything of the similar negotiations of Cooke and Wheatstone during the same period, will find some startling revelations.

At one time his inventions were on the point of being adopted by more than one English railway, and, had he stood his ground but six months longer, there can be no doubt that it would have gone hard with his rivals, Messrs. Cooke and Wheatstone. But alas! just as his labours seemed on the point of fruition, private affairs, which we can never cease to deplore, drove him from England, and, of course, left them an easy triumph. Davy sailed from the Thames for

Australia, on April 15, 1839, and, amid the new cares of a somewhat unsettled Colonial life, soon forgot all about the telegraph. Indeed, we believe that nobody will read these pages with more surprise than the old man himself who is the subject of them.

The first extracts that we shall give have reference to the Exhibition at Exeter Hall, described on pp. 374–78. In a letter to his father, dated January 23, 1838, he says :—

"I write you a few lines in haste, upon a different subject from the last. By the advice of several friends, whom I have deemed trustworthy counsellors in such matters, I have been induced to open an exhibition of my electrical telegraph, accompanied with electrical and galvanic experiments of a some-what novel nature to illustrate its principle.* You will observe that the present apparatus is, in appearance and effect, totally different from what you have seen, though founded on similar elementary principles.

"The degree of success of the last three days has been sufficient to encourage me in the correctness of what I have done. I have had Captain Beaufort from the Admiralty to look at it, as well as Mr. Jay, who is superintendent of the Government telegraphs,

* "This exhibition is accompanied with a variety of interesting experiments, the room *lighted by an enormous galvanic battery*, and, altogether, I have seldom passed an hour more amused."—Extract from letter in *Mechanics' Magazine*, for February 3, 1838, p. 296.

and who invited me to the Admiralty to-morrow, to examine the telegraphic arrangements, and furnish me with an exact estimate of the expenses of the present system for the sake of comparison. This I think a good introduction. To-day I have had eighteen persons, paying their 1*s.* each, and yesterday twelve, to see it, several expressing themselves gratified, and saying that they should bring their friends. An old gentleman came yesterday, and to-day he came again with four ladies. He says he is coming again to-morrow with some male friends and others.

"On the principle, *parvis componere magna*, I am led to presume that if the thing were generally known (instead of being merely left to the attraction of a board or two at the door) a great many persons would come to see it, paying their 1*s.* each, and that thus I might realise a considerable sum [which would be] very acceptable. To make it pretty generally known is impossible without some expense, which, at present, it is out of my power sufficiently to compass. And yet the thing appears to me so promising in success, that I would not willingly lose the chance, after having bestowed so much care, anxiety, and labour on the invention, and having, as I have now the best reason to believe, brought it to greater perfection than any other person. It is my anxious wish, now that every principal expense has already been met, immediately to advertise the exhibition, once or more, in every principal newspaper, and to take other necessary means of making it public. From present experience I believe the returns will be speedy, and in any case the prospect of indirect advantage to me is sufficient to justify so doing. If I neglect, or am unable to avail myself of, the present opportunity, there are others ready who will instantly take it up.

"Clarke, Palmer, and Cooke himself have been to see it at the private exhibition on Thursday last, and though they could not *immediately* make out the principle on which the effects were produced, yet it is all come-at-able by dint of pondering and patient experiment by such long-headed persons.

* * * * * *

"From 11 to 5, exhibition hours, I have scarcely had time to warm my fingers in the late bitter weather, from the all-sorts of questions, explanations, illustrations, demonstrations, &c., I have had to deal forth to the learned and unlearned—the former being the least troublesome."

A few days later he wrote to the same address :—
" The exhibition to-day had about the same number of visitors within two or three, which, all things considered, is pretty well, and, if continued, would set aside all apprehension of losing by it.

* * * * * *

"Among the visitors were Lord Euston and his son, who were pointed out to me by a gentleman

2 E

present. Mr. James Wheeler, my old master's brother, was there. He was at a lecture at the Royal Institution last week, when Cooke and Wheatstone's telegraph was exhibited, and said that, on comparison of action and effect, he much preferred mine. He also said that theirs would be rather advantageous to me than otherwise, as the public would soon draw the parallel.

"It is my earnest desire now to make the thing promptly known in every direction [by advertising largely].

*   *   *   *   *   *

I calculate that by the time 1000 persons have been to see the telegraph their retail conversation will be enough to dispense with other advertisements than rare and occasional ones, because, out of 1000 persons on an average computation, 100, by their gossiping propensities, will act as walking advertisements.

"I have with me a boy who is remarkably sharp and handy at repeating the experiments. * * * The little fellow appears to be able to understand anything he has once seen, and has, moreover, a very good address, asking for the One Shilling, Sir, or Madam, very genteely, &c.

"You did not expect to have a son turn showman, but I trust I am merely instrumental in promulgating a useful discovery, and that you will live to see it established, generally, throughout the country. I must *endeavour* to persuade the Admiralty to lay it down from London to Chelsea, or Putney, for experiment, this being the most foggy part of the line towards Portsmouth ; but I fear they are too stingy of the revenues of the nation. I rather expect that some enterprising individuals will take it up for public use. Time will show.

"P.S.—Receipts to-day about 25s. Among the visitors was pointed out, after he had left the room, Earl Grosvenor."

Towards the end of February, 1838, he wrote:—

"My dear Father,—My business with Mr. Welch is concluded—my lease cancelled—and I am no longer the occupier of the house 390, Strand. Please, therefore, address to me at Mr. Smith's, 199, Fleet Street.

*　　*　　*　　*　　*　　*

"As we some months ago prognosticated, the telegraph, being once promulgated, has interested the public, and is in a fair way to be generally adopted. The Great Western Railway have decided upon laying it down upon their line, and the only question, both in this, and in all subsequent cases, will be, whether my plan, or that of Cooke and Wheatstone, be preferred.

"Mr. Brunel, Junior, Engineer to the Great Western Railway, with Mr. Tite, and other Directors of the company, came to see my apparatus, and wished

2 E 2

me distinctly to point out the advantages which it possessed over the rival scheme. Mr. Brunel, being on intimate terms with Mr. Cooke, was somewhat inclined to lean the other way, but the principal difficulty under which I laboured was the impossibility of rendering manifest all the advantages of my mechanism, without entering, more or less, into such explanations as would, more or less, betray my secret —as yet unpatented. When, therefore, I stated that I could effect such and such objects he could not see how it was possible—thought the attempt would be dangerous, or precarious. Seeing also that I employed six wires, he could not conceive but that my plan must be an infringement upon the patent of Cooke and Wheatstone, and that the company could not safely carry it into execution without risk of action for damages, &c.

"Moreover, that, as I was not prepared fully to develop my plans, I could not be considered in a condition to treat with them, for they would have to buy of me what he designated 'a pig in a poke,' which, though it might produce very pretty effects, yet, as the *rationale* was not open for canvass, its practicability could not fairly be judged of, nor could he confidently assure the company but that it might prove to be an infringement on the others' patent. Mr. Brunel is a particularly sharp, intelligent man, capable of comprehending anything in all its bearings, and of improving the barest hint. I had, of course, to be on the alert to divulge nothing that would impair the security of a future patent-right. I could not fail to learn something from him, and the result of this interview has been to prove to me the necessity of ascertaining, with the greatest care, the precise footing upon which I stand, before taking any further steps. I have endeavoured to persuade the company to delay a week, or two, before they ultimately decide on adopting any plan.

" In the meantime, my first object will be to obtain the opinion of the most eminent lawyer in patent affairs, and I have been nearly all this day engaged in conference with Mr. Carpmael upon the subject. This may cost me two or three guineas, but will be infinitely cheaper than a blindfold course of proceeding. To-morrow I shall get his opinion. Should this be favourable to my views, I shall almost think it right to obtain a second opinion of some eminent barrister, or of the Attorney, or Solicitor-General, before venturing to act upon it. But if fully confirmed as to my right to secure as exclusive, and to act upon, or license others to act upon, my own invention, there can be little question as to the peremptory necessity for immediately raising funds to take out a patent, which will place me on a par, or more than a par, with Cooke and Wheatstone. The time has now arrived when the thing is on the point of being acted upon throughout Europe.

" As to the particulars of my mechanism, there are

guesses enough at it, but, though it is simple as can be, the guesses are as far wide of the actual truth as need be. Mr. Cooke himself is in perfect ignorance of it.*

"I hope, in a postscript, to subjoin Mr. Carpmael's opinion. He told me this evening that, though he would not record it on paper until he had investigated the matter fully, yet his present impression was that the two inventions [*i. e.*, Cooke and Wheatstone's and his own] differed most essentially in all main points, and that a separate patent might be obtained and maintained without hazard of litigation. He has appointed to-morrow morning to inspect my mechanism (of which as yet he has seen the description only) at 10 o'clock, at Exeter Hall.

"28 February, 1838: I enclose a copy of Mr. Carpmael's opinion.† I am now passing the patent through the first stage, which will cost about 12*l.*, but beyond this, unassisted, I shall not be able to go. Mr. Carpmael thinks it may not be difficult to get some one to advance money for future patents, if I can only place myself in a condition to explain, by

---

* "I have sufficient reason to know that the true principle of [my apparatus] has not been discovered by any one, not even by Mr. Wheatstone. I have purposely, and for a veil, allowed it to be supposed that the principle is the same as that in Mr. Cooke's invention, which, as I designed, is taken for granted."—Davy MSS., No. 10.

† This document, copied by Davy himself, is preserved amongst his MSS., No. 11. It bears the date February 24, 1838.

securing the English patent first, after which it will be just as desirable to do the same thing in Belgium, America, and other places.

* * * * * *

" Your ever affectionate Son,

"E. DAVY."

" May 30, 1838.

" My dear Father,—This long-pending decision upon my application for a patent has at length been given. I believe I told you that, owing to the Solicitor-General not being able fully to comprehend some points, it had been agreed to call in the assistance of some eminent scientific man, and, accordingly, Mr. Faraday was referred to as being the highest electrical authority in the kingdom, and he was kind enough to undertake [the examination of the points in question]. The result has been in my favour, *i. e.*, I am entitled to the patent I am applying for with the retention of every point of the least value. The Solicitor-General's report will be ready for delivery to-morrow, Thursday, and then all that will be wanted to proceed with the patent will be the money. It will then take about ten days to pass the Great Seal, and until that there is no security for it, and I will still labour under the difficulty of not being able to explain its nature, or advantages, to any one so as to get it taken up. Besides, there is every day the risk of persons finding out the particulars for themselves.

"Once the patent secured, I think it not improbable that it may end in a compromise with Cooke and Co., for when I have the patent I must get connected with some one possessed of capital. They have, I understand, already laid out 2000*l.* upon their telegraph, and are very anxious at present, as Mr. Wheatstone told me they were in treaty with some of the great railway companies, but that the latter delayed coming to a decision, understanding that there might probably be another patent in the market. So, if I pass my patent, they will either have to wait six months to see the specification, or else offer me terms at once.

"Whether the Great Western is the company alluded to I know not, but I had previously been given to understand by Mr. Gibbs that they had already contracted with them, and were going on with the preparations (as I was told by a different party) of coating an immense quantity of copper wire with india-rubber. It may, therefore, or may not, be some other great company.

"I am happy in being able to communicate the intelligence contained in this note, for, from the long and vexatious delay, I have been not without apprehension that the decision would be against me. The circumstance of Mr. Faraday having been called in will also render the patent safer, as his opinions on such matters would naturally be looked upon by the public with some confidence.

*   *   *   *   *   *

"With kindest loves, believe me,

"My dear Father,

"Your ever affectionate Son,

"E. DAVY.

"P.S.—I enclose a copy of the claims upon which Mr. Faraday advised that my patent might be granted. You will perceive that it contains the most important points." *

"June 16, 1838.

"My dear Father,—I have only time to say that I received from Messrs. Gibbs 130*l.*, and Mr. Carpmael informs me that the patent will be sealed early in next week. I must write you again to explain what I purpose doing as soon as that is accomplished, *viz.*, to send circulars immediately to all the Boards of Directors of Railway Companies, and to give one, or more public lectures on the subject, inviting as many influential people as possible to attend. It must now

---

* This document is preserved amongst the Davy MSS., No. 11.

be pushed forward with all our might and main, and I hope it will not be long before it does some good.

"You will soon hear from me again, and believe me,

"My dear Father,

"Your ever affectionate Son,

"E. DAVY."

"June 23, 1838.

"My dear Father,—I think that I ought to give you notice from time to time of my moves with the telegraph, in order that, in case of any sudden accident to me, and the concern being in a promising state, my successors might know better where to take it up, and what I had been doing.

"The patent has not yet passed the Seal. I expect that it will about Wednesday, or Thursday next.

"I have been endeavouring to make connections with some business men, to assist me in making negotiations with the railway companies, or in getting up a general telegraph company.* The principle on which I endeavour to engage their services is that of percentage on whatever money I may obtain for licenses under my patent, through *their particular influence, or interference.* The amount I have fixed upon is 10 per cent., which will, perhaps, be liable to deviations in some cases. The present difficulty is in getting the thing *started.* When known practically

---

* A few letters to and from business men and Railway Boards on this subject are preserved amongst the Davy MSS., No. 11.

and appreciated, it may be that the companies will come to me, instead of my having to seek after them.

" The best business man I have at present retained is Mr. P—— * * * I requested him to apply first to the Birmingham Railway Company, and the subject has been brought before the directors. The only answer obtained is that, if ever the directors should deem it necessary to adopt any electrical telegraph, they will make the most minute and careful examination into the comparative merits and advantages of each plan before deciding on either. I saw Mr. Creed, secretary to, and original getter-up of, the Birmingham Railway Company, who told me only that he would be happy to receive any memorial from me on the subject of my invention in order to lay it before the directors. Mr. P—— is to introduce me to their domestic engineer in about a week.

* * * * * *

" Mr. P—— is next about to apply to the Southampton Railway, and I am now preparing letters* for him to make use of, setting forth that, when once laid down, the Admiralty will, no doubt, be glad to make advantageous contracts with them for the use of it for Portsmouth, which is at no great distance.

" We must, of course, rake our brains to find out all the inducements we can to tempt people to these speculations.

* Original drafts of these preserved.—MSS., No. 11.

" That will be the next move. Then there will be the grand junction from Birmingham to Liverpool and Manchester.

" The next business man I hope to retain, and have partly, is Captain B——. He is intimate with the engineer of the Birmingham and Gloucester Railway, and has influence with the Midland Counties Railway, either of which would be a good step.

" Another is Mr. B——, of whom you have heard before.

  *   *   *   *   *   *

" I have an appointment to meet a capitalist, name as yet unknown, at three o'clock on Monday about money for taking out the foreign patents, all which may, or may not, come to nothing. Another appointment with a broker, named L——, to aid in getting up a company, at four o'clock the same day. There are many of these appointments for the one that leads to any result. Therefore, do not be on the look-out for such results, I will be sure to tell you if anything good comes. It is no use to be either sanguine, or easily put out of one's opinions.

" Believe me, my dear Father,

" Your ever affectionate Son,

" E. DAVY.

" P.S.—My impression at this moment is that it will be better, if possible, to get up a general company,

and sell the patent out and out, particularly as the
Birmingham directors scarcely appear to comprehend
the advantages of the system further than for mere
railway uses. It will, I know, be a very difficult
matter to get the proper people in the mind for
entering into such a scheme. Mr. Hesseldine appears
to listen to the proposition, but has some objections
of which I cannot clearly see the drift, unless it be
this—that the Government could scarcely allow such
a powerful instrument to be in the hands of individuals,
or a private company, and would either prohibit it, or
else take it under their own management ; and, therefore, that the best possible parliamentary, or government, influence ought to be made in order to secure
the probability that such future arrangements with
the Government may be advantageous to us.—I know
very well that the French Government would not
permit it except in their own hands ; but though I
think our Government ought, *and, perhaps, will eventually take it upon themselves as a branch of the Post
Office system*, yet I can scarcely imagine that there
would be such absurd illiberality as to prohibit, or
appropriate it, without compensation.

"There is, however, prudence in what he suggests
as to making friends in high places, if it can only be
done.

"Are there any of the directors of the Bristol and
Exeter Railway with whom interest could be made?
They are, I believe, in great part Exeter people."

"July 4, 1838.

"My dear Father,—It was not until this morning that my patent actually passed the Great Seal. It is now secure for England and Wales, and you will see it in the list in the next *Gazette*.

"The enclosed was written some time back. It may be well to preserve whatever details I send you with regard to the telegraph.

"My object now is to get a company formed to take my patent off my hands, and, either pay me a large sum down for it out and out, or else a smaller sum down, and an agreement for a further remuneration hereafter, and proportioned to the success of the scheme, such as a percentage on dividends, &c.

* * * * * *

"There is plenty of money in the market, and plenty of people ready to vest it in such schemes, if they can only be satisfied that they will pay more than 5 per cent. interest. All I have to do is to make people believe this, and the money will come without any pressing on my part. But this *is* the difficulty, and one which I have now to make every possible exertion to overcome. The practicability of the plans will, I believe, not be much longer doubted. I have several persons at work to get some influential names sufficient to head a prospectus* as Directors, &c., and find

* A draft of such a prospectus, headed "Voltaic Telegraph Company," is preserved in the Davy MSS, No 9. It is a powerfully-written and exhaustive document, and will well repay perusal. *À propos*

current expenses of printing, advertising, journeys, models, &c.

*   *   *   *   *   *

" Mr. P.—— says that ' before forming a company, we must first secure the consent of some railway company to the laying down of the wires upon their line on certain terms. If one railroad will do this, we may afterwards reason that others will agree to the same ; otherwise people will say, ' How are you going to en- force permission from the railways, or turnpike trusts, without an Act of Parliament'?' Now, I don't see that the want of previous agreement with a railway should at all deter us from endeavouring to form a company, but it is clear enough that such an agree- ment, previously obtained, would be a step gained, and an argument in our favour. With this view an appointment is now pending with the domestic, or resident, engineer of the Birmingham Railway.

*   *   *   *   *   *

"I have had notice of another application for a patent by a person named *Morse*. Messrs. Cooke and Wheatstone have entered an opposition to this ap- plication, and I shall have to do the same, so that one, or other, of us may be able to stop it. We are now

---

of the name, the following memorandum may be quoted :—" A satis- factory name is not yet decided on. It might be called the ' Oerstedian,' after Oersted, the Danish philosopher, who first discovered the magnetic powers of electricity ; or the ' Instanterian ' ; but a better name may turn up."—MSS., No. 11. In another place Davy speaks of a system of *Electroloquism !* (MSS., No. 7).

both equally interested in keeping a third rival out of
the field, and it may save much after trouble and
competition. * * *

"Your ever affectionate Son,

"E. DAVY."

In a letter to the same address, dated July 21, 1838,
the following passage occurs :—" I find the people who
undertake to make appointments about the telegraph
very dilatory in so doing, which prevents my making
progress as fast as I could wish.   I trust the prospectus
will be a help.

" It is not every one who is willing to be a director
that will suit, as I wish to confine it to the highest
respectability, and avoid all poison.   Mr. E—— has
evidently his enemies, but if I can find that he has also
his friends, he will be a valuable acquisition, having
much money and connections, and I believe he would
liberally support me, or lend his voice to pay me a
large sum."

Two days later he wrote :—" I had a further conversation with Mr. P—— on Saturday.   He expects
an appointment with the engineer in question
[? Mr. Fox] on Tuesday, and entertains a hope that
we may also secure Mr. ——, the chairman of the
Southampton Railway Company, than whom, for one,
we need not have a better.   Mr. P—— having considered the prospectus, and suggested some slight
alterations, said that, now the thing was distinctly laid

out, his views had quite altered, and he should have a difficulty in seeing how the thing should do otherwise than 'pay' to the shareholders. He would be the managing director, or whipper-in, as he is in the A——— Mining Company. As I have but slender acquaintance among the great commercial people, I am obliged to apply to, and make use of, persons of this kind, and I believe he is well known and knows many, and that his persuasion may have some effect, where I should not be listened to. The difficulty is to get him to stir himself sufficiently. I do not anticipate much good from either Captain B———, or Mr. B———, but perhaps they may be of some aid."

"July 30, 1838.

"My dear Father,— * * * I had an interview on Wednesday last with Mr. Fox, resident engineer of the Birmingham Railway, the whole particulars of which I can scarcely enter upon at this moment, except that it was quite satisfactory and friendly, as far as it went. The main purport of it was that if we had a company who would go to the expense of laying down the wires, &c., the railway directors would willingly grant the use of their line and afford every facility and protection, on condition only of a license under the patent as far as relating to railway purposes only.

"This has been one of the problems, 'How is the Telegraph Company, without an Act of Parliament, to lay down the wires?' He says there is no doubt

2 F

that few of the railway companies will object to these terms. Mr. P—— has promised to-morrow to see the Southampton Railway people, and I shall have another interview with Mr. Fox for further explanations, so that I trust we shall soon be enabled to come before the public ; but it is a tedious business.

"31st July.—Since writing the above I have received M. A.'s letter and enclosure, to which I shall give the earliest attention. You may presume, if you do not hear from me for some little time together, that there is nothing particular going forward. I understand the directors of the Great Western Railway are under discussion to ascertain the nature of my patent, but I shall take no notice at present.—With kindest love, believe me,

"My dear Father,

"Your ever affectionate Son,

"E. DAVY."

In a postscript to a letter dated Aug. 1, 1838, he writes :—

"I spent yesterday evening with Mr. Fox, explaining my inventions to him. He expressed the most favourable opinions of them, and, if he does not alter his mind, we may consider the Birmingham line as secured, or nearly so. He is the sharpest and quickest man I have yet had to talk to, comprehending everything before it was half explained, and suggesting improvements, or remedies for difficulties, &c. We are in other quarters making slow but, I trust,

effective progress, and the chances against eventual
success appear daily diminishing."

" August 17, 1838.

"My dear Father,—Mr. P—— has had interviews
with Mr. Easthope and his son. Mr. E—— seemed
to take very enlarged views of the applications of the
telegraph, and the revenues to be derived from it,
and promised his strenuous influence for its immediate adoption on the Southampton and Portsmouth
line, of which he is chairman of the directors. The
conversation, as repeated to me, coincides with my
own (perhaps sanguine) idea to an extent which I
have never had the satisfaction of meeting with before.
Mr. P—— came home quite red-hot upon the matter,
sees it in a new light, and has this morning agreed to
take out the principal foreign patents (to find the
money and have half), on condition that I would also
give him a further interest in the English patent
already obtained. I have, consequently, agreed that
he shall pay me 150*l.*, and have one-fourth of the
patent right. This I have done, not but that the
patent may be worth far more than four times 150*l.*,
but with ulterior objects, to secure his interest and
exertion to help it on, for I could do little by myself.
And it appears to me the remaining three-quarters
may be thereby increased in value more than they
will have lost. I do not suppose that Mr. P—— will
go from this arrangement, which yet remains to be
executed. He says, that, if sufficiently interested, he
will devote nearly all his time to it.

2 F 2

"I spent yesterday evening and took tea with Mr. Fox, resident engineer of the Birmingham Railway. I have his decided approbation of my plans, in preference to those of the other party, and, therefore, a powerful voice is secured on that line. We had much conversation on various details of the subject, but it takes time to work people into an acting humour. Until now Mr. P—— has been lukewarm, or, at least, tardy, in his movements. The encouragement, which Mr. Easthope has given, has put a new life into the thing.

"I believe I told you that Mr. Easthope is a Member of Parliament, proprietor of the *Morning Chronicle*, and a large shareholder in the Southampton, and in the Havre and Paris Railways. He is, perhaps, as good a patron as could be obtained for one. He said that the subject was not new to him, and that it had been frequently under discussion in society where he had been.

"You will perceive that if I have been in error as to the prospects of this invention, I have now some people of high standing to keep me in countenance in the 'moonshine.' Mr. Easthope speaks of its opening communications between London, or Liverpool, and the Mediterranean! With kindest love to all the circle, believe me, my dear Father,

"Your affectionate Son,

"E. DAVY."

About the end of August, or beginning of September, 1838, he wrote :—

"My dear Brother,—I have just received a packet enclosing, among others, a letter from you. You will perceive by my communications to Father, &c., that I am trying hard to dispose of my telegraph. I wish to get it clean off my hands, and, if possible, an employment in laying it down at an annual salary. I believe I may now almost calculate on the Birmingham Railway, the best line in the kingdom; there is little now to fear from the rivals, Cooke and Wheatstone, and there are no others. My object is to form a company of affluent people, who will purchase my patent right, and, if this succeeds, it will produce a large sum of money, as 10*l.* or 20,000*l.*, just as easily as so many hundreds. The value of the invention has very greatly increased since what it was six months ago, and I would not now sell it to Cooke and Co. for any sum they would be likely to offer, and which I would gladly have accepted once.* I have every assurance that I shall get together a set of wealthy directors, and that the shares will be taken up. We have, as you will perceive, some first-rate people already engaged, and much interested in it. Mr. Wright could, if he chose, advance 200,000*l.* We

* From a letter of Wheatstone to Cooke, in Mr. Latimer Clark's possession, dated July 18, 1839, it seems that they then contemplated buying up Davy's patent. Ultimately it was bought for 600*l.* by the old Electric Telegraph Company, and—smothered, like a good many others. See letter of May 12, 1847, amongst the Davy MSS., No. 12.

are promised the Marquis of Douro and Lord Sandon for trustees, through the interest of Messrs. Mac-Dougall, the solicitors in Parliament Street. There is no present reason to apprehend but that I shall get my price for it by persevering and securing influence step by step. But for all this my presence here is indispensable. It would come to nothing if I left London at this juncture which would be madness. There is no one able, or willing, to push it forward for me, and, if allowed to sleep, the patent would not be worth a rush. I am now anxious to connect with some sharp, wary solicitor, not too young, whom I can engage to protect my own personal interests in driving the bargain with the directors, which will be very essential—one is so apt to be talked over by these keen monied men.

*　　*　　*　　*　　*　　*

" The enclosed piece of paper contains a statement of the progress made in organising our company. Only the names with asterisks are fully secured, but the others we have not much reason to doubt of.

" When a meeting of the directors can be called together, I shall propose that as soon as the deposits are paid up they give me 10,000*l.* in money and one or two thousand shares ; in fact, the best bargain I can make. Something will come of it.

" Your ever affectionate Brother,

" E. DAVY.

"P.S. Exeter Hall cannot be said to pay at present. It is kept open rather to answer a purpose in getting up the company."

The above letter, as appears from another to his father of September 9, 1838, was written to his brother, Henry Davy, about the end of the previous month.

The piece of paper referred to contains the following :—

TRUSTEES.

Marquis of Douro, and Lord Sandon.

DIRECTORS.

*Sir F. Knowles, Bart., F.R.S.	Mr. Bagge (of Norfolk), MP.
*John Wright, Esq., Banker.	Mr. Harrison (Chairman of the
Em Tennant, Esq., M P.	Southampton Railway).

ENGINEER.

Mr. Fox.

SUPERINTENDENT OF MACHINERY.

E[dward] D[avy].

SOLICITORS.

*Messrs. M'Dougall and Co.

Capital £500,000 in 10,000 shares, £50 each, deposit £5.

In a letter to his sister, dated September 22, 1838, he says :—

"I am obliged to remain in, or near, London, on account of the telegraph, as there is a probability that the arrangements with the Southampton Railway will soon be completed. Nothing is certain as yet, but the directors appear decided upon having one on their line. The Birmingham line also is in prospect.

" I have sent a circular to all the principal railway companies. The Grand Junction say they have no intention of adopting any telegraph at present, and the Birmingham and Derby seem to imply, in their answer, that they have it in contemplation, and will take it into consideration as soon as they are prepared to do so, &c. All that could be expected from the circular was to prevent them from engaging with the other party before they knew anything about me.

" The idea of forming a company is suspended for the present, on account of the opinion of the Southampton Railway directors, that all the railway companies would eventually take it upon themselves, and find the capital. If so, it is all that is wanted.

" I presume that if terms are made with the Southampton, which is the first company that is likely (being on the line of the Government telegraph to Portsmouth, and having a probability of a contract with the Admiralty), they will require me to superintend the laying it down at a salary, independent of the remuneration for license under the patent."

On October 8, 1838, he writes to his father :—" I have done all I can to bring my patent before the Southampton Railway Company, and have received every assurance as to their intention of adopting it. I must now wait their final decision if they do so as to the precise terms, and also as to the time when they will be ready to commence operations. What I have proposed to them is, that they shall be at all the expenses, pay me one-third of the net profits, and employ me, at a reasonable salary, to lay down the telegraph and keep it in repair.

"I will shortly make an attempt to urge forward the Birmingham Company, where, I believe, I have sufficiently secured the preference. As soon as one of these companies brings the telegraph into operation along the entire line, and it is found to answer, the others will quickly follow.

"The Polytechnic Institution, after giving much trouble, declined purchasing the Exeter Hall model, and I am now in treaty for it with a Mr. Coombes, an American, who proposes to take it home with him, and open an exhibition, with some other models, in his own country. I have closed the Exeter Hall room, and paid off Mr. Spicer, my assistant, and also Downy, the other man who used to attend there."

Again on November 17, 1838, he says:—"I have made some arrangements with Mr. Watson, rather on the principle of co-operation than of actual partnership, and have pretty well explained to him how I am situated. * * *

"You will perceive I am anxious to be doing something, independent of my expectations from the telegraph. I have put it into as good a position as possible, and have no doubt of its final success; and must now wait the answer from the Southampton directors, who have verbally promised its adoption, as well as from other quarters. I shall not let the matter go to sleep, but as it is in a channel, and as my interference will not hasten it, there is little to be gained by thinking of it ; so I must be doing other things.

"There is no relying on Mr. P——. He agreed to purchase a fourth of the patent for 150*l.*, and was redhot to conclude the bargain, but after a few days he told me that he could not at present provide the money. I cannot help these disappointments.

"Mr. Fox is steadily friendly to me as yet."

"November 29, 1838.

"My dear Father,— * * * I have got rid of my room at Exeter Hall, which, now it is no longer required, is a saving of 14*s.* a week. Altogether it has somewhat more than paid its expenses, or thereabouts. It would doubtless have done better, but I was driven from personally attending to it by incessant annoyances. It has, however, answered more effectually by the notoriety which it has given to the telegraph. You will perceive by a number of the *Railway Times* which I shall enclose, a reason for the delay in the decisions of the Southampton Company, *viz.*, the question whether they would obtain the branch line to Portsmouth. Mr. Easthope is a spirited man, by whom many other monied persons are guided, and he has influence with the present Government. I have as yet no reason to doubt that he will keep his word with me, that my telegraph shall be adopted immediately they are prepared to commence operations with it.* The bringing this to bear may be a work of some little time, as such things usually are, but I am sure you will not regret the attention I have paid to it, nor even the manner in which it has diverted me from my other business; nor do I think you need feel doubtful as to its eventual success. The next month will be occupied with completing the specification (due Jan. 4), much of which will have to be remodelled by Mr. Carpmael's direction. Everything depends upon this being as perfect as it can be, and I wish I were more fit for the task, by having a mind more at ease than is at present possible.

*　　*　　*　　*　　*　　*

* The following extract is *à propos*. It is from a letter of Wheatstone to Cooke, now in Mr. Latimer Clark's possession, dated July 18, 1839:—
"Now on another subject, Mr. Easthope, the chairman of the board of directors of the Southampton Railway, wishes to see the telegraph at work. Will Saturday next be a convenient day for the purpose? If so, I will bring him with Mr. Irving and Mr. Wright, the banker. This visit will be an important one. He is fully impressed with the advantages which may result from the invention, and I think would not be disinclined to encourage it. I need not say that he is a person of influence and wealth.

"One great difficulty with respect to him and the railway with which he is connected is now obviated, for I understand he gave considerable encouragement to Davy so long as he thought his plans likely to succeed."

It will be remembered that at this date Davy was in Australia, and "his plans" in the hands of people who did not understand them. Naturally, then, Mr. Easthope turned to Cooke and Wheatstone. It is also interesting to note that Mr. Wright was one of the fully secured directors of Davy's proposed Telegraph Company. See p. 439.

" I have to-day been informed that the Brighton Railway Company are about to adopt an electrical telegraph, which is a quarter in which I scarcely expected it. I must look after them to ascertain if it is correct, for Mr. Cooke is making all his interest. I think the London and Dover will be a better line, but will not be complete for a long while. I am obliged to employ a good deal of Nickols' time on the working model, which will be the principal thing in the specification, in order that it may be ready to show at work. The month after the specification is enrolled it will appear, at length, in the *Repertory of Arts,* a number of which I shall purchase and forward to you. With kindest loves to mother, brothers, and sisters, believe me, your ever affectionate Son,

" E. DAVY."

" December 12, 1838.

" My dear Mother,— * * * My specification must be enrolled by the 4th January, and afterwards will be published, gratuitously, by the patent agents, in the *Repertory of Inventions.* This also I shall have to look to, procure a few copies, and send them to the parties most likely to serve us. After this, for aught I know at present, there is little more I can do to forward the matter, and it must wait the good time of the railway directors to take it up, which there is no reason to doubt nearly all of them will do eventually.

" I believe I have effectually barred any hasty

adoption of Cooke and Wheatstone's telegraph, which has made no further progress.

"You must be aware that, although it may come into operation almost immediately, yet it may possibly not until some little time hence ; but this is a question which now rests with others, and not so much with me, and we must not, therefore, be disappointed at some delay.

*   *   *   *   *   *

"Your ever affectionate Son,

"E. Davy."

We have arrived at the end of our MSS., and, consequently, at the end of our task, which, we need hardly say, has been to us a labour of love ; but before dismissing the subject we cannot resist the pleasure of quoting two short passages,* which will serve to show in what estimation Mr. Davy was held by those most capable of comprehending his character. The first is from Mr. Thomas Watson, Dentist, London, a gentleman of great scientific attainments, to Davy's father :—

"May 20, 1839.

"My dear Sir,—I have to apologise for not acknowledging the receipt of your kind present ere this, an exceeding pressure of business must plead my excuse.

"Permit me to say that any service I may have

---

* Davy MSS., No. 12.

rendered your son has been to me a source of much gratification. I much regret, upon private grounds, that by his absence I lose an acquaintance which I highly prized, while, upon public grounds, science has lost an adjunct as talented, as zealous, and, without flattery, I must add *his pursuits would have so enlightened and benefited his countrymen that his secession to the primitive shores of South Australia must be deplored as a national calamity.*

*   *   *   *   *   *

"Believe me, yours sincerely,

"THOS. WATSON."

The next extract is from a letter (October 21, 1839), of Charles Pain, the family solicitor, of Surrey Street, Strand, to the same address :—

"Mr. Carpmael passed some high encomiums on your son's talents in matters of science, and said he considered *his leaving England a great loss to the country*, and he particularly regrets his absence on account of the telegraph, which, had he been present, he would have had no difficulty in disposing of to the Great Western Railway Company, who are now adopting that of Messrs. Cooke and Wheatstone, and to whose, he says, your son's is very superior."

The rest can be told in a few words. For a year or two after Davy's departure for Australia, his father and one or two friends tried, but in a half-hearted way, to carry on the negotiations from the point where he himself had left them. Another exhibition of both the screen and recording telegraphs was opened in Exeter Hall for a few months in 1839-40, but, as those in charge of the instruments did not thoroughly understand them, and could not always get them to work satisfactorily, no good came of it.

The machines were sent down to Ottery St. Mary at the end of 1840, and were stowed away in an out-house as so much rubbish. In the hope of rescuing them we lately paid a visit to Davy's native place, but found, to our grief, that only three years before, on a change of residence, they were broken up and sold as old metal! Our informant, the family gardener, added " 'twas such a pity, as there was as much mechanism about them as would fit up a hundred clocks!"

In a field we found some pieces of cotton-covered iron and copper wire, and six of the Daniell cells—huge things of three or four gallon-capacity.* The outer jars are of glazed earthenware about eighteen inches high, and the porous pots are more than half an inch thick !   These relics will now be carefully preserved.

And so ends the story of a *magnificent failure.*†

* Two of these are now in the Library of the Society of Telegraph-Engineers and Electricians. Nov. 15, 1883.

† In the belief that our readers will now be interested in everything relating to Mr. Davy, we have collected a few biographical notes of the venerable pioneer, which will be found in the Appendix B, to this volume.

CHAPTER XVI.

TELEGRAPHS BASED ON ELECTRO-MAGNETISM AND
MAGNETO-ELECTRICITY (*continued*).

1837.—Alexander's Telegraph.

IN May 1837, William Alexander, of Edinburgh, published a scheme for telegraphic communication, which was the realisation of Ampère's and Ritchie's ideas. It was widely noticed at the time, having appeared in the following amongst other journals :— Edinburgh *Scotsman*, July 1, and November 18; Edinburgh *Evening Courant*, July 3; London *Times*, July 8; and London *Mechanics' Magazine*, August 12, and November 25, 1837. In this paper he showed the practicability of his project; estimated its cost; and pointed out its utility as well to the public as to the state.

After a brief reference to the then existing system of semaphoric signalling, he says :*—"The plan of a telegraph underground, by means of electric or voltaic currents, transmitted by metallic conductors, was some time ago devised, and its practicability supported by electricians of eminence; but their ideas on the subject have not hitherto been matured,

---

* We quote from his *Plan and Description of the Original Electro-Magnetic Telegraph*, &c., 8vo., 30 pp., London, 1851.

or carried into actual practice upon the scale which is now contemplated.

"It has been found by experiments made with a view to ascertaining the velocity of electricity, that it is transmitted instantaneously, by means of a common iron wire, a distance of eight miles; and electricians of the first eminence have declared their opinion that, judging from all scientific experience, the electric or galvanic influence would be almost instantaneously transmitted from one end to the other of a metallic conductor, such as ordinary copper wire of moderate thickness, of some hundred miles in length.

"If this scientific theory is correct, it follows that a wire, secured by a coating of non-conductors, and protected from external influence or injury, and laid under the turnpike road between Edinburgh and London, could be the means of distinctly indicating to a person stationed in London that such wire had been electrified or galvanised in Edinburgh — the transmission of the electric or galvanic influence being clearly discernible by various well-known means.

"How, then, is this scientific fact to be applied to purposes of practical and general utility? Simply by laying as many wires, separated from each other, as will correspond to the letters of the alphabet, and preconcerting between the persons stationed at two extremities of the line of communication that each individual wire is to represent a particular letter;

2 G

because, if the person stationed in Edinburgh can, by applying the electric influence to any one wire, instantaneously apprise another person stationed in London that a particular letter of the alphabet is thereby indicated, words and sentences *ad infinitum* may be communicated, and the idea of a perfect telegraph would be realised.

" Without experience it is impossible to say with what rapidity this electro-magnetic telegraph could be worked, but in all probability intelligence could be conveyed by such a medium as quickly as it is possible to write, or at least to print ; and apparatus could be constructed somewhat resembling the keys of an organ, by which the letters of the telegraph could be touched with the most perfect ease and regularity.

" It has been mentioned that the transmission of the electricity or galvanism could be discernible by various means well known. If any indication, however slight, is made, that is enough—all that is wanted being that it should be perceivable by the person placed to watch the telegraph.

" It has been assumed that the electric current is capable of transmission by means of a single impulse from Edinburgh to London. But it is not indispensable that so great a distance should be accomplished at once. Intermediate stations for supplying the telegraph with new galvanic influence could be resorted to, and its perfect efficiency still be preserved.*

---

* Manual retransmission, not automatic translation, is here meant.

"The best mode of troughing or protecting the metallic conductors, and separating them both from each other, and from the surrounding substances by which the electric or galvanic influence might be diverted, would of course require considerable scientific and mechanical skill ; but the object appears perfectly attainable. Insulating or non-conducting substances, as gumlac, sulphur, resin, baked wood, &c., are cheap, and the insulation might be accomplished in many ways. For example, by laying the wires, after coating them with some non-conducting substances, in layers betwixt thin slips of baked wood, similarly coated, the whole properly fastened together and coated externally. These slips might be perhaps ten yards long, and at the joinings precautions for the expansion and contraction of the wire, by the change of temperature, might be adopted. The whole might be enclosed in a strong oblong trough of wood, coated within and pitched without, and buried two or three feet under the turnpike road.

"The expense of making the telegraph proposed, is of course an important element in the consideration of its practicability and utility.

"The chief material necessary, *viz.*, copper wire, is by no means expensive. It is sold at 1*s.* 6*d.* per pound, of sixty yards in length. The cost of a wire from Edinburgh to London, say 400 miles, would thus be about 900*l.*—but say for solderings, &c.,

100*l*. additional ; or that each copper wire, laid from Edinburgh to London, would cost 1000*l*. sterling, and that the total expense for the wires necessary to indicate separately each letter of the alphabet, would be 25,000*l*. The purchase of so large a quantity would of course be made at a considerably less price ; but probably one or two additional wires might be needed, and the circuit of the electrical influence must be provided for by one or more return wires.

" The coating, separating, and troughing of the wires can be accomplished by low-priced materials, and the total expense of the whole work (except the price of the wires), allowing a large sum for incidental expenditure, has been roughly estimated at 75,000*l*. ; making a maximum expenditure of, say 100,000*l*., for the completion of the telegraph.   For a proportionately additional sum it might be extended to Glasgow.

" The average of the parliamentary estimates for railways is about 15,700*l*. per mile, so that the whole cost of the electro-magnetic telegraph proposed would only amount to as much as the construction of a railway of between six and seven miles in length. Were the details of this plan decided on by competent scientific and practical persons, the cost would be accurately estimated with unusually few sources of error.   Here are no levels to adjust—no viaducts to erect—no morasses to cross—no property to purchase. Buried under the public road to the depth of two or three feet, the machine would be amply protected against injury, as well as from any atmospheric influence. For change, or damage occasioned by changes in the road, it would be easy to provide. Damage by mischievous persons is quite unlikely, as is shown by the safety of water-pipes, gas-pipes, and railroads. But it would be quite easy to arrange a system for immediately detecting the seat of any damage, and its repair would be perfectly easy.

" As to the working of the telegraph, it is apprehendéd that even if the speed of writing were not attained, there could at least be no difficulty in indicating one letter per second. At this rate, a communication which would contain sixty-five words would occupy about five minutes. This is supposing the vowels to be all indicated. But abbreviations in this, and many other respects, would no doubt be contrived ; and the number of words in the communication supposed, are greater than necessary for an ordinary banking or commercial letter, or for friendly inquiries and responses. Supposing, however, that each communication was to occupy five minutes, and to be charged five shillings—if the telegraph was worked twelve hours a day (that is, six hours from each end), it would produce a revenue of 36*l.* daily, or 10,800*l.* per annum, supposing there were to be 300 working days in the year. If, however, the plan is practicable, the public intelligence that would no doubt be transmitted by the telegraph, would be sufficient to keep it in operation night and day.

"No one can doubt that there would be a very great demand for the services of such a perfect telegraph as is here supposed capable of being constructed. In every department of commerce, in shipping, in banking, and all money transactions, in the communication of public and political intelligence, in the law, and in family and friendly intercourse, the utility of the telegraph would be immense. By coming at the same time to the two ends of the telegraph, parties might almost enjoy all the advantages of a personal interview, at a trifling expense. The consequence of such a machine being established, would be to bring, as it were, the cities of London and Edinburgh into the immediate neighbourhood of each other, and to produce transactions and communications of kinds not hitherto known or practised—communications which do not at present pass through Edinburgh or London would be brought to these points for the sake of rapid transmission—communications might be made to intermediate points, and public intelligence could be disseminated all along the line. Were the example followed all over the kingdom, it would create perhaps one of the greatest changes in human affairs, called into operation by the ingenuity of man.

"After the uses to which the power of steam and coal-gas has been so successfully and wonderfully applied, the telegraph now proposed may not be an unworthy follower in the march of discovery and improvement.

"The present sketch is submitted for the private consideration of a limited number of scientific and influential gentlemen, of whom a meeting will soon be convened, to give their opinion of the practicability and utility of the plan here generally developed."

The following is a description of the apparatus as given by the author at p. 19 of his pamphlet :—

"The model is contained in a mahogany case, or frame, 6 feet long, 2 feet wide, and 3½ feet high.

Fig. 30.

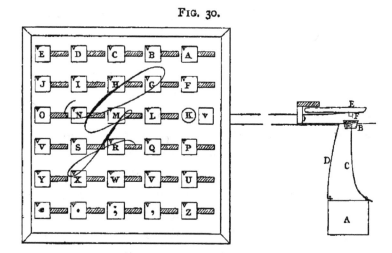

"The end of the case, intended to face the north, is composed of a wooden board or tablet coloured black, with the twenty-six letters of the alphabet, a comma, a semicolon, a full point, and an asterisk, shown on white enamel, at equal distances, in six rows or tiers. The tablet is protected by a sheet of plate glass, and the top or lid of the case is also of glass, for more easy inspection of the interior.

"Behind the tablet are placed (also in six rows or tiers) thirty steel magnets, about two inches long, poised on their centres, so as to admit of their assuming their natural position in the magnetic meridian, and thus having their north poles pointed to the back of the tablet.* On the north pole of each of the thirty magnets a small piece of brass wire is fixed, protruding through a slit or aperture in the tablet; and from the point of this brass wire a thin piece of brass of about one-half inch square, coloured black outside, is suspended.

"Each of these thirty pieces of brass, when the needles are in their natural direction of north and south, conceal or veil one of the letters or points marked on the tablet; and in this position the observer of the tablet perceives nothing but one uniform black surface.

"Each of the magnets is poised within a coil of several convolutions of copper wire, and a galvanometer is thus formed.

"At the other, or south end of the model, is a horizontal line of thirty wooden keys, resembling the keys of a pianoforte, and on these keys are marked the twenty-six letters of the alphabet, a comma, a semicolon, a full point, and asterisk, in the same

---

* Many writers, as Moigno and Shaffner, describe and illustrate these needles as suspended *vertically*, which is a mistake.

manner as on the tablet. Thirty insulated copper wires traverse the model from the keys to the galvanometers, with both of which they are connected.

"Each galvanometer is also connected by an insulated wire, about three inches in length, with a transverse copper rod, extending from one side of the model to the other. There are six such transverse rods, placed horizontally and at right angles with the six rows or tiers of galvanometers. These copper rods are connected by wires with each other—and a thick copper wire traverses the model from the undermost rod to the south end of the model, and is there connected with the copper plate or positive pole of a small galvanic battery.

"In a small trough or reservoir, extending under the whole length of the line of keys, a small quantity of mercury is deposited, and the zinc plate or negative pole of the galvanic battery is connected by a wire with the mercury in the trough.

"It must be here noticed that the two poles of the galvanic battery are thus connected together by the wires and metallic conductors above described, *except in the space that intervenes between the keys and the trough of mercury placed beneath them.*

"It has, therefore, in the next place to be remarked, that thirty pendant platinum wires are attached to the under part of the thirty keys of the model, and that when any key is pressed down with the finger, the pendant platinum wire is immersed in the mercury,

and the galvanic circuit, by means of metallic conductors, between the two poles (copper and zinc) of
the battery completed.

" *The instantaneous effect of the galvanic circuit being
so completed is to cause one of the magnets to deflect
towards the west, carrying the small brass veil along
with it, and thereby exhibiting on the tablet the same
letter of the alphabet or point that is marked on the
key pressed down.*

"When the finger is taken off the key it rises, by
means of a spring underneath, to its former position
on a level with the other keys ; and the pendant
platinum wire ceasing to be dipped in the mercury,
the galvanic circuit is again broken, and the magnet
returns to its natural position, and veils the letter
that was shown on the tablet.

"Hence it follows, that by simply pressing down
with the finger any of the keys (precisely in the same
manner as the keys of a pianoforte are touched), the
same letter that is marked on the key is shown on the
tablet for a sufficient length of time to allow it to be
observed by any person watching the motions of the
veils on the tablet ; and words are thus communicated
in rapid succession from the one terminus of the telegraph to the other.

"When, in the course of a communication, it is
wished to indicate a comma, semicolon, or full period,
these will be disclosed on the tablet on the corresponding key being pressed down ; and in order to

indicate that the spelling of a word is finished, the key marked with the *asterisk* may be pressed down, and the asterisk being at the same instant exhibited on the tablet, will show the observer that the word is completed, and that a new one is about to be spelled.

"In order either to send or receive a communication by a telegraph of the simple construction proposed, no greater learning would be required than is necessary in reading a common book ; and the rapidity with which a communication could be made, would be as great as that with which most persons are able to write, or as a compositor is able to set up types.

"In telegraphing between distant points, the connecting wires would be made to traverse the intermediate space through a tube of wood, or some other material that would protect the wires from external injury ; and the wires would of course be separated from each other by laying them in separate grooves in the tube, or by coating them with some non-conducting substance. The diameter of the tube might be very small ; and in order to protect the wires from any atmospheric influence, and the tube itself from violence, it would be best placed under ground.

"Following out the scientific principles that have been explained, and taking advantage of the mechanical contrivances illustrated by the model now exhibited,

it appears perfectly practicable to construct an electro-
magnetic telegraph surpassing all other kinds of
telegraphs in respect to the rapidity, facility, and
certainty with which every species of communication
can be made between points, however distant."

Having perfected his plans, Alexander submitted
them to the Government in the following letter which
he addressed to Lord John Russell, Home Secretary,
on June 12, 1837, the date, by the way, of Cooke and
Wheatstone's first patent :—

> "Edinburgh, 19 Windsor Street,
> "12th June, 1837.

"My Lord,—I have the honour to enclose for your
Lordship's consideration a plan for an electro-magnetic
telegraph between Edinburgh and London, and capable of being adopted all over the kingdom with the
most important national advantages.

"I have had the honour of submitting the plan to
some of the most eminent scientific and influential
persons here, and have met with the most flattering
approbation from them.

"In order to test the practicability of the plan,
experiments have, by the obliging permission of Dr.
Hope, the professor of chemistry, been made in his
class-room in our University, by his very able practical demonstrator Mr. Kemp, both upon a small and
pretty extensive scale—a metallic conductor of about
four miles in circuit having been operated upon both

by mechanical electricity and by galvanism. These experiments have been performed in the presence of Professor Hope himself, and also of the Lord Provost; the Solicitor-General; the Master of the Merchant Company; Sir Charles Gordon, the secretary of the Highland Society; Sir John Hall; Professor Jameson; Professor Traill; Mr. Patrick Robertson; Mr. George Monro; Mr. Hamilton, architect; and other scientific and literary gentlemen, and have proved most satisfactory. A plate of copper and a plate of zinc (about the size of a crown-piece each) immersed in a little acid water, were found sufficient to move an ordinary magnetic needle at the termination of a copper wire of four miles in length, notwithstanding numerous joinings and sinuosities in the conductor.

"Should the telegraph projected ultimately prove capable of the general utility and application contemplated, I humbly think, from its affinity to the Post Office Department, it may prove worthy of the attention of Government, in place of being left to individual enterprise.

"In either case, I venture to hope that sufficient has been already demonstrated both of its practicability, and national importance and utility, to excuse my bringing the plan under your Lordship's notice, and to request your distinguished patronage.

"It has been suggested to me that the fund, amounting to upwards of 6000*l.* per annum, administered, subject to the orders of the Lords of the Treasury, by the trustees for the encouragement of arts and manufactures in Scotland, could not be better applied to the extent of a few hundred pounds than by defraying the cost of additional experiments under the superintendence of the gentlemen I have named, or such others as may be selected, upon a still larger scale than it can be expected that individuals should supply for a national object.

" If upon an extent of 50 or 100 miles of metallic conductors, the same instantaneous and perfect indication of the passage of the electric or galvanic fluid is found, as has been in the case of our recent experiments at the University, the triumph of the scheme would be complete.

" May I request the favour of your Lordship's views.

" The Solicitor-General transmitted a print of the plan about a week ago to the Lord Advocate, and I have since made a communication to his Lordship (to whom I have the honour of being known) on the subject.

" I have the honour, &c.,

" W. ALEXANDER."

The reply was as brief and as little satisfactory as those vouchsafed to Wedgwood (1814), Ronalds (1816), Porter (1825), and " Corpusculum " (1832). It ran as follows :—

" Whitehall, 15th June, 1837.

" Sir,—I am directed by Lord John Russell to acknowledge the receipt of your letter of the 12th instant, enclosing a Plan for an Electro-Magnetic Telegraph.

" I am, Sir,

"Your obedient servant,

"S. M. Phillips."

In December 1837, Alexander memorialised the Lords of the Treasury on the subject, but with no better fortune.   The memorial set forth

" That the attention of your memorialist has been for some time past directed towards effecting an instantaneous telegraphic communication between London and Edinburgh by means of electro-magnetism; and he now respectfully submits to your Lordships a print of the plan respecting this object, that was circulated by him last summer for the consideration of certain scientific and official gentlemen in Scotland.

"The result of the investigation instituted was, that the heads of the public bodies in Edinburgh, and the most scientific persons resident in that city, concurred in the accompanying testimonials of their belief of the success of the plan, and an expression of their readiness to act as a committee to direct and superintend experiments further to test the practicability of the proposed undertaking.

"That in order further to illustrate your memorialist's

views, he caused to be prepared a model of the projected telegraph, and exhibited the same to the first meeting of the Society of Arts in Edinburgh, and refers to a copy of the *Scotsman* newspaper of the 18th November, 1837, as containing a description of the model, and its powers in effecting instantaneous telegraphic communication between distant places.

"That your memorialist, from a strong conviction of the vast national importance attached to the completion of an instantaneous telegraphic communication between all parts of the kingdom, at a moderate expense, feels himself warranted in laying the above-mentioned documents before your Lordships, and in praying that such provision may be made by a grant of money, reference to scientific persons or otherwise, as to your Lordships may seem meet.

"That the model above referred to has been conveyed from Edinburgh to London ; and your memorialist will have much satisfaction in exhibiting it in complete operation to your Lordships, and to give any further explanations that may be thought proper.

"All which is respectfully submitted by

"W. ALEXANDER."

The following is the paragraph in the *Scotsman* referred to in the memorial :—

"The telegraph thus constructed operates with

ease and accuracy, as many gentlemen can witness. The term model, which we have employed, is in some respects a misnomer. It is the actual machine, with all its essential parts, and merely circumscribed as to *length*, by the necessity of keeping it in a room of limited dimensions. While many are laying claim to the invention, to Mr. Alexander belongs the honour of first following out the principle into all its details, meeting every difficulty, completing a definite plan, and showing it in operation. About twenty gentlemen, including some of the most eminent men of science in Edinburgh, have subscribed a memorial, stating their high opinion of the merits of the invention, and expressing their readiness to act as a committee for conducting experiments on a greater scale, in order fully to test its practicability. This ought to be a public concern. A machine which would repeat in Edinburgh words spoken in London three or four minutes after they were uttered, and continue the communication for any length of time, by night or by day, and with the rapidity which has been described—such a machine reveals a new power whose stupendous effects upon society no effort of the most vigorous imagination can anticipate."

Alexander opposed the application of Cooke and Wheatstone for a Scotch patent in 1837, but ultimately withdrew his opposition under circumstances which are thus described in a letter of Miss

2 H

Wheatley (now Lady Cooke), dated December 2, 1837: *—

"I have the pleasure to tell you that the Scotch patent is now free from all opposition, and will be obtained immediately. William had an interview with Mr. Alexander on Thursday at the Lord Advocate's (of Scotland) office, and he agreed to accompany the Judge and Lord Lansdowne to see some experiments yesterday at Euston Square terminus. These proved so satisfactory that Alexander at once acknowledged the superiority of William's and Mr. Wheatstone's plans, and gave up his own. This was an agreeable way of arranging the matter, and William was pleased with Alexander's manner of yielding the point; though, of course, he saw he had no chance of succeeding."

Alexander did not, however, cease to advertise his own invention. Wheatstone, writing to Cooke, December 15, 1837, says :—" Alexander continues to make a great noise about his invention. A few days ago he took it to Kensington Palace for the inspection of the Duke of Sussex; and last night he was at the Royal Society." Early in 1838 he placed it on exhibition at the Royal Gallery of Practical

* Extracted by kind permission of Mr. Latimer Clark. See also Cooke's evidence in *The Electric Telegraph Company* versus *Nott and others*, Chancery Proceedings, p. 49. A printed copy of the evidence and affidavits in this celebrated case is preserved in Mr Latimer Clark's magnificent collection of books on electricity and magnetism—a collection which rivals and, in some respects, excels that of the late Sir Francis Ronalds.

Science, Adelaide Street, Strand ; * at the Polytechnic Institution, London, in 1839 ; at the Glasgow meeting of the British Association, in 1840 ; and finally at the Great Exhibition (Hyde Park), in 1851.

In answer to some inquiries of ours, Mr. George P. Johnston, the well-known bookseller, of 21, Hanover Street, Edinburgh, has kindly sent us the following letter :—

" Edinburgh, 5th May, 1883.

" Dear Sir,—I fear you will think I have forgotten about your queries as to Mr. Alexander, but the delay has been caused by the difficulty of finding people in, &c.

" I can obtain no information regarding his family whatever, and he seems to have passed out of the remembrance of Edinburgh people. All I can gather is that he was considered by some 'a clever man always inventing,' by others as 'half-crazed on the subject of inventions.' I am sure he is the same W. Alexander who is author of several treatises on Scotch Bankruptcy Acts, &c., published between 1847 and 1859. It is curious that one philosophical instrument maker here—the only one old enough to have known him—never heard of him.

" I am, yours respectfully,

" GEO. P. JOHNSTON.

" To J. J. Fahie, Esq."

* See p. 381, *ante.* He also brought it before the Society of Arts, which, however, decided that it was not new, and, on that account, unworthy of attention.—Letter, Wheatstone to Cooke, of 24th March, 1838.

2 H 2

Dr. Edward Sang, Secretary of the Royal Scottish Society of Arts, informs us that Alexander proposed a bridge over the Forth at Inch-Garvie, just where one is now about to be built.

Alexander's death is mentioned in the *Transactions of the Royal Society of Edinburgh* as having occurred during the session of 1859-60; but there is no obituary notice.

1837-8.—*Mungo Ponton's Telegraph.*

On the 15th November, 1837, a working model of the apparatus last described was exhibited at the Society of Arts, Edinburgh, and excited the keenest interest amongst all the members present. One of these, the late Mr. Mungo Ponton, was so impressed with the subject that he set about at once to devise a telegraph of his own, which should be free from the imperfections with which he saw that Alexander's was hampered; and so rapidly did execution follow upon the heels of design, that in five days, *i e.*, on the 20th November, 1837, he forwarded to the Society a "Model and Description of an Improved Electric Telegraph." The paper was read at the meeting of the 10th January, 1838; and again, on the 20th June following, a supplement was presented, together with a model of the telegraph in an improved form.*

* Both these documents are now before us, having been lately discovered after a long search which was kindly made for us by Dr. George Macdonald, of Edinburgh. The model was for many years in the museum of the Society, but on a change of offices it was put away in a damp cellar, where it soon fell to pieces.

In the first communication Ponton begins by saying :
" When, on a former evening, Mr. Alexander's electric
telegraph was exhibited, I took occasion to point out
one or two defects under which it appeared to me to
labour. These were—1st, the weak and vacillating
character of the force employed to discover the letters ;
2nd, the great size of the reading-board ; and 3rd, the
very unnecessary multiplication of the lines of wire.
Had Mr. Alexander had a just claim to the original
invention of an electric telegraph, I should have considered it only fair to have pointed out to him the
remedies which had occurred to me for the removal of
those defects ; but as I am led to understand that his
claim to originality extends no further than to the
mode of constructing the telegraph, I felt myself at
liberty to follow out my own ideas.

"The objects I have had in view in the construction of the models now submitted to the Society are
—1st, to show how a powerful and decided force may
be developed at the reading end of the telegraph ;
2nd, how the reading surface may be reduced within
very narrow limits ; 3rd, how the quantity of motion
required for the display of the characters may be made
very minute, only a quarter of an inch ; 4th, how that
motion may be rendered independent of the swing of
the needle ; and 5th, how with only eight lines of wire
we may exhibit all the letters of the alphabet, all the
figures, a variety of points and signs, and a considerable number of combinations of letters.

"The first model to which I would call attention is that which shows the method of developing a powerful and steady force at the reading end of the telegraph. This consists of a dipping needle, delicately poised, and furnished with a galvanometer coil. From either side of the centre of the needle is suspended a small slip of wood, from which project downwards the four ends of two bent wires, which dip into mercury cups. The mercury in the cups is connected with the opposite ends of the wires of a pair of common electro-magnets, whose poles are opposed to each other. When one end of the dipping needle is down, and the wires on one side touching the mercury in the cup, the effect is to make the two electro-magnets attract each other. When the other end of the needle is sent down by means of the electric current passing through the galvanometer coil, the effect is to make the two electro-magnets repel each other. In this manner a very considerable and a perfectly steady force is produced, which may be employed for the raising and depressing of levers for the display of the characters.

"Although I should consider this a very decided improvement, yet it is not necessary for the other improvements which I have suggested, and accordingly I have not used it in the construction of the model telegraph now before the Society. To that model I would now direct attention.

"It will be seen to contain eight galvanometers,

having their needles suspended vertically like a dip-ping needle. These are placed in two sets of four, piled one above another, each lower needle pro-jecting beyond the one immediately above it. To the end of each needle is attached a fine thread, which is stretched upwards and attached to the bottom of a card. There are eight of these cards placed one before another. They are suspended by threads to the end of eight levers placed one above another. When the needles are in their natural position they hold down the cards against two bars which are placed so as to support them. When the needles are affected by the electric current passing through their coils, they swing upwards, and slacken the thread by which the cards are held down, and the cards are then pulled upwards by the levers attached to their upper ends. Their motion, however, is checked after they have moved a quarter of an inch, by a piece of wood placed for that purpose. Thus the motion of the cards is rendered in a great measure independent of the swing of the needle.

" The ends of the galvanometer wires are passed onwards to a key-board at the working end of the telegraph. This key-board contains eight keys, a pair being attached to each wire, the one passing a negative, the other a positive electric influence through the line of wire. The needles are so adjusted that the one is affected by the positive, the other by the

negative current, so that each wire works a pair
of needles.

" Behind the cards at the reading end of the telegraph is a permanent back, with characters upon it,
disposed in four columns and nine lines. The two
backmost cards have also characters upon them,
disposed in two columns and nine lines. There are
thus in all eight columns and nine lines, or seventy-
two distinct signals. The cards have a variety of
openings cut in them for displaying the characters, so
arranged that only one character is displayed at a
time. The four hinder cards are cut so as to display
in succession the eight columns, and the four front
ones so as to display eight of the nine lines, also in
succession, the uppermost line being seen when the
four front cards are at rest.

" Thus when any one of the four hinder cards is
touched alone, it displays a character on the first line.
The outermost displays the character situated on the
extreme left column of the permanent back, and the
innermost that situated on the extreme right. When
the first and third keys are touched, the characters on
the left column of the backmost card are discovered,
while the first and fourth display those in the right-
hand column. When the second and third keys are
touched, those in the left-hand column of the second
backmost card are displayed ; the second and fourth
display those on the right. It will be thus seen how
the whole eight columns are displayed. The inner-

most of the four front cards, corresponding to No. 5 of the keys, when raised alone, uncovers the second line; the next the third, the next the fourth, and the outermost the fifth. When the fifth and seventh keys are touched, the sixth line is displayed; the fifth and eighth display the seventh line. The sixth and seventh keys display the eighth line; the sixth and eighth display the ninth.

"From this explanation it is easy to see how any one of the seventy-two signals can be made to appear.

"The following is the arrangement adopted in the present model :—

Signs.	Keys.	Signs.	Keys.	Signs.	Keys.
A	1	I	3	Q	3·5
B	2·7	J	1·3·6	R	4·8
C	4·5	K	1·5	S	1·6
D	4·6	L	3·8	T	3·6
E	2	M	1·8	U	1·3
F	3·7	N	2·8	V	4·7
G	2·5	O	4	W	1·4
H	2·4	P	1·7	X	1·4·6
		Y	2·3	Z	2·6

The remaining signals are dedicated to the exhibition of figures, points, arithmetical signs, and combinations of letters; but it is unnecessary to particularise the arrangement. Of course they are all produced by touching either three or four keys together.

"It will be observed that any one of the four front cards may be touched alone without producing any signal. This might be taken advantage of to extend the number of signals to 120, by employing the needles which move the front cards to shift the permanent back."

In the improved model of June 20, 1838, Ponton had so contrived matters as to be able to reduce the number of line wires to four, by the various combinations of which into metallic circuits as in Cooke and Wheatstone's five-needle telegraph, and by using positive and negative currents, he was able to show forty-eight different signals. The work of the electric currents was also reduced to a minimum, slight deviations of the needles being all that was necessary.

Proceeding on the principle that in all correct telegraphing every signal made from one end should be repeated back from the other, he adopted the plan of exhibiting the signals, not in their direct transmission, but on their repetition, and by an arrangement which allowed the signal-indicating apparatus to be worked, not by the needles as before, but by the hand of the receiver in the act of repetition.

The galvanometer needles were eight in number (two in the circuit of each wire), and bore marks corresponding to keys, which, on being depressed, sent positive or negative currents into their respective line wires, and, at the same time, uncovered the letters, figures, or signs with which they were in train. Whenever, therefore, any of the needles were deflected, the receiver had only to depress the corresponding keys, by which act he repeated the signal back to the sending station, and uncovered the letter, figure, or sign which that signal was intended to express.*

The following table shows all these signs and the numbers of the keys which entered into their formation :—

Signs.	Keys.	Signs.	Keys.	Signs.	Keys.
A	1·4	I	4·5	Q	1·4·7
B	1·6	J	4·7	R	1·4·8
C	1·8	K	5·8	S	1·6·8
D	2·3	L	6·7	T	2·3·5
E	2·5	M	1·3·6	U	2·3·6
F	2·7	N	1·3·8	V	2·3·7
G	3 6	O	1·4·5	W	2·3·8
H	3·8	P	1·4·6	X	2·4·5
		Y	2·4·7	Z	2·5·7

Signs.	Keys.	Signs.	Keys.	Signs.	Keys.
,	2·5·8	1	1·3·5·8	8	2·3·5·7
;	2·6·7	2	1·3·6 7	9	2·3·5·8
.	3·5·8	3	1·3·6·8	0	2·3·6·7
&	3·6·7	4	1·4·5·7	+	2·3·6·8
Sig.	3·6·8	5	1·4·5·8	−	2·4·5·7
E R.	4·5·7	6	1·4·6·7	×	2·4·5·8
R E.	4·5·8	7	1·4·6·8	÷	2·4·6·7
E N D.	4·6·7				

Ponton added an alarum which was a very simple affair, and recalls a somewhat similar one of Edward Davy described on p. 357. An extra galvanometer was placed in the circuit of one of the line wires, and across its needle was placed a fine platinum wire,

* With a little practice this would have been found to be a work of supererogation, for the operators would soon have come to know the values of the deflections, without the need of reproducing them in black and white. Again, it would soon have been found that four galvanometers would suffice ; and thus the system would resolve itself into a four-needle telegraph.

which, normally, rested against a lucifer or other quick match. Above the match was a thread holding back the hammer of a bell or the detent of an alarum. To sound the alarum the proper keys were held down for a few seconds, the needle of the galvanometer was deflected and carried the piece of platinum wire into the flame of a spirit lamp ; then on reversing the direction of the current in the line the red-hot wire was suddenly brought in contact with the match, which, igniting, burnt the string above it and so released the hammer of the bell or the detent of the alarum.*

On the report of a committee an honorary silver medal was awarded in December 1838 to Mr. Ponton "for the ingenuity of his plans as manifested in the working model which had been presented to the Society."

---

* A supply of matches and strings would of course be necessary. In the ordinary course of signalling, the platinum wire would often be in the flame of the lamp, but never sufficiently long to be heated to the point of igniting the match ; so that, as Ponton points out, there would be small risk of unseasonable alarums. Ponton also suggested an alarum by the direct action of a rather heavy galvanometer needle striking against a bell when deflected by the current.

CHAPTER XVII.

TELEGRAPHS BASED ON ELECTRO-MAGNETISM AND
MAGNETO-ELECTRICITY (*continued*).

1837.—" *Corpusculum's* " *Telegraph.*

WE copy the following letter from the *Mechanics'*
Magazine, for December 30, 1837, p. 219—a most
valuable work for all engaged in scientific research,
and to which we gratefully acknowledge ourselves
indebted :—

"Sir,—I first met with an account of Alexander's
telegraph last night in the *Mechanics' Magazine,* and
a very important improvement suggested itself, which
will render fifteen of the thirty-one wires unnecessary.
I see no reason why each of fifteen wires should not
represent two letters, thus; let each of the letter
screens affixed to the movable magnets be wide
enough to cover two letters. Then the positive end
of the galvanic battery being connected with the
conducting wire, by a touch of the keys, the magnet
an1 screen will move in one direction and discover
one letter. The negative end of the battery being
then connected with the same wire, the magnet will
move in the contrary direction and discover the other
letter. There must of course be something fixed to
prevent the magnet going so far in either direction as

to discover both letters.   The returning wire connected with all the other thirty (? fifteen), must, of course, have its connection with the battery poles reversed, at the same time as the lettered wire.

" Not having seen a model of the instrument, I am in doubt whether the magnet would not, on returning to its stationary position, want a contrivance to prevent its oscillation ; I have, therefore, devised the following plan which would perhaps be the best of the two .—Let each wire act upon two magnets and screens, one magnet and screen moving in one direction, but prevented from moving in the other as now. The current of electricity if reversed, would, on account of this prevention, not move *this* magnet and screen in the opposite direction, but it would the *other* magnet and screen, having a similar stop or prevention, but placed on the other side of the pole.

" It seems many persons have formed designs for telegraphs, I, too, formed mine, and prepared a specification of it five years ago, and that included the plan of making one wire only serve for the returning wire for all the rest, as in Alexander's telegraph ; *but even that might, I think, be dispensed with where a good discharging train, as gas or water, pipes, at each end of the telegraph could be obtained.*   I wrote to the Admiralty at the time I mention on the subject of my invention, but facilitating commercial correspondence it seems was too contemptible a subject for state philanthropy.

"My telegraph was designed to print off its own communications (and I think might be made to convey hundreds in a minute) by means of a machine I invented for rapidly writing in the common printing characters, and which I wished to get some one to join me in perfecting and patenting, but was unsuccessful, as I have been in two or three other instances in which others are now reaping the advantages which I should have done myself, but for that infamous plundering incubus upon talent, the English Patent Laws. I think the following method might serve to secure the poor man's patent without interfering with the legal right of plunder. Let him have the liberty of filing a specification (? sealed or open), which should have the effect of preventing every other person, and also himself, from deriving any benefit from his invention, till the plunderers by legal authority should have their ' pound of flesh.'

"This preliminary specification should enable the inventor to take a patent for anything coming fairly within its scope and spirit. It would enable him to enter into a contract on fair terms with any person able to bear the expense of a patent, which he cannot now do without risk of being victimised, as I have lately had reason to know to my cost. There is also a valuable protection that I think might be extended to scientific, as it is to literary, inventions. A man's play cannot be exhibited to others without his sanction. There is just as good reason why a working model of either a useful or pleasing piece of mechanism should be protected from piracy. I may mention as an instance (of which there are many others), that I am constructing a model twenty inches broad by twenty-four long and twenty-four high, for a machine to produce light by a succession of electric sparks. I have completed one element which is, of itself, all but sufficient to read by, and when the other elements are complete, it will be certainly capable of being thirty-two times as powerful, and not improbably sixty or one hundred times. A larger one which I had commenced (but which I fear will be too expensive for me to complete, in the present unprotected state of science) is eight feet, by four feet eight inches high, and I calculate it would produce a million, and not improbably many millions, of sparks of various colours in a minute, and would give 100,000 moderate shocks, or by (combination) 4000 or 5000 far too intense for endurance, in the same short period.

" This would doubtless form a very excellent subject for exhibition ; but as any blockhead may imitate, I have given up the thought, at least for the present.

"CORPUSCULUM.

" December 8, 1837."

We have copied this letter *in extenso*, with all its ambiguities, for two reasons, 1st, because it is intrinsically interesting and valuable, and 2nd, in the hope that our doing so will afford a clue to some of our readers who may wish to discover the true name of the writer. The points of interest in this letter are:—

(1) The suggestion of a telegraph like Davy's.

(2) The suggestion of the earth circuit seven months before Steinheil's *accidental* discovery of it, and exactly as we use it to-day.

(3) The construction of a Roman type printing telegraph in 1832.

(4) The suggestion of a patent law which was subsequently passed, and of a law applicable to instruments, as copyright is to literary productions.

(5) A system of electric lighting—the light-giving part to consist apparently of one or more vacuum tubes, guardedly called "elements," no doubt, with the object of misleading "pirates and blockheads."

1837.—*Magrini's Telegraph.*

The proposal that we have now to notice is one of great merit, and resembles in some respects Cooke and Wheatstone's five-needle, or *Hatchment*, telegraph of 1837. It is the invention of Professor Luigi Magrini, of Venice, and is described by him, at length, in a *brochure*, which he published, at Venice, in 1838, entitled *Telegrafo Elettro-Magnetico, Praticabile a Grandi Distanze.* From an Appendix on pp. 85-6, it appears that the first published account of this telegraph is that contained in the *Gazzetta Privilegiata di*

Venezia, No. 189, of 23rd August, 1837; * but, as far as we can discover, it was never tried on any extensive scale. Had this been done, there can be no doubt that it would have succeeded as well as the English one, and we should have had the curious result of seeing the simultaneous and independent establishment in Italy and in England of electric telegraphs, which are not only based on the same principles, but, in some respects, are almost identical.

The signal apparatus consisted of a horizontal table, one metre long, and sixty centimetres broad, into which fitted three galvanometers as shown in the Fig. 31. By means of two batteries of different strengths, and a commutator, each needle was susceptible of four movements, one weak and one strong to the right, and one weak and one strong to the left. These four positions indicated for each needle a different letter which was suitably inscribed on the board, or table. Thus, the letters appertaining to the first galvanometer were A, B, C, D; those of the second, I, L, M, N; and those of the third, S, T, U, V.

In order to indicate all the other letters, the needles were employed, two and two at a time; F, for example, corresponded to *weak*, right-handed deflections of needles 1 and 2; H, to *strong* deflections of the same two needles, and in the same direction; O, to weak,

---

* In the *Annales Télégraphiques*, for March–April 1882, p. 140, it is said to date back to 1832; but this is probably a misprint.

left-handed deflections of the second and third needles ; R, to strong deflections of the same needles, but in the other, or right-handed direction ; and so on for the rest.

FIG. 31.

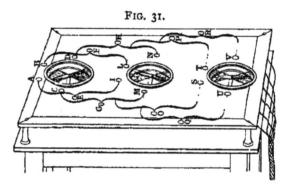

Magrini employed six line wires, forming three metallic circuits. At the sending station these dipped into troughs of mercury placed on a table, and a little above which was laid the commutating board on short supports. This board, which for clearness sake is shown in the Fig. 32, in a raised, or vertical, position, carried, underneath, twenty-four glass rods, in three rows of eight rods each. To the ends of each rod were attached elastic strips of brass, terminating in projecting pins of the same material, which could be pushed downwards (by means of a handle affixed to the centre of the rod and projecting through the top face of the board) so as to dip into the mercury troughs. The other ends of the elastic strips were permanently connected, the one with the positive,

2 I 2

and the other with the negative pole of one or other of the batteries E, and E'. Taking, for example, the first row of keys, or rods, on the left of the figure, which, we will suppose, was connected to the first

FIG. 32.

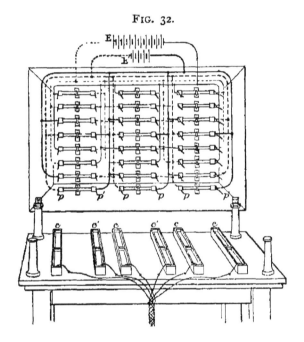

galvanometer at the distant station, then the first rod, at the top, was in connection with the poles of the strong battery; the second rod was connected to the same battery, but in the reverse way to the first; the third and fourth rods were connected to the weak battery in the same manner that the first and second were to the strong. The remaining four rods were connected, rod for rod, like the last, that is to say, the fifth was connected to the same battery as the first, and in the same manner, the sixth like the second, and so on.

Whenever, then, the first rod was depressed, a current from the strong battery E, flowed out to line, and circulating through the coils of the first galvanometer, produced a strong deflection of the needle (say, to the left), and so pointed to the letter C. Depressing the second rod produced a strong deflection of the same needle to the right, and so indicated D; and so on for all the rest. With regard to the last four rods of each row they were used in pairs, one from each row; thus, when the fifth rods in the first and second rows were depressed, the needles of the first and second galvanometers were strongly deflected to the left, and indicated the letter G; while depressing the last rods of the second and third rows produced feeble deflections to the right of the second and third galvanometers, and indicated P.

It is easy to see that all these combinations could be obtained by making use of the first four rods of each row, but it was no doubt in order to avoid all chance of confusion that the inventor introduced special ones for this purpose.

Magrini added an alarum whose construction will be seen from the accompanying Fig. 33; the bar *m, o, n,* was so balanced that in its normal state its hammer *m,* rested against the bell. When it was required to

attract attention a current was set up in the electro-
magnet *a*, *b*, *c*, which brought down the soft iron
armature F, G; then by means of a pole-reversing
arrangement Q, S, the direction of the current in

FIG. 33.

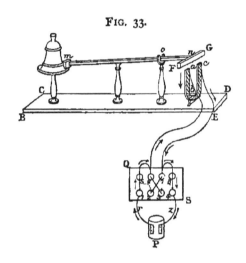

a, b, c, was altered. Owing to the residual magnetism
in F, G, the first effect of this inversion of current was
to repel the armature, then immediately after to
attract it afresh. At each reversal, therefore, of the
current the hammer *m*, clicked against the bell and
produced a tinkling sound.

1837.—*Stratingh's Telegraphs.*

These aim no higher than to be lecture-room
demonstrations of the possibility of an electric telegraph, and coming as they do at a time when not

only the *possibility*, but the *practicability* of this mode of communication was completely established, they would not deserve notice, were it not that they contain the suggestion of a contrivance which, we believe, to be of great practical utility in the construction of relays, and electro-magnets generally, and which, in this conviction, we utilised in our patent of February 3, 1876.* On p. 3 of our Provisional Specification we say :—

" I will now describe the second part of the invention, which relates to improvements whereby the ordinary unpolarised relay, or electro-magnet, is rendered susceptible of great sensibility both for duplex and for single transmission. This is done for duplex by so utilising the local current, which the working of the relay brings into play at certain times, *that the armature, at these times, is itself an electro-magnet and opposes the attraction subsisting between its poles and those of the coils to the pulling force of the spring, so that, if need be, the magnetism of the coils has only to overcome the inertia of the lever, and not the force of the spring as well ;* and for single working by making the counteracting springs parts of the local battery circuits and placing within them cylindrical bar-magnets or bars of soft iron."

As will presently be seen (p. 490), the words that

---

* No. 433. "Improvements in Electric Telegraphs, comprising an Improved System of Duplex Working, and an Improved Relay or Electro-Magnet, the principle of which may also be used in any Instrument or Contrivance where Relays or Electro-Magnets with Counteracting Springs are employed."

we have italicised in the above extract are but the (of course unconscious) realisation of a suggestion made by Professor Stratingh nearly forty years before.

For the following account of Stratingh's telegraphs we are indebted to Mr. J. M. Collette, engineer of the Netherlands telegraphs.

"To Mr. J. J. Fahie, London.

"The Hague, April 28, 1883.

"Sir,—In reply to your letter of 17th instant, I have the honour to inform you that the late Mr. Stratingh, professor in the University of Groninque, published his article, "Iets over eenen Electro-Magnetischen Klokken-Telegraaf" (On an Electro-Magnetic Acoustic Telegraph), in the *Journal for the Encouragement of Industry*,* for 1838.

"I send you a copy of the woodcut, Fig. 34, which accompanied the description of his apparatus. The latter consisted of two electro-magnets of the horse-shoe form, two levers, each having at one end an armature, and at the other a small hammer, and two bells, or gongs, of different tones. It is evident that when a current passed, for example, through the coil *i*, the armature *e*, would be attracted, and the hammer attached to the other end of the lever would descend and strike the bell *h*. To prevent the *sticking* of the armature it was provided on its under surface with a thin plate of ivory. The ordinary clockwork alarum *o, o', o'', o'''*, was intended to warn the attendant of the

* *Tydschrift ter bevordering van Nyverheid*, vol. v. part 2.

coming of a despatch. The armature e, in ascending released the detent q, and so set the wheel work in motion.

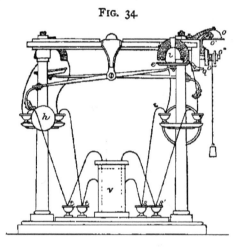

FIG. 34.

" The current was produced by a pair v, composed of a copper cylinder containing acidulated water, in which was plunged when required another cylinder of zinc. Wires from the copper and zinc poles dipped into little cups of mercury, into which were also plunged as required the terminal wires of the electro-magnets.

" In the above-mentioned paper Professor Stratingh stated that experiments made with this apparatus before the Physical Society of Groninque succeeded perfectly through twenty metres of line wire, but that when the distance was increased to one hundred metres the results were not so good, the current then proving to be insufficient.

"It would seem that the Professor did not continue his experiments. He was content with describing his plans 'for what they were worth,' and added that better results would probably be obtained (1), by increasing the force of the battery; (2), by employing insulated wires; (3), by the use of thicker wires; (4), by more delicately suspending the levers, &c. He even remarked that he had surrounded the armatures with covered wire in such a way that the current in circulating through this wire and through the coil *i*, should produce opposite poles in the contiguous parts of the armature and electro-magnet, which would make the attraction stronger.

FIG. 35.

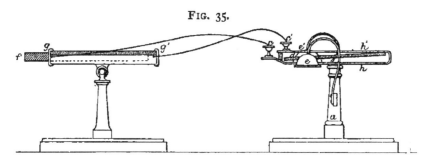

"In continuation of these experiments the Professor made and tried an acoustic apparatus of a simpler construction. This, as represented in Fig. 35, consisted of a stand *a*, supporting a copper band *h*, *h'*, bent on itself and holding cups of mercury into which dipped the wires coming from the electromotor. Within the band *h*, *h'*, was pivoted a bar magnet *d*, which, when deflected to one side or the other, struck one of the bells *e*, or *e'*. Instead of the simple band of copper, Schweigger's multiplier coil could be used.

"As electromotor Stratingh employed a magneto-electric arrangement, consisting of a helix *g*, *g'*, and a bar magnet *f*. By introducing *f*, into *g*, *g'*, first from one side and then from the other, he caused the needle *d*, to be deflected, and so to strike the bell *e*, or *e'*, as required.

"The above, dear Sir, is the substance of Mr. Stratingh's paper, and I hope it will be sufficient for your purpose.

"Receive, &c.,

"COLLETTE."

1837.—*Amyot's Telegraph*.

For much the same reason that we have noticed Stratingh's crude proposals, we must say a few words on another plan which dates from about the same time, and which, if we comprehend it rightly, must be regarded as the first automatic telegraph. Unfortunately, we know very little of Amyot's plans—no more, in fact, than is contained in the following paragraph which we extract from his *Note Historique*, in the *Compte Rendu* * :—

"As for myself, after having studied the problem [of

* For July 9, 1838, pp. 80-3. The *Journal des Travaux de l'Académie de l'Industrie Française*, for March 1839, p. 43, says that Amyot's *note* was addressed to the Academy of Sciences in April 1838 ; but the date of his telegraph appears to be still earlier, for it is referred to in the *Compte Rendu*, for December 26, 1837, p. 909.

electric telegraphy] as thoroughly as I could, I contrived an apparatus, with only one current and one
needle, which itself wrote down on paper and with
mathematical precision whatever a simple wheel
[drum] at the distant end of the line transmitted.
The signals were previously arranged on the wheel by
means of points differently spaced, as on the wheels
of our Barbary organs, and, as in these, the motion of
the wheel was obtained from an ordinary clock
spring.   To transmit a despatch it was only necessary to set it up on the wheel by means of movable
characters [types], and to deposit it in a box, and
immediately it would be reproduced at the distant
station on paper which was moved along regularly by
a machine.   The attendant had only to collect the
paper and hand it to an *employé* who was specially
charged with the interpretation of the ciphers.   With
such an apparatus no errors could possibly occur, for
everything went like clockwork.

"As regards the conducting wires, it would suffice
to put them out of the way of oxydation, by burying
them in the earth, having previously coated them with
a simple varnish of mineral pitch.

"I have communicated all my ideas on this subject
to M. Savary, who has not only encouraged me in my
experiments, but has assisted me with his great
scientific knowledge."

, M. Guerout says that Amyot made a model of his
machine at the request of Baron de Meyendorff, who

sent it to St. Petersburg,* and that he vainly urged its adoption in his own country. M. Foy, the Sir John Barrow of the French semaphores, decided that the invention was public property, and that his department would make the instruments for itself when it was deemed necessary to do so.†

* From a passage in Vail's *American Electro-Magnetic Telegraph,* p. 91, it would seem that in 1838 Amyot joined Morse in an attempt to introduce the latter's invention into Russia. Everything had been settled with Baron de Meyendorff, but at the last moment the Emperor refused his sanction. Why ? See note p. 317.

† *La Lumière Électrique,* March 24, 1883, p. 364. In the *Compte Rendu,* for December 31, 1838, p. 1162, we find the following paragraph :—"M. Amyot, who had presented in the month of June [? December 1837] a note on a plan of correspondence by means of electric telegraphs, addressed to-day tables on a language and a system of signals which he proposed to be used in connection with this correspondence."

APPENDIX A.

—••—

WE make the following extracts from the *Smithsonian Reports*, for 1857 and 1858 :—

Communication from Professor Joseph Henry, Secretary of the Smithsonian Institution, relative to a Publication by S. F. B. Morse.

" Gentlemen,—In the discharge of the important and responsible duties which devolve upon me as Secretary of the Smithsonian Institution, I have found myself exposed, like other men in public positions, to unprovoked attack and injurious misrepresentation. Many instances of this, it may be remembered, occurred about two years ago, during the discussions relative to the organic policy of the Institution ; but, though very unjust, they were suffered to pass unnoticed, and generally made, I presume, no lasting impression on the public mind.

" During the same controversy, however, there was one attack made upon me of such a nature, so elaborately prepared and widely circulated, by my opponents, that, though I have not yet publicly noticed it, I have, from the first, thought it my duty not to allow it to go unanswered. I allude to an article in a periodical entitled ' Shaffner's Telegraph Companion,' from the pen of Professor S. F. B. Morse, the celebrated inventor of the American electro-magnetic telegraph. In this, not my scientific reputation merely, but my moral character was pointedly assailed; indeed, nothing less was attempted than to prove that in the testimony which I had given in a case where I was at most but a reluctant witness, I had consciously and wilfully deviated from the truth, and this, too, from unworthy and dishonourable motives.

" Such a charge, coming from such a quarter, appeared to me then, as it appears now, of too grave a character and too serious a consequence to be withheld from the notice of the Board of Regents. I, therefore, presented the matter unofficially to the Chancellor of the Institution, Chief Justice Taney, and was advised by him to allow the matter to rest until the then existing excitement with respect to the organisation of the Institution should subside, and that in the meantime the materials for a refutation of the charge might be collected and prepared, to be brought forward at the proper time, if I should think it necessary.

"The article of Mr. Morse was published in 1855, but at the session of the Board in 1856 I was not prepared to present the case properly to your consideration, and I now (1857) embrace the first opportunity of bringing the subject officially to your notice, and asking from you an investigation into the justice of the charges alleged against me. And this I do most earnestly, with the desire that when we shall all have passed from this stage of being, no imputation of having attempted to evade in silence so grave a charge shall rest on *me;* nor on *you*, of having continued to devolve upon me duties of the highest responsibility, after that was known to some of you individually, which, if true, should render me entirely unworthy of your confidence. Duty to the Board of Regents, as well as regard to my own memory, to my family, and to the truth of history, demands that I should lay this matter before you, and place in your hands the documents necessary to establish the veracity of my testimony, so falsely impeached, and the integrity of my motives, so wantonly assailed.

" My life, as is known to you, has been principally devoted to science, and my investigations in different branches of physics have given me some reputation in the line of original discovery. I have sought, however, no patent for inventions, and solicited no remuneration for my labours, but have freely given their results to the world, expecting only, in return, to enjoy the consciousness of having added, by my investigations, to the sum of human knowledge, and to receive the credit to which they might justly entitle me.

" I commenced my scientific career about the year 1828, with a series of experiments in electricity, which were continued at intervals up to the period of my being honoured by election to the office of Secretary of this Institution. The object of my researches was the advancement of science, without any special or immediate reference to its application to the wants of life or useful purposes in the arts. It is true, nevertheless, that some of my earlier investigations had an important bearing on the electro-magnetic telegraph, and brought the science to that point of development at which it was immediately applicable to Mr. Morse's particular invention.

" In 1831 I published a brief account of these researches, in which I drew attention to the fact of their applicability to the telegraph; and in 1832, and subsequently, I exhibited experiments illustrative of the application of the electro-magnet to the transmission of power to a distance, for producing telegraphic and other effects. The results I had published were communicated to Mr. Morse, by his scientific assistant, Dr. Gale, as will be shown on the evidence of the latter; and the facts which I had discovered were promptly applied in rendering effective the operation of his machine.

" In the latter part of 1837 I became personally acquainted with Mr. Morse, and at that time, and afterwards, freely gave him information in regard to the scientific principles which had been the subject of my investigations. After his return from Europe, in 1839, our intercourse was renewed, and continued uninterrupted till 1845. In that year, Mr. Vail, a partner and assistant of Mr. Morse, published a work purporting to be a history of the Telegraph, in which I conceived manifest injustice was done me. I complained of this to a mutual friend, and subsequently received an assurance from Mr. Morse that if another edition were published, all just ground of complaint should be removed. A new emission of the work, however, shortly afterwards appeared, without change in this respect, or

2 K

further reference to my labours. Still I made no public complaint, and set up no claims on account of the telegraph. I was content that my published researches should remain as material for the history of science, and be pronounced upon, according to their true value, by the scientific world.

" After this, a series of controversies and lawsuits having arisen between rival claimants for telegraphic patents, I was repeatedly appealed to, to act as *expert* and witness in such cases. This I uniformly declined to do, not wishing to be in any manner involved in these litigations, but was finally compelled, under legal process, to return to Boston from Maine, whither I had gone on a visit, and to give evidence on the subject. My testimony was given with the statement that I was not a willing witness and that I laboured under the disadvantage of not having access to my notes and papers, which were in Washington. That testimony, however, I now reaffirm to be true in every essential particular. It was unimpeached before the court, and exercised an influence on the final decision of the question at issue.

" I was called upon on that occasion to state, not only what I had published, but what I had done, and what I had shown to others in regard to the telegraph. It was my wish, in every statement, to render Mr. Morse full and scrupulous justice. While I was constrained, therefore, to state that he had made no discoveries in science, I distinctly declared that he was entitled to the merit of combining and applying the discoveries of others, in the invention of the best practical form of the magnetic telegraph. My testimony tended to establish the fact that, though not entitled to the exclusive use of the electro-magnet for telegraphic purposes, he was entitled to his particular machine, register, alphabet, &c. As this, however, did not meet the full requirements of Mr. Morse's comprehensive claim, I could not but be aware that, while aiming to depose nothing but truth and the whole truth, and while so doing being obliged to speak of my own discoveries, and to allude to the omissions in Mr. Vail's book, I might expose myself to the possible, and, as it has proved, the actual, danger of having my motives misconstrued and my testimony misrepresented. But I can truly aver, in accordance with the statement of the counsel, Mr. Chase (now Governor of Ohio), that I had no desire to arrogate to myself undue merit, or to detract from the just claims of Mr. Morse.

"I have the honour to be, your obedient servant,

"JOSEPH HENRY.

"To the Board of Regents."

Report of the Special Committee of the Board of Regents on the Communication of Professor Henry.

"Washington, May 19, 1858.

"Professor Henry laid before the Board of Regents of the Smithsonian Institution a communication relative to an article in Shaffner's Telegraph Companion, bearing the signature of Samuel F. B. Morse, the inventor of the American electro-magnetic telegraph. In this article serious charges are brought against Professor Henry, bearing upon his scientific reputation and his moral character. The whole matter having been referred to a committee of the Board, with instructions to report on the same, the committee have attended to the duty assigned to them, and now submit the following brief report, with resolutions accompanying it.

"The committee have carefully examined the documents relating to the subject, and especially the article to which the communication of Professor Henry refers. This article occupies over ninety pages, filling an entire number of Shaffner's Journal, and purports to be 'a defence against the injurious deductions drawn from the deposition of Professor Joseph Henry (in the several telegraph suits), with a critical review of said deposition, and an examination of Professor Henry's alleged discoveries bearing upon the electro-magnetic telegraph.'

"The first thing which strikes the reader of this article is, that its title is a misnomer. It is simply an assault upon Professor Henry; an attempt to disparage his character; to deprive him of his honours as a scientific discoverer; to

2 K 2

impeach his credibility as a witness and his integrity as a man. It is a disingenuous piece of sophistical argument, such as an unscrupulous advocate might employ to pervert the truth, misrepresent the facts, and misinterpret the language in which the facts belonging to the other side of the case are stated.

"Mr. Morse charges that the deposition of Professor Henry 'contains imputations against his (Morse's) personal character,' which it does not, and assumes it as a duty 'to expose the utter non-reliability of Professor Henry's testimony;' that testimony being supported by the most competent authorities, and by the history of scientific discovery. He asserts that he 'is not indebted to him (Professor Henry) for any discovery in science bearing on the telegraph,' he having himself acknowledged such indebtedness in the most unequivocal manner, and the fact being independently substantiated by the testimony of Sears C. Walker, and the statement of Mr. Morse's own associate, Dr. Gale. Mr. Morse further maintains, that all discoveries bearing upon the telegraph, were made, not by Professor Henry, but by others, and prior to any experiments of Professor Henry in the science of electro-magnetism; contradicting in this proposition the facts in the history of scientific discovery perfectly established and recognised throughout the scientific world.

"The essence of the charges against Professor Henry is, that he gave false testimony in his deposition in the telegraph cases, and that he has claimed the credit of discoveries in the sciences bearing upon the electro-magnetic telegraph which were made by previous investigators; in other words, that he has falsely claimed what does not belong to him, but *does* belong to others.

"Professor Henry, as a private man, might safely have allowed such charges to pass in silence. But standing in the important position which he occupies, as the chief executive officer of the Smithsonian Institution; and regarding the charges as un- doubtedly containing an impeachment of his moral character, as well as of his scientific reputation; and justly sensitive, not only for his own honour, but for the honour of the Institution, he has a right to ask this Board to consider the subject, and to make their conclusions a matter of record, which may be appealed to hereafter should any question arise with regard to his conduct in the premises.

"Your committee do not conceive it to be necessary to follow Mr. Morse through all the details of his elaborate attack. Fortunately, a plain statement of a few leading facts will be sufficient to place the essential points of the case in a clear light.

"The deposition already referred to was reluctantly given, and under the compulsion of legal process, by Professor Henry, before the Hon. George S. Hillard, United States Commissioner, on the 7th of September, 1849.

*   *   *   *   *   *

"Previous to this deposition, Mr. Morse, as appears from his own letters and statements, entertained for Professor Henry the warmest feelings of personal regard, and the highest esteem for his character as a scientific man. In a letter, dated April 24, 1839, he thanks Professor Henry for a copy of his 'valuable contributions,' and says, 'I perceive many things (in the contributions) of great interest to me in my telegraphic enterprise.' Again, in the same letter, speaking of an intended visit to the Professor at Princeton, he says : 'I should come as a learner, and could bring no 'contributions' to your stock of experiments of any value.' And still further : 'I think that you have pursued an original course of experiments, and discovered facts more immediately bearing upon my invention than any that have been published abroad.'

"It appears from Mr. Morse's own statement, that he had at least two interviews with Professor Henry—one in May 1839, when he passed the afternoon and night with him, at Princeton ; and another in February 1844—both of them for the purpose of conferring with him on subjects relating to the telegraph, and evidently with the conviction, on Mr. Morse's part, that Professor Henry's investigations were of great importance to the success of the telegraph.

"As late as 1846, after Mr. Morse had learned that some dissatisfaction existed in Professor Henry's mind in regard to the manner in which his researches in electricity had been passed over by Mr. Vail, an assistant of Mr. Morse, and the author of a history of the American magnetic telegraph, Mr. Morse, in an interview with Professor Henry, at Washington, said, according to his own account, ' Well, Professor Henry, I will take the earliest opportunity that is afforded me in anything I may publish, to have justice done to your labours ; for I do not think that justice has been done you, either in Europe or this country.'

" Again, in 1848, when Professor Walker, of the Coast Survey, made his report on the theory of Morse's electro-magnetic telegraph, in which the expression occurred, ' the helix of a soft iron magnet, prepared after the manner first pointed out by Professor Henry,' Mr. Morse, to whom the report was submitted, said : ' I have now the long-wished-for opportunity to do justice publicly to Henry's discovery bearing on the telegraph.' And in a note prepared by him, and intended to be printed with Professor Walker's report, he says : ' The allusion you make to the helix of a soft iron magnet, prepared after the manner first pointed out by Professor Henry, gives me an opportunity, of which I gladly avail myself, to say that I think that justice has not yet been done to Professor Henry, either in Europe or in this country, for the discovery of a scientific fact, which, in its bearing on telegraphs, whether of the magnetic needle or electro-magnet order, is of the greatest importance.'

" He then proceeds to give an historical synopsis, showing that, although suggestions had been made and plans devised by Soemmering, in 1811, and by Ampère, in 1820, yet that the experiments of Barlow, in 1824, had led that investigator to pronounce ' the idea of an electric telegraph to be chimerical '—an opinion that was, for the time, acquiesced in by scientific men. He shows that, in the interval between 1824 and 1829, no further suggestions were made on the subject of electric telegraphs. Then he proceeds : ' In 1830, Professor Henry, assisted by Dr. Ten Eyck, while engaged in experiments on the application of the principle of the galvanic multiplier to the development of great magnetic power in soft iron, made the important discovery that a battery of intensity overcame that resistance in a long wire which Barlow had announced as an insuperable bar to the construction of electric telegraphs. Thus was opened the way for fresh efforts in devising a practicable electric telegraph ; and Baron Schilling, in 1832, and Professors Gauss and Weber, in 1833, had ample opportunity to learn of Henry's discovery, and avail themselves of it, before they constructed their needle telegraphs.' And, while claiming for himself that he was 'the first to propose the use of the electro-magnet for telegraphic purposes, and the first to construct a telegraph on the basis of the electro-magnet,' yet he adds, '*to Professor Henry is unquestionably due the honour of the discovery of a principle which proves the practicability of exciting magnetism through a long coil, or at a distance, either to deflect a needle or to magnetise soft iron.*'

"What Mr. Morse here describes as a 'principle,' the discovery of which is unquestionably due to Professor Henry, is the law which first made it possible to work the telegraphic machine invented by Mr. Morse, and for the knowledge of which Mr. Morse was indebted to Professor Henry, as is positively asserted by his associate, Dr. Gale. This gentleman, in a letter, dated Washington, April 7, 1856, makes the following conclusive statement :—

"'Washington, D. C., April 7, 1856.

"' Sir,—In reply to your note of the 3rd instant, respecting the Morse telegraph, asking me to state definitely the condition of the invention when I first saw the apparatus in the winter of 1836, I answer : This apparatus was Morse's original instrument, usually known as the type apparatus, in which the types, set up in a composing stick, were run through a circuit breaker, and in which the battery was the cylinder battery, with a single pair of plates. This arrangement also had another peculiarity, namely, it was the electro-magnet used by Moll, and shown in drawings of the older works on that subject, having only a few turns of wire in the coil which surrounded the poles or arms of the magnet. The sparseness of the wires in the magnet coils

and the use of the single cup battery were to me, on the first
look at the instrument, obvious marks of defect, and I accordingly suggested to the Professor, without giving my reasons for
so doing, that a battery of many pairs should be substituted for
that of a single pair, and that the coil on each arm of the
magnet should be increased to many hundred turns each ;
which experiment, if I remember aright, was made on the same
day with a battery and wire on hand, furnished I believe by
myself, and it was found that while the original arrangement
would only send the electric current through a few feet of wire, say
fifteen to forty, the modified arrangement would send it through
as many hundred. Although I gave no reasons at the time to
Professor Morse for the suggestions I had proposed in modify-
ing the arrangement of the machine, I did so afterwards, and
referred in my explanations to the paper of Professor Henry, in
the nineteenth volume of the American Journal of Science, p. 400
and onward. It was to these suggestions of mine that Professor
Morse alludes in his testimony before the Circuit Court for the
eastern district of Pennsylvania, in the trial of B. B. French and
others *v.* Rogers and others.—See printed copy of Com-
plainant's Evidence, p. 168, beginning with the words ' Early
in 1836 I procured 40 feet of wire,' &c., and p. 169, where
Professor Morse alludes to myself and compensation for services
rendered to him, &c.

 " ' At the time I gave the suggestions above named, Professor Morse was not familiar with the then existing state
of the science of electro-magnetism. Had he been so, or had
he read and appreciated the paper of Henry, the suggestions
made by me would naturally have occurred to his mind as they
did to my own. But the principal part of Morse's great
invention lay in the mechanical adaptation of a power to
produce motion, and to increase or relax at will. It was only
necessary for him to know that such a power existed for him
to adapt mechanism to direct and control it.

 " ' My suggestions were made to Professor Morse from
inferences drawn by reading Professor Henry's paper above
alluded to. Professor Morse professed great surprise at the

contents of the paper when I showed it to him, but especially at
the remarks on Dr. Barlow's results respecting telegraphing,
which were new to him, and he stated at the time that he was
not aware that any one had even conceived the idea of using the
magnet for such purposes.

        " ' With sentiments of esteem, I remain, yours truly,

                                  "' L. D. Dale.
" ' Prof. Jos. Henry,
      "' Secretary of the Smithsonian Institution.'

      *   *   *   *   *   *

    "It thus appears, both from Mr. Morse's own admission
down to 1848, and from the testimony of others most familiar
with the facts, that Professor Henry discovered the law, or
' principle,' as Mr. Morse designates it, which was necessary to
make the practical working of the electro-magnetic telegraph at
considerable distances possible; that Mr. Morse was first
informed of this discovery by Dr. Gale ; that he availed himself
of it at once, and that it never occurred to Mr. Morse to deny
this fact until after 1848. He had steadily and fully acknowledged the merits and genius of Mr. Henry, as the discoverer
of facts and laws in science of the highest importance in the
success of his long-cherished invention of a magnetic telegraph.
Mr. Henry was the discoverer of a principle, Mr. Morse was
the inventor of a machine, the object of which was to record
characters at a distance, to convey intelligence, in other words,
to carry into execution the idea of an electric telegraph. But
there were obstacles in the way which he could not overcome
until he learned the discoveries of Professor Henry, and applied
them to his machine. These facts are undeniable. They
constitute a part of the history of science and invention. They
were true in 1848, they were equally true in 1855, when Professor
Morse's article was published.

      *   *   *   *   *   *

    "What changed Mr. Morse's opinion of Professor Henry, not
only as a scientific investigator, but as a man of integrity, after
the admissions of his indebtedness to his researches, and the oft-

repeated expressions of warm personal regard? It appears that Mr. Morse was involved in a number of lawsuits, growing out of contested claims to the right of using electricity for telegraphic purposes. The circumstances under which Professor Henry, as a well-known investigator in this department of physics, was summoned by one of the parties to testify have already been stated. The testimony of Mr. Henry, while supporting the claims of Mr. Morse as the inventor of an admirable invention, denied to him the additional merit of being a discoverer of new facts or laws of nature, and to this extent, perhaps, was considered unfavourable to some part of the claim of Mr. Morse to an *exclusive* right to employ the electro-magnet for telegraphic purposes. Professor Henry's deposition consists of a series of answers to verbal, as well as written, interrogatories propounded to him, which were not limited to his published writings, or the subject of electricity, but extended to investigations and discoveries in general having a bearing upon the electric telegraph. He gave his testimony at a distance from his notes and manuscripts, and it would not have been surprising if inaccuracies had occurred in some parts of his statement; but all the material points in it are sustained by independent testimony, and that portion which relates directly to Mr. Morse agrees entirely with the statement of his own assistant, Dr. Gale. Had his deposition been objectionable, it ought to have been impeached before the Court; but this was not attempted; and the following tribute to Professor Henry by the Judge, in delivering the opinion of the Supreme Court of the United States, indicates the impression made upon the Court itself by all the testimony in the case: ' It is due to him to say that no one has contributed more to enlarge the knowledge of electro-magnetism, and to lay the foundations of the great inventions of which we are speaking, than the Professor himself.'

" Professor Henry's answers to the first and second interrogatories present a condensed history of the progress of the science of electro-magnetism, as connected with telegraphic communication, embracing an account of the discoveries of Oersted, Arago, Davy, Ampère; of the investigations by Barlow and

Sturgeon; of his own researches, commenced in 1828, and continued in 1829, 1830, and subsequently. The details of his experiments and their results, though brief, are very precise. There is abundant evidence to show that Professor Henry's experiments and illustrations at Albany [in 1831], and subsequently at Princeton, proved, and were declared at the time by him to prove, that the electric telegraph was now practicable ; that the electro-magnet might be used to produce mechanical effects at a distance adequate to making signals of various kinds, such as ringing bells, which he practically illustrated. In proof of this, we quote a letter to Professor Henry, from Professor James Hall, of Albany, late president of the American Association for the Advancement of Science.

"'January 19, 1856.

"'Dear Sir,—While a student of the Rensselaer School, in Troy, New York, in August 1832, I visited Albany with a friend, having a letter of introduction to you from Professor Eaton. Our principal object was to see your electro-magnetic apparatus, of which we had heard much, and at the same time the library and collections of the Albany Institute.

"'You showed us your laboratory in a lower story or basement of the building, and in a larger room in an upper story some electric and galvanic apparatus, with various philosophical instruments. In this room, and extending around the same, was a circuit of wire stretched along the wall, and at one termination of this, in the recess of a window, a bell was fixed, while the other extremity was connected with a galvanic apparatus.

"'You showed us the manner in which the bell could be made to ring by a current of electricity, transmitted through this wire, and you remarked that this method might be adopted for giving signals, by the ringing of a bell at the distance of many miles from the point of its connection with the galvanic apparatus.

"'All the circumstances attending this visit to Albany are fresh in my recollection, and during the past years, while so much has been said respecting the invention of electric telegraphs, I have often had occasion to mention the exhibition of your electric telegraph in the Albany Academy, in 1832.

" ' If at any time or under any circumstances this statement can be of service to you in substantiating your claim to such a discovery at the period named, you are at liberty to use it in any manner you please, and I shall be ready at all times to repeat and sustain what I have here stated, with many other attendant circumstances, should they prove of any importance.

" ' I remain, very sincerely and respectfully, yours,

" ' JAMES HALL.[*]

" ' Professor Joseph Henry.'

" In his deposition, Professor Henry's statements are within what he might fairly have claimed. But he is a man of science, looking for no other reward than the consciousness of having done something for its promotion, and the reputation which the successful prosecution of scientific investigations and discoveries may justly be expected to give. In his public lectures and published writings he has often pointed out incidentally the possibility of applying the facts and laws of nature discovered by him to practical purposes ; he has freely communicated information to those who have sought it from him, among whom

---

[*] In the American telegraph suit, Smith *v.* Downing, Oliver Byrne gave evidence as follows :—

" In the year 1830, I attended the public lectures of Abraham Booth (afterward scientific reporter for *The Times* newspaper, and who became Dr. Booth), delivered in Dublin, among other subjects, on electricity and electro-magnetism. In said lectures, the said Booth, in my presence, used in combination a long circuit of insulated wire conductors, a galvanic battery, an electro-magnet with an armature and mercury cups to join and disjoin the circuit, with which he magnetised and demagnetised the iron of the electro-magnet, causing it to attract the armature when the circuit was joined, and to recede from it [allow it to fall away] when disjoined. Mr. Booth, at that time, stated to his audiences that that power could be produced and used at distant places, as signs of information ; and he repeatedly illustrated what he meant, by causing the armature to approach the magnet, and then to fall from it on the floor, stating at the same time that it made marks by so falling."—Jones' *Historical Sketch of the Electric Telegraph*, &c., New York, 1852, p. 32.

has been Mr. Morse himself, as appears by his own acknowledg-
ments. But he has never applied his scientific discoveries to
practical ends for his own pecuniary benefit. It was natural,
therefore, that he should feel a repugnance to taking any part in
the litigation between rival inventors, and it was inevitable that,
when forced to give his testimony, he should distinctly point out
what was so clear in his own mind and is so fundamental a fact
in the history of human progress, the distinctive functions of
the discoverer, and the inventor who applies discoveries to
practical purposes in the business of life.

" Mr. Henry has always done full justice to the invention of
Mr. Morse. While he could not sanction the claim of Mr.
Morse to the *exclusive* use of the electro-magnet, he has given
him full credit for the mechanical contrivances adapted to the
application of his invention. In proof of this we refer to his
deposition, and present also the following statement of Hon.
Charles Mason, Commissioner of Patents, taken from a letter
addressed by him to Professor Henry, dated March 31, 1856:—

" ' U.S. Patent Office, March 31, 1856.

" ' Sir,—Agreeably to your request, I now make the following
statement :

" ' Some two years since, when an application was made for
an extension of Professor Morse's patent, I was for some time in
doubt as to the propriety of making that extension. Under
these circumstances I consulted with several persons, and
among others with yourself, with a view particularly to ascertain
the amount of invention fairly due to Professor Morse.

" ' The result of my inquiries was such as to induce me to grant
the extension. I will further say that this was in accordance
with your express recommendation, and that I was probably
more influenced by this recommendation and the information I
obtained from you, than by any other circumstance, in coming
to that conclusion.

" ' I am, Sir, yours very respectfully,

" ' CHARLES MASON.

" ' Professor J. Henry.'

" To sum up the result of the preceding investigation in a few words.

" We have shown that Mr. Morse himself has acknowledged the value of the discoveries of Professor Henry to his electric telegraph ; that his associate and scientific assistant, Dr. Gale, has distinctly affirmed that these discoveries were applied to his telegraph, and that previous to such application it was impossible for Mr. Morse to operate his instrument at a distance ; that Professor Henry's experiments were witnessed by Professor Hall and others in 1832, and that these experiments showed the possibility of transmitting to a distance a force capable of producing mechanical effects adequate to making telegraphic signals ; that Mr. Henry's deposition of 1849, which evidently furnished the motive for Mr. Morse's attack upon him, is strictly correct in all the historical details, and that, so far as it relates to Mr. Henry's own claim as a discoverer, is within what he might have claimed with entire justice ; that he gave the deposition reluctantly, and in no spirit of hostility to Mr. Morse ; that on that and other occasions he fully admitted the merit of Mr. Morse as an inventor ; and that Mr. Morse's patent was extended through the influence of the favourable opinion expressed by Professor Henry.

" Your committee come unhesitatingly to the conclusion that Mr. Morse has failed to substantiate any one of the charges he has made against Professor Henry, although the burden of proof lay upon him ; and that all the evidence, including the unbiassed admissions of Mr. Morse himself, is on the other side. Mr. Morse's charges not only remain unproved, but they are positively disproved."

Extract from Professor Henry's evidence in the Telegraph suit of Morse *v.* O'Reilly, Boston, September, 1849 :—

" In February 1837, I went to Europe ; and early in April of that year Professor Wheatstone, of London, in the course of a visit to him in King's College, London, with Professor Bache, now of the Coast Survey, explained to us his plans of an electro-magnetic telegraph; and, among other things, exhibited to us his method of bringing into action a second galvanic circuit. This consisted in closing the second circuit by the deflection of a needle, so placed that the two ends of the open circuit projecting upwards would be united by the contact of the end of the needle when deflected, and of opening or breaking the circuit so closed by opening the first circuit and thus interrupting the current, when the needle would resume its ordinary position under the influence of the magnetism of the earth. I informed him that I had devised another method of producing effects somewhat similar. This consisted in opening the circuit of my large quantity magnet at Princeton, when loaded with many hundred pounds weight, by attracting upward a small piece of movable wire, with a small intensity magnet, connected with a long wire circuit. When the circuit of the large battery was thus broken by an action from a distance, the weights would fall, and great mechanical effect could thus be produced, such as the ringing of church bells at a distance of a hundred miles or more, an illustration which I had previously given to my class at Princeton. My impression is strong, that I had explained the precise process to my class before I went to Europe, but testifying now without the opportunity of reference to my notes, I cannot speak positively. I am, however, certain of having mentioned in my lectures every year previously, at Princeton, the project of ringing bells at a distance, by the use of the electro-magnet, and of having frequently illustrated the principle of transmitting power to a distance to my class, by causing in some cases a thousand pounds to fall on the floor, by merely lifting a piece of wire from two cups of mercury closing the circuit.

" The object of Professor Wheatstone, as I understood it, in bringing into action a second circuit, was to provide a remedy for the diminution of force in a long circuit. My object, in the process described by me, was to bring into operation a large quantity magnet, connected with a quantity battery in a local circuit, by means of a small intensity magnet, and an intensity battery at a distance."

Up to the date of Henry's visit to Wheatstone in February 1837, the latter did not know how to construct an "intensity" electro-magnet. It will be remembered by all readers of Mr. Latimer Clark's interesting biography of Sir W. F. Cooke,* that it was a difficulty of this kind that first brought Cooke and Wheatstone together. Cooke had contrived a telegraph and alarum, to be operated by clockwork mechanism, the detents of which were to be released, as occasion required, by electro-magnets. The apparatus worked well enough on short circuit, but when he came to try it through such lengths as a mile of wire, the electro-magnets were so enfeebled that they could not withdraw the detents. In this difficulty Cooke sought the advice of Roget, Faraday, Clarke, and Wheatstone.

The latter's opinion was very unfavourable. "Relying," he says, "on my former experience, I at once told Mr. Cooke that his plan would not and could not act as a telegraph, because sufficient attractive power could not be imparted to an electro-magnet interposed in a long circuit; and to convince him of the truth of this assertion, I invited him to King's College to see the repetition of the experiments on which my conclusion was founded. He came, and after seeing a variety of voltaic magnets, which even with powerful batteries exhibited only slight adhesive attraction, he expressed his disappointment."

* *Journal of the Soc. of Tel. Engs.*, vol. viii. p. 374.

And again :—" When I endeavoured to ascertain how a bell might be more efficiently rung, the attractive power obtained by temporarily magnetising soft iron first suggested itself to me. The experiments I made with the long circuit at King's College, however, led me to conclude that the attraction of a piece of soft iron by an electro-magnet could not be made available in circuits of very great length, and, therefore, I had no hopes of being able to discharge an alarum by this means." *

In reference to these experiments, Cooke wrote on March 4, 1837 :—

" Mr. Wheatstone called on Monday evening, and postponed our meeting at King's College till Wednesday. The result was nearly what I had anticipated, the electric fluid losing its magnetising quality in a lengthened course. An idea, however, suggested itself to Mr. Wheatstone, which I prepared to experiment on last Saturday, but again failed in producing any effect. I gave up my object for the time, and proposed explaining the nature of my discomfited instrument to the Professor. He, in return, imparted his to me. He handsomely acknowledged the advantage of mine, had it acted ; his are ingenious, but not practicable. His favourite is the same as mine, made at Heidelberg, and now in one of my boxes at Berne, requiring six wires, and a very delicate arrangement. He proposed that we should meet again next Saturday, and make further experiments. For a time I felt relieved at having decided the fate of my own plan, but my mind returned to the subject with more perseverance than ever, and before three o'clock the next morning I had re-arranged my unfortunate machine under a new shape.

---

* *The Electric Telegraph, was it invented by Professor Wheatstone?* by W. F. Cooke, London, 1856–57. See part ii. pp. 87 and 93.

2 L

" I now use a true [permanent] magnet of considerable power, with the poles about four inches apart, and a slender armature, four and a half inches long, covered with several hundred coils of insulated copper wire, and suspended like a mariner's compass in the plane of the poles of the magnet. Whenever the galvanic circuit is completed, the ends of the armature are respectively attracted by the poles of the magnet with a force sufficient to overcome the opposition of a feeble spring, the movement not exceeding one-twentieth of an inch. A lever forming part of the detent of my fan is moved by a projecting pin, and liberates the clockwork. I have seen an arrangement of this sort in the Adelaide Gallery, but used there merely as a toy."*

Even this arrangement, which was obviously capable of good results, and in which our practical readers will recognise the germ of the Brown and Allan relay, was not approved by Professor Wheatstone. Cooke writes :—" On many occasions during the months of March and April 1837, we tried experiments together upon the electro-magnet ; our object being to make it act efficiently at long distances in its office of removing the detent. The result of our experiments confirmed my apprehension that I was still without the power of exciting magnetism at long distances. * * * In this difficulty we adopted the expedient of a secondary circuit, which was used *for some time* in connection with my alarum."†

"From all this," says Mr. Latimer Clark, "it is evident that Professor Wheatstone at this time [April 1837] did not appreciate the importance of using fine

* *Jour. Soc. Tel. Engs.*, vol. viii. p. 378.
† The italics are our own. See *The Electric Telegraph, was it invented by Professor Wheatstone?* part ii. p. 27.

wire, and that he had not studied Professor Henry's paper on electro-magnets, in the twentieth volume of *Silliman's Journal*, for January 1831, in which he so clearly shows the advantage of using long fine wires [in the coils] and numerous elements for long circuits." *

It is equally evident that it was not until *after* the interview with Henry that Wheatstone recognised the applicability of Ohm's laws to telegraphic circuits, the study of which would, likewise, have enabled him to ascertain the best proportions between the length, thickness, &c., of the coils, as compared with the other resistances in the circuit, and to determine the number and size of the elements of the battery necessary to produce a maximum effect.†

* *Jour. Soc. Tel. Engs.*, vol. viii. p. 381.

† As soon as Professor Wheatstone had thus learnt how to construct an "intensity" electro-magnet, the use of the secondary or relay circuit referred to on pp. 511 and 514 was abandoned. "These secondary circuits," says Wheatstone, "have lost nearly all their importance, and are scarcely worth contending about, since my discovery that electro-magnets may be so constructed as to produce the required effects by means of the direct current, even in very long circuits. Previously, however, to this discovery they appeared to be of great importance to both of us—to me, as the means of ringing the alarum connected with my telegraph ; to Mr. Cooke, as the only means of enabling him to work his instrument."—*The Electric Telegraph, was it invented by Professor Wheatstone ?* part ii. pp. 95-6.

As these words were written in the winter of 1840-41, they must be taken to represent Professor Wheatstone's estimate of the relay at that date. It was not thus that Edward Davy appraised it. From *March 1837 onward* he steadily regarded it to be what it is—one of the key-stones of electric telegraphy. See pp. 359 and 366, *ante.*

APPENDIX B.

———◦◦◦———

Short Memoir of Edward Davy, M.R.C.S., M.S.A. By HENRY DAVY, M.D. (Lond.), M.R.C.P., Physician to the Devon and Exeter Hospital. *Reprinted from " The Electrician,"* No. 11, vol. xi., 1883.

The following short biographical sketch of my uncle has been written by me at the request of Mr. Fahie in order to complete the history of the MSS. which he has lately published :—

Edward Davy's family originally settled near the coast in Dorsetshire, where, about the year 1616, they were living on their own estates. Unfortunately, they took an active part against the king in the Monmouth Rebellion, and, when Judge Jefferys commenced the " Bloody Assize," they found it convenient to migrate into Devonshire, where they commenced life afresh, mostly pursuing the occupation of farmers. His grandfather was a farmer, partly owning, and partly renting, his estates in the neighbourhood of Exeter ; while his father was Thomas Davy, who resided at Ottery St. Mary, and had an extensive medical practice in Ottery and the neighbourhood. Thomas Davy was educated at Ottery and Guy's Hospital, at the latter, being a pupil of Sir Astley Cooper, but from the time he left London it is much to be questioned whether he was ever out of Devonshire more than five or six times in his life.

Edward Davy was born on June 6, 1806, and was educated at a school kept by his maternal uncle, Mr. Boutflower, in Tower Street, London. Subsequently he was apprenticed to Mr. Wheeler, house surgeon at St. Bartholomew's Hospital, and, about the year 1828, he became a " Member of the Royal

College of Surgeons," and soon after a " Member of the Society of Apothecaries." Shortly after this he bought a business at 390, Strand. I have always heard it stated that some eight or nine hundred pounds were advanced by his father to buy a medical practice, but that he was taken in, and found that the so-called practice was that of a dispensing chemist. However this may be, he soon began to trade as an operative chemist, under the name of Davy and Co., and in 1836 he published a small work, termed " Experimental Guide to Chemistry," at the end of which is a catalogue of the instruments, &c., supplied by his firm. This guide book might even now serve as a useful text-book for a beginner in experimental work, whilst in the catalogue at the end he mentions several of his original modifications of instruments, such as " Davy's Blow-pipe," " Davy's Improved Mercurial Trough," &c., proving how completely he had given himself up to his favourite pursuit.

About this time, 1835, he invented and patented a cement for mending broken china and glass, which for many years brought him in a small income, and was well known as " Davy's Diamond Cement," and it was during these years that he first commenced to experiment on the Electric Telegraph ; but this part of his history up to the time of his leaving England has already been told by Mr. Fahie. One question which will be asked by all the readers of Mr. Fahie's narrative is : Why did Edward Davy fail in his attempt to get his system of telegraphy adopted? It seems certain that at the time of his leaving England his system was in a more perfect state than that of Cooke and Wheatstone. It is seen, too, that Edward Davy was and had been negotiating with some of the leading engineers, railway companies, and railway directors, and that many of them had promised to adopt his system. Why, then, did he fail? Chiefly because he left England just at the wrong moment. Into the reasons of his leaving there is no occasion to enter. Suffice it to say he had been contemplating doing so all through the end of 1837 and 1838, and that his reasons were entirely of a private kind. But in leaving England when he did, he struck the death-blow to all his hopes ; and had he remained, his system, as Mr. Fahie says, would probably have been adopted. It is important to note that he himself did not realise the fact that his leaving England would ruin his invention, and that had he realised his true position he would probably have stayed on until his negotiations with the railway companies were concluded. In a letter to his father early in 1839, in which he announces his final decision to leave, he says :—

"I have perfected, as far as I can, secured, and made public the telegraph. What remains, *i. e.*, to make the bargain with the companies when they are ready and willing, can be managed by an agent or attorney as well as if I were present."

How entirely wrong this opinion was subsequent events soon proved, for the directors, having no one to deal with who thoroughly understood his instruments, adopted those of his rivals, Cooke and Wheatstone.

But other causes greatly contributed to his failure. The first is brought out in the sketch I have given of his family. His father was, for his day, a well-qualified medical man, but he was quite destitute of any scientific training, whilst his close residence in Devonshire had prevented his seeing the direction in which the thought of the day was moving. To most people in 1837 the idea even of a railway was new ; and when Edward Davy talked of " *sending messages along a wire for hundreds of miles*," when he predicted the use of "marine cables," hinted at the " *telephone*," and prophesied that the " *Government would adopt the telegraph as part of their postal system*," it is excusable that his father should regard him as a visionary, and should tell him that his plans were all " *moonshine*." When, too, he found that he had to pay for this " *moonshine*" by constant remittances, one can easily forgive his anxiety that his son should go back to the regular practice of his profession ; especially when to his old-fashioned ideas it was almost a disgrace for a medical man to have a son an operative chemist. Not only did his father discourage Edward Davy in his pursuits, but he took no pains to bring his invention to the notice of many influential friends, who might have helped him in his endeavours to make it known to the public.

From Mr. Fahie's narrative, it seems evident that had any well-known firm taken up the invention, and pressed its advantages on the public, the railway directors would have adopted it. Thomas Davy's brother was a merchant of wide local reputation, who at one time had been one of the Government's largest contractors for building wooden frigates, while his nephew was a rising partner in one of the best-known and largest mercantile houses of the day (Anthony Gibbs and Co.), and yet neither of these was in the least made acquainted with the patent of the electric telegraph. This was the more deplorable, since Edward Davy was evidently unfortunate in his choice of business men. Repeatedly in his letters to his father he states that neither Mr. P—— nor Captain B—— nor Mr. B—— was assisting him as he should wish, and had he been assisted by any energetic man of business it is probable that Mr. Fahie's history would have ended very differently. Like most other geniuses, Edward Davy had no marked business capacity; he could invent an original machine, he was not able to hold his own with far-seeing men of the world. He had no friend to advise him, and he was unfortunate in the agents whose assistance he obtained. Had he offered 50 per cent. of his eventual gains he would have attracted the service of men of acknowledged position. As it was, his offer of 10 per cent. did not attract these, and even when he offered 25 per cent. to Mr. P——, the latter did not keep his part of the undertaking, so that the compact broke through. After careful perusal of the MSS., I am convinced that Edward Davy did everything in his own power to make his invention succeed; he failed because he had little business capacity, and his father's line of action prevented his getting even friendly advice from quarters where it might have been obtained. But the questions will be asked, Why has Edward Davy allowed his claims as a pioneer in telegraphy to be so completely ignored? And why have not his family published these MSS. before?

The answer to the latter question is simple. I very much doubt if the family knew anything about these MSS. They were collected and labelled by another uncle of mine long since dead, and it was only after the death of my father last year that they fell into my hands, after having narrowly escaped being burned as rubbish. They came into my possession last March [1883], and when by chance Mr. Fahie in April 1883 wrote to ask me for information as to my uncle, I readily placed them at his disposal, only stipulating that, as he was quite a stranger to me, he should publish nothing without my permission. Until I saw his annotations I was not aware of the extent of Edward Davy's inventions, and I am quite sure my father had no idea as to the value of these MSS., for his training as a solicitor had not taught him any science.

I do not know why Edward Davy himself allowed his claims to be ignored. Probably he did not know that these MSS. had been preserved, and without them he would have no proof with which to support his claims.* After leaving England in 1839 he threw all his energies into the colonial life he had adopted. His letters to his father, mother, and brothers are full of references to this new mode of life. He busied himself in acclimatising trees, grasses, &c., the seeds of which he obtained from England. His leisure he filled up with writing newspaper articles on hygiene and other subjects. He also pursued his favourite subject, chemistry, and patented a " plan for saving fuel during the process of smelting ores," which he had invented in 1838. For many years he was assayer to the Mint at Melbourne, while for

---

* In a letter, dated October 10, 1883, and received since the above was written, Mr. Edward Davy himself says, in answer to our inquiry, " How is it that, being alive, you have never asserted your claims?"— " When the Solicitor-General passed Cooke and Wheatstone's first patent in the face of my opposition and of the grounds thereof, how could I say that he had not done rightly? Again, when my father sold the patent for so insufficient a sum, I looked upon it that all hope of pecuniary benefit to myself was gone. I might still have fought for the credit of the invention, but I *was not aware* that the documents, which you have unearthed, had been so carefully preserved ; besides, being at such a distance, I should have had to carry on a controversy at a great disadvantage, with the risk of being considered an impostor. I had friends in England ; but none able, if ever so willing, to defend the claim. The other party was in the midst of friends, and in possession of the field."

the past twenty-five years he has carried on a medical practice, latterly in partnership with one of his sons. In this busy colonial life he has, no doubt, found more happiness than in brooding over his disappointments. Only in one letter to his family do I find any reference to the telegraph. Writing to one of his sisters in 1841, after saying that a storm is preventing him from going to sleep, he adds :—" I shall therefore enter into some conversation with you, although, from there being no electro-telegraph, it may be five months ere my voice reaches you." Whatever be the cause, it is certain that he has never once referred to this period of his life, and, as Mr. Fahie says, he will be quite as surprised as any one at finding that his labours of forty-five years ago have now been made public.

Mr. Fahie concludes the narrative of Edward Davy's inventions and negotiations by terming it a magnificent failure, and I think no one will deny this who has read how nearly he obtained complete success. It was, however, only a failure as far as he, himself, was concerned. His labours were, in reality, most useful. His experiments in Regent's Park, his exhibition of his instrument in Exeter Hall, brought electric telegraphy before the public in a way which was done by no other person. His correspondence with, and the constant advocacy of his invention to, such men as Brunel, Fox, and Easthope forced the electric telegraph on their notice with a double force, and, no doubt, did much to cause its early adoption. I do not here enter into the question as to how far his ideas were adopted by others. It is certain that, with a strange lack of business-like foresight, he exhibited his machine before it was patented, and that his exhibition was visited by Cooke, Wheatstone, &c. Probably his work has assisted many of his successors in working out improvements ; but quite apart from this his labours were useful, and are well worthy of recognition. It is interesting to note that he spent some thousands in experimenting and making his experiments public. Mr. Fahie has shown me a letter printed in *The Electrician* (October 11, 1879) from the late Dr. Cornish, vicar of Ottery St. Mary. This letter says Edward Davy's family spent thirty thousand pounds on the electric telegraph. I do not know whether there is a misprint here, but if the thirty be divided by ten I think the resulting three thousand would be more near the mark. It is also interesting to note that Edward Davy is still alive, and well appreciated by his fellow-townsmen in his colonial home. Almost the last paper I got from him contained an account of an entertainment given in his honour, at which, he having refused any more substantial acknowledgment, he was presented with an illuminated address in recognition of his having been for many years a magistrate and on three occasions mayor of his town, and for having for twenty-five years gratuitously held the office of medical officer of health to the district.

It is to me a satisfaction that these MSS. have come to light during his lifetime. He is now seventy-seven years of age, and for the past forty-five years his claims have been quite ignored. He, I am sure, would be the last to claim any position which was undeserved, but it cannot but be a pleasure to him to see the real value of his work recognised, and his name rescued from an unmerited oblivion, and placed in its proper position as one of the very first pioneers of electric telegraphy.

A writer in *The Exeter and Plymouth Gazette*, for September 25, 1883, supplies the following additional information :—

"My attention having been called to the series in *The Electrician* by a recent article from the pen of the ubiquitous Harry Hems, I made inquiries of an old and respected inhabitant of Ottery (Mr. Jeffrey, solicitor), who knew the Davy family well. He says that at the end of the last century Thomas Davy, surgeon, commenced practice at Ottery St. Mary. Soon after, he married the daughter of a literary gentleman of Exeter, named Boutflower. Trade in the town of Ottery at that time was brisk. In the year 1782 a large woollen manufactory was completed, at a ruinous cost to Sir George Younge, Bart., the then lord of the manor. The manufactory was conducted by Messrs. Ball and Fowell, a daughter of the former of whom is still alive. It was always said that Mr. Davy was of the same family as Sir Humphry Davy, who was born at Penzance in the year 1778. He also was intended for the medical profession, and served under an apothecary in order to study chemistry—a circumstance that resulted in his invention of the Davy safety lamp, the metallic bases of the alkalies and earths, and of the principles of electro-chemistry. Mr. Thomas Davy had possessions in the West Indies, and held them until his death, in the year 1852. At the end of the last century a large fire occurred in Mill Street, Ottery, in a butcher's shop occupied by a person named James. Mr. Thomas Davy became the purchaser of the ruins, and erected on the site a mansion, where he resided until his death. He was blessed with four sons and two daughters. Edward Davy (the inventor of telegraphy) was the eldest son. At an early age he showed precocity, and having his father's surgery at command, he for a time quite gave himself up to the study of chemistry, particularly electro-chemistry. He was convinced that the time would arrive when communication would be made by wire round the world. Later on in life he left Ottery for London, where he married the daughter of a London magistrate named Minshell [see p. 404]. He then again followed his old hobby, and expressed himself confident of discovering the secret of communication by telegraphy. At last he accomplished his end, and took a room in the Exeter Hall, Strand, where his instrument, not unlike a piano, was exhibited. He cut letters from a printed bill, and gummed them on to the keys. Wires were placed round the room, which he said were one mile long. Attached were a number of small jets like lamps, placed at intervals, which he instantaneously lighted through the wire. He said to his audience, ' The most extraordinary part of it is, that if I touch one of these letters it moves a corresponding letter instantaneously at the end of the wire.' A person asked if distance made any difference, and Mr. Davy replied, ' No ; if I had these wires 3000 miles long, or even round the world, one could not discover the difference of time.' Strange to say, he ultimately sold his discovery and right for 600*l.*, and left for Australia. As a young man he took a great interest also in geology, and frequently started off from Ottery to examine the hills in the district. On reaching Melbourne he pursued his study of geology, and in due time appeared an article from his pen in the Melbourne *Argus* (which by-the-by was edited by a gentleman who had been educated at the King's Grammar School, when kept by the Vicar of Ottery, Dr. Cornish), showing that he was of opinion that the country, or certain parts of it, was auriferous. This article Mr. Thomas Davy retained up to the time of his death, and often read it to his friends. Mr. Thomas Davy had an extensive practice at Ottery. He had more than an ordinary share of common sense, and was amiable and kind, particularly to the poor. For many years he studied the art of producing fat stock, and was the only person who then grew mangel-wurzel in the parish, which was called at that period by the farmers 'A gentleman's crop.' He was brother to the late—I had almost said centenarian of Topsham—James Davy, whose remarkable career is worthy of notice, and who possessed a successful business for over sixty years in a ship-building, coal, and lime trade. Although deprived of sight for a great number of years, his energy of mind and body was not impaired. He was benevolent and much respected. The only representative of Mr. Thomas Davy left in the county is Dr. Davy, of Exeter."

We make the following extract from the Melbourne *Argus*, for November 16, 1883 :—

Royal Society of Victoria.

The ordinary monthly meeting of the Royal Society was held on Thursday evening ; Mr. R. L. J. Ellery (president) in the chair. Mr. Ellery read the following paper describing an interesting fact in connection with the early history of the electric telegraph :—

" It is no new thing to say that the one who by intellectual process, or rational experiment, makes a discovery seldom reaps the benefit, either as regards reputation, or more substantial results. The man of science, or the patient investigator, is nowhere in the race, as compared with the man of business, and so it often, almost always, happens that the discoverer is forgotten, while those who, ghoul-like, turn his brains to account are the only ones who reap the reward and are remembered. This is because men like Faraday, and many more, are not business men, their life is spent in inquiring of nature's forces and nature's laws, and giving the results for the benefit of mankind, and not in learning and following the more popular ways of money-making. The instance I am about to refer to is a case in point. Let us think for a moment what a mess we should be in if we were suddenly deprived of the electric telegraph, or electricity as a means of communication at a distance, and we may perhaps form some sort of an idea of what we owe to those early workers who laid the foundation-stones of this great and universal benefit. Nevertheless one, and, as it now seems likely, the first who by his discoveries made the electric telegraph a fact has been hidden among us for over thirty years, scarcely known except as a country surgeon, and certainly never till now recognised as one to whom the gratitude, if nothing else, of the whole civilised world belongs for his investigations into the applications of electricity and magnetism, which are now considered by competent authorities to have constituted those first important steps which rendered all subsequent details of the electric telegraph an easy task. From some articles in *The Electrician*, it is pretty clear that Dr. Edward Davy (who was known by some of us thirty years ago, as superintendent of the Assay Office in Melbourne, was one of the founders of the Philosophical Institute, the parent of this Royal Society, and now resides at Malmesbury, following his profession as a medical man), must be regarded in virtue of his most important discoveries, exhibitions of working models at Exeter Hall, and his invitation to carry out his electric telegraph on the Great Western line in England, the real first inventor of the electric telegraph. The history in brief seems to be this: As early as 1836 Davy conceived the possibility of an electric telegraph, and appears to have had an excellent knowledge and thorough grasp of the properties of electricity.

He had been educated for the medical profession, and took his diploma at the Royal College of Surgeons in 1828. He then seems to have taken up the business of an operative or analytical chemist, and we have heard of several chemical instruments invented or improved by him. During this time (about 1835) he seems to have made some investigations into electricity, and in 1836 the possibility of using the electric current for telegraphic purposes suggested itself to him, and he matured a method, which he patented in 1838, as already stated. Cooke and Wheatstone patented in 1837, and afterwards actually carried their needle telegraph into operation, and obtained its adoption on the railway lines of Great Britain. Davy, who had matured his plan and exhibited working models before this time, contested unsuccessfully the granting of the patent. Perhaps from the want of means, or perhaps for lack of the commercial afflatus, so often absent in scientific men, yet so essential to the substantial success of a discovery or invention, Davy failed to carry his telegraph into practical use, and eventually we hear of his having come to Australia in 1839 ; his connection with the early discovery of the electric telegraph was forgotten, nor does he ever seem to have in any way resuscitated the matter until his work is referred to in Mr. Fahie's papers on the early history of the electric telegraph, published lately in *The Electrician*. Although Cooke and Wheatstone succeeded, and Dr. Davy did not, this does not alter the fact that to the latter we are decidedly indebted for discoveries which eventually resulted in the perfection of both what are known as the needle and Morse systems. To those interested in this subject, I may state that copies of Dr. Davy's work and inventions can be seen in *The Electrician*, vol. xi., Nos. 8, 9, 10, and 11 of this year. There is, however, one paragraph taken from his letters and communications which is interesting and prophetic ; it is in a postscript to a letter to his father, dated July 1838. Speaking of a suggestion that had been made, to the effect that Government would scarcely allow such a powerful instrument to be in the hands of individuals, he says —

" ' I know very well the French Government would not permit it except in their own hands, but though I think our Government ought, and perhaps will eventually take it upon themselves as a branch of the Post Office ; yet I can scarcely imagine that there would be such absurd illiberality as to prohibit or appropriate it without compensation.'

" Again in 1838 Davy wrote :—

" ' I cannot, however, avoid looking at the system of electrical communication between distant places, in a more enlarged way, as a system which will one of these days become an especial element in social intercourse. As railways are already doing, it will tend still further to bring remote places, in effect, near together. If the one may be said to diminish distance, the other may be said to annihilate it altogether, being instantaneous.'

" There is a ring of prescience in these words, uttered as they were forty-five years ago, before a mile of telegraph wire had been erected except the single mile he constructed himself for experimental purposes in Regent's Park ; and although, so far as is known, the idea of submarine communication was at that time scarcely dreamt of, Davy, in his ' Outline description of his improved electrical telegraph,' refers to and describes an insulated conductor or 'cable' for such a purpose. This Society will, I am sure, feel proud to know that it may rank among its founders the name of Edward Davy, the almost forgotten pioneer and inventor of the electric telegraph, and at the eleventh hour to do what honour to him it may be within its province and power to do."

Mr. Ellery intimated, amidst applause, his intention to move at next meeting that Dr. Davy be elected a life honorary member of the Society.

Professor Kernot thought a much higher honour was due to Dr. Davy, if he was really the actual first discoverer of the electric telegraph.

Mr. C. R. Blackett suggested the appointment of a sub-committee to look into the matter, and report as to the best means of recognising the scientific services of Dr. Davy.

The suggestion was adopted, the sub-committee appointed being Messrs. S. W. M'Gowan, J. Cosmo Newbery, C. R. Blackett, Professor Kernot, and Dr. Wilkie.

The following paragraph appeared in *The Electrician*, for January 12, 1884 :—

"*Edward Davy.*—The graphic and interesting accounts of Edward Davy's telegraphic inventions, which have appeared in these pages from the pen of Mr. J. J. Fahie, have aroused considerable interest in the colony of Victoria, Dr. Davy's adopted home. We notice that Mr. Ellery recently read a paper before the Royal Society of Victoria in which he concurred with Mr. Fahie in thinking that, if Davy had stayed in the Strand instead of emigrating to Australia in disgust, he would have succeeded in establishing his claim to be the first inventor of the electric telegraph. At the conclusion of his paper, Mr. Ellery proposed that the Society should do him such honour as lay in their power by electing him a life honorary member. The meeting, however, were of opinion that Dr. Davy deserved some still better recognition of his so long neglected genius, and appointed a sub-committee to report upon the best means of doing him public honour. We hope that our colonial kinsfolk will not be allowed to entirely show us the way in this matter. Davy lived in London, and exhibited his apparatus here long before he went to live in Australia. Who can tell how much modern telegraphy is indebted to the inventive genius of the chemist who resided at 390, Strand? Had he stayed here he would no doubt have exercised a very powerful influence on the history of the telegraph, for, as Mr. Fahie has so ably shown, he was far in advance of his contemporaries both in theory and practice. Circumstances caused him to leave us, and he was for a time forgotten. Do not let us forget how much he is deserving of honour at our hands, however—not a mere empty, formal, and official recognition of his services, but something substantial, and that may prove of benefit to the man himself in his declining years. Why not place his name on the Civil List, as has been already suggested? There are many far less deserving than Dr. Davy to be counted in this list. We feel sure that the many telegraph engineers and electricians of the present day who know how to appreciate Davy's genius will not allow their colonial brethren to out-do them in honouring him, nor let the matter sleep for want of a little energy. It is curious to notice that Dr. Davy was not altogether fortunate in his Australian career, and that misfortune came upon him through no fault of his. It appears that Dr. Davy was Assay Master at the Melbourne Mint, from 1853 to 1855, and enjoyed a salary of 1500*l.* This was when Mr. La Trobe was governor. Dr. Davy had been specially invited to accept this post whilst occupying another similar, but less lucrative position in Adelaide. A succeeding governor, however, Sir Charles Hotham, abolished the office, giving Dr. Davy, by way of compensation, six months' salary. Here, then, was Davy once more turned adrift by Fortune's wheel. At the time that he was at Melbourne, Mr. Childers, the present Chancellor of the Exchequer, was Auditor-General there. After this Davy tried his hand at farming, with but indifferent success, and finally settled down in Malmesbury to practise his profession of surgeon. Here he gradually rose high in popular esteem, has been several times mayor, and has been prime mover in several local public works of great benefit to the town. Dr. Davy is now in his seventy-eighth year, and is not so well able to carry on his profession as in his younger days, and from various causes his practice is not so good as it used to be, consequently a grant from the Civil List would be as good a mode of honouring him as any, and we think it ought to be made."

BIBLIOGRAPHY.

———✕———

[See also Catalogue of Works in which the Sympathetic Telegraph is referred to, p. 20.]

A.

ABOUT, EDMOND, *Le Nez d'un Notaire*, 20

Addison, *Spectator*, No 241 (1711), *Guardian*, No. 119 (1713), 9

Akenside, *The Pleasures of Imagination* (1744), 9

Aldini, *Essai théorique et expérimental sur le Galvanisme* (Paris, 1804), 264, 343

Alexander, W., *Plan and Description of the Original Electro-Magnetic Telegraph*, 448

Alibert, *Éloges historiques de Galvani* (Paris, 1802), 183

American Polytechnic Review (1881), 263

Annales de Chim. et de Phys., 194, &c.

Annales Télégraphiques (1859), 109

Anon [Hamilton Walker ?], *Notes to Assist the Memory* (Lond., 1835), 306

Aristotle, *History of Animals* (ix. 37), 27, 170

Augustine, St., *De Civitate Dei*, 413

B.

Bacon, Lord, *Advancement and Proficience of Learning*, 311

Bakewell, *Manual of Electricity* (1857), 274

Beraud, *Dissertation sur le rapport qui se trouve entre la cause des effets de l'Aiman, et celles des phénomènes de l'Électricité* (Bordeaux, 1748), 251

Berio, *Ephemerides of the Lecture Society, Genoa* (1872), 122

Birch, *History of the Royal Society*, 36

Blasius de Vigenere, *Les cinq premiers livres de Tite Live* (1576), 5

Bockmann, *Versuch uber Telegraphie und Telegraphen* (1794), 98

Bologna Academy Transactions, for Galvani's papers, 180

Bostock's *History of Galvanism*, 214, 270, 297

Boyle, Robert, *Experiments and Notes about the Mechanical Origine or Production of Electricity* (1675), 32

Brewster, Sir D., *Letters on Natural Magic*, 33

—— *Life of Sir Isaac Newton*, 36

—— *Edinburgh Encyclopædia*, 48

Browne, Sir Thos., *Pseudodoxia Epidemica* (London, 1646), 13

Bryant, *Transactions American Soc.*, vol. ii., 172

K.

Kirby and Spence, *Introduction to Entomology*, 169

Kircher, A., *Magnes, sive de arte magnetica* (1641), 18

Komaroff's *La Presse Scientifique des Deux Mondes*, 317

L

La Lumière Électrique (Mar. 1883, Guerout), 119, 162

Lardner, *Manual of Electricity*, 27, &c.

La Rive, De, *Treatise on Electricity* (1853-58), 345

Larrey, Baron, *Clinique Chirurgicale* and *Bulletin de la Société Médicale d'Émulation*, 239

Le Correspondant (1867), 82

Lehot, *Observations sur le Galvanisme et le Magnétisme* (Paris, 1806), 256

Le Journal des Sçavans, 89

Le Mercure de France (1782), 87

Leonardus, Camillus, *Speculum Lapidum* (1502), 4

Les Mondes (1867), 82

Linguet's *Mémoires sur la Bastille*, 88

Livy, Blasius de Vigenere, 5

M.

Madrid, Gaceta de (1796), 107

Magazine of Popular Science, 328

Magrini, *Telegrafo Elettro - Magnetico* (Venezia, 1838), 167, 481

Maimbourg, *Hist. de l'Arianisme* (1686), 2

Marana, G. P. (or the Turkish Spy), Letters of (Paris, 1639), 12

Martyn and Chambers, *The Phil. Hist and Mems. of the Roy. Acad. of Sciences at Paris* (1742), 176

Mechanics' Magazine, 148, &c.

Melbourne Argus, for E Davy, 524

Memorials Scientific and Literary of Andrew Crosse, the Electrician, 137

Metra, *Correspondance Secrète* (1788), 87

"Misographos," *The Student, or the Oxford and Cambridge Miscellany* (1750), 9

Moigno, *Traité de Télégraphie Électrique* (Paris, 1852), 91, 101, 150

Monthly Magazine, The, 107, 268

Morning Herald, for 1837, 152

N.

Nicholson's *Journal of Natural Philosophy*, 194, &c.

Noad's *Manual of Electricity*, 256

Notes and Queries, 32, 74, 306

O.

Oersted, *Recherches sur l'Identité des forces chimiques et électriques* (Paris, 1813), 271

O'Shaughnessy's *Electric Telegraph in British India*, 348

P.

Paris, *Life of Sir H. Davy*, 213

Parthenius, J. M, *Electricorum* (1767), 77

Philosophical Magazine, 211, &c.

Philosophical Transactions, 33, &c.

Pliny, *Nat. Hist.* xxxii. 2, 170; xxxvii. 3, 27

Plutarch, *Life of Timoleon*, 28

INDEX.

———•◦•———

LONDON: PRINTED BY WILLIAM CLOWES AND SONS, LIMITED, STAMFORD STREET
AND CHARING CROSS

LIST OF BOOKS

RELATING TO

ELECTRICITY, TELEGRAPHY,
THE ELECTRIC LIGHT, &c.

PUBLISHED AND SOLD BY

E. & F. N. SPON, 16, CHARING CROSS, LONDON.

18mo, boards, 2s.

A Handbook of the Electro-Magnetic Telegraph. By A. E. LORING, a Practical Telegrapher.

With numerous Illustrations, 8vo, cloth, 21s

Electricity and the Electric Telegraph. By G. B. PRESCOTT.

With numerous Illustrations, royal 8vo, cloth, 10s 6d.

Electricity in Theory and Practice, or the Elements of ELECTRICAL ENGINEERING. By Lieut BRADLEY A. FISKE, U S N.

248 pp. and 7 Plates, 8vo, cloth, 9s.

Reports of the Committee on Electrical Standards appointed by the BRITISH ASSOCIATION Revised by Sir W. THOMSON, Dr. J. P. JOULE, Professors CLERK MAXWELL and FLEEMING JENKIN With a Report to the Royal Society on Units of Electrical Resistance, by Professor F. Jenkin. Edited by Professor FLEEMING JENKIN, F R S

With Illustrations, crown 8vo, cloth, 12s. 6d.

Elementary Treatise on Electric Batteries. From the French of ALFRED NIAUDET Translated by L M FISHBACK.

Oblong 8vo, cloth, 10s. 6d

Handbook of Electrical Diagrams and Connections. By C. H. DAVIS and FRANK B. RAE

With Illustrations, 8vo, sewed, 1s 6d

The Telephone. A Lecture, entitled " Researches in Electro-Telephony," by Professor ALEXANDER GRAHAM BELL, delivered before the Society of Telegraph Engineers.

Crown 8vo, cloth, 4s 6d.

Elements of Construction for Electro-Magnets. By COUNT TH. DU MONCEL, Membre de l'Institut de France. Translated from the French by C. J. WHARTON.

8vo, sewed, 1s.

Electric Lighting : its State and Progress, and its Probable INFLUENCE ON THE GAS INTERESTS. By J. T. SPRAGUE, Mem. Soc. Tel. Eng.

With 113 Engravings and Diagrams, crown 8vo, cloth, 7s. 6d.

Electricity as a Motive Power. By COUNT TH. DU MONCEL, Membre de l'Institut de France, and FRANK GERALDY, Ingénieur des Ponts et Chaussées. Translated and edited by C J WHARTON, Assoc Soc. of Telegraph Engineers and of Electricians.

With Woodcut Illustrations, crown 8vo, cloth, 4s.

Useful Information on Electric Lighting. Embodying the Rules of the Fire Risk Committee and the Electric Lighting Bill · also plain directions for the working of Electric Lamps and Dynamo-Machines. By KILLINGWORTH HEDGES, F.C.S., M. Soc. T.E., Assoc. Mem. Inst C.E. Fourth Edition, revised and enlarged.

8vo, sewed, 1s.

The Supply of Electricity by Local Authorities. By KILLINGWORTH HEDGES, M. Inst. M.E.

8vo, sewed, 1s.

The Electric Light for Industrial Uses. By R. E. CROMPTON.

8vo, cloth, 10s 6d

Electric Lighting by Incandescence, and its Application to INTERIOR ILLUMINATION. A Practical Treatise, with 96 Illustrations. By W. E. SAWYER.

285 pp., with numerous Diagrams and Tables, crown 8vo, cloth, 12s. 6d.

Electrical Tables and Formulæ for the use of Telegraph INSPECTORS AND OPERATORS. Compiled by LATIMER CLARK and ROBERT SABINE.

18mo, cloth, 1s. 6d.

Practical Electrical Units popularly explained. By JAMES SWINBURNE, late of J. N. Swan and Co., Paris, late of Brush-Swan E.L Co., U.S.A.

126 pp., with 36 Illustrations, fcap. 8vo, cloth, 1s.; sewed, 6d.

Electro-Telegraphy. A Book for Beginners. By FREDERICK S. BEECHEY, Telegraph Engineer

With Portrait and Wood Engravings, 8vo, cloth, 7s. 6d

Philipp Reis, Inventor of the Telephone: a Biographical Sketch, with documentary testimony, translations of the original papers of the Inventor and contemporary publications. By SILVANUS P. THOMPSON, B.A., D.Sc., Professor of Experimental Physics in University College, Bristol.

8vo, sewed, 8s., large paper, 12s. 6d.

Catalogue of Books and Papers relating to Electricity, MAGNETISM, AND THE ELECTRIC TELEGRAPH, &c Including the Ronalds Library, compiled by Sir FRANCIS RONALDS, F.R S, together with a Biographical Memoir of the Author. Edited by ALFRED J. FROST, Assoc. Soc. T.E, Member of the Library Association of the United Kingdom.

8vo, sewed, 1s.

Electric Lighting by Water Power. By THOS. B. GRIERSON, Assoc. Inst. C E.

With 62 Illustrations, crown 8vo, cloth, 4s. 6d.

Practical Electric Lighting. By A. BROMLEY HOLMES, Assoc. M. Inst C.E

With numerous Diagrams and Tables, crown 8vo, cloth, 3s 6d.

The Laying and Repairing of Electric Telegraph Cables. By Capt. V. HOSKIER, Royal Danish Engineers

Second Edition, crown 8vo, cloth, 4s. 6d

A Guide for the Electric Testing of Telegraph Cables. By Capt. V. HOSKIER, Royal Danish Engineers.

8vo, sewed, 4s.

Journal of the Society of Telegraph Engineers and of Elec-TRICIANS. Edited by Professor W. E. AYRTON, F.R.S. No. 50.

Royal 8vo, cloth, 16s.

Lightning Conductors: their History, Nature, and Mode of APPLICATION. By R. ANDERSON, F C.S., F.G S., Member of the Society of Telegraph Engineers.

8vo, cloth, 7s. 6d

Lightning-Rod Conference. Report of the Delegates from the following Societies, viz.: the Meteorological Society, the Royal Institute of British Architects, the Society of Telegraph Engineers and Electricians, the Physical Society, and co-opted Members. Edited by G. J. SYMONDS, F.R.S.

Royal 8vo, cloth, 15s.

The Speaking Telephone, Electric Light, and other Elec-TRICAL INVENTIONS. By G. B. PRESCOTT. Fully illustrated.

Crown 8vo, cloth, 5s.

Practical Information for Telephonists. By T. D. LOCKWOOD.

8vo, sewed, 1s.

Electro-Motors and their Government. By Professors W. E. AYRTON, F R.S., and JOHN PERRY, M.E.

18mo, boards, 2s.

Incandescent Electric Lights. With particular reference to the Edison Lamps at the Paris Exhibition By COUNT DU MONCEL and W. H. PREECE.

London: E. & F. N. SPON, 16, Charing Cross.
New York: 35, Murray Street.

1891.

BOOKS RELATING

TO

APPLIED SCIENCE

PUBLISHED BY

E. & F. N. SPON,

LONDON: 125, STRAND.

NEW YORK: 12, CORTLANDT STREET.

The Engineers' Sketch-Book of Mechanical Move-
ments, Devices, Appliances, Contrivances, Details employed in the Design
and Construction of Machinery for every purpose. Collected from
numerous Sources and from Actual Work. Classified and Arranged for
Reference. *Nearly* 2000 *Illustrations.* By T. B. BARBER, Engineer.
8vo, cloth, 7s. 6d.

A Pocket-Book for Chemists, Chemical Manufacturers,
Metallurgists, Dyers, Distillers, Brewers, Sugar Refiners, Photographers,
Students, etc., etc. By THOMAS BAYLEY, Assoc. R.C. Sc. Ireland, Analytical and Consulting Chemist and Assayer. Fourth edition, with
additions, 437 pp., royal 32mo, roan, gilt edges, 5s.

SYNOPSIS OF CONTENTS :

Atomic Weights and Factors—Useful Data—Chemical Calculations—Rules for Indirect
Analysis—Weights and Measures—Thermometers and Barometers—Chemical Physics—
Boiling Points, etc.—Solubility of Substances—Methods of Obtaining Specific Gravity—Con-
version of Hydrometers—Strength of Solutions by Specific Gravity—Analysis—Gas Analysis—
Water Analysis—Qualitative Analysis and Reactions—Volumetric Analysis—Manipulation—
Mineralogy — Assaying — Alcohol — Beer — Sugar — Miscellaneous Technological matter
relating to Potash, Soda, Sulphuric Acid, Chlorine, Tar Products, Petroleum, Milk, Tallow,
Photography, Prices, Wages, Appendix, etc., etc.

The Mechanician : A Treatise on the Construction
and Manipulation of Tools, for the use and instruction of Young Engineers
and Scientific Amateurs, comprising the Arts of Blacksmithing and Forg-
ing ; the Construction and Manufacture of Hand Tools, and the various
Methods of Using and Grinding them ; description of Hand and Machine
Processes ; Turning and Screw Cutting. By CAMERON KNIGHT,
Engineer. *Containing* 1147 *illustrations,* and 397 pages of letter-press.
Fourth edition, 4to, cloth, 18s.

B

Just Published, in Demy 8vo, cloth, containing 975 pages and 250 Illustrations, price 7s. 6d.

SPONS' HOUSEHOLD MANUAL:

A Treasury of Domestic Receipts and Guide for Home Management.

PRINCIPAL CONTENTS.

Hints for selecting a good House, pointing out the essential requirements for a good house as to the Site, Soil, Trees, Aspect, Construction, and General Arrangement, with instructions for Reducing Echoes, Waterproofing Damp Walls, Curing Damp Cellars.

Sanitation.—What should constitute a good Sanitary Arrangement; Examples (with Illustrations) of Well- and Ill-drained Houses; How to Test Drains; Ventilating Pipes, etc.

Water Supply.—Care of Cisterns; Sources of Supply; Pipes; Pumps; Purification and Filtration of Water.

Ventilation and Warming—Methods of Ventilating without causing cold draughts, by various means; Principles of Warming; Health Questions; Combustion; Open Grates; Open Stoves; Fuel Economisers; Varieties of Grates; Close-Fire Stoves; Hot-air Furnaces; Gas Heating; Oil Stoves, Steam Heating; Chemical Heaters; Management of Flues; and Cure of Smoky Chimneys.

Lighting.—The best methods of Lighting; Candles, Oil Lamps, Gas, Incandescent Gas, Electric Light; How to test Gas Pipes; Management of Gas.

Furniture and Decoration.—Hints on the Selection of Furniture; on the most approved methods of Modern Decoration; on the best methods of arranging Bells and Calls, How to Construct an Electric Bell.

Thieves and Fire.—Precautions against Thieves and Fire; Methods of Detection; Domestic Fire Escapes; Fireproofing Clothes, etc.

The Larder—Keeping Food fresh for a limited time; Storing Food without change, such as Fruits, Vegetables, Eggs, Honey, etc.

Curing Foods for lengthened Preservation, as Smoking, Salting, Canning, Potting, Pickling, Bottling Fruits, etc., Jams, Jellies, Marmalade, etc.

The Dairy.—The Building and Fitting of Dairies in the most approved modern style; Butter-making, Cheesemaking and Curing.

The Cellar.—Building and Fitting; Cleaning Casks and Bottles; Corks and Corking; Aërated Drinks; Syrups for Drinks; Beers, Bitters; Cordials and Liqueurs, Wines; Miscellaneous Drinks.

The Pantry.—Bread-making; Ovens and Pyrometers; Yeast; German Yeast; Biscuits; Cakes; Fancy Breads; Buns.

The Kitchen.—On Fitting Kitchens; a description of the best Cooking Ranges, close and open; the Management and Care of Hot Plates, Baking Ovens, Dampers, Flues, and Chimneys; Cooking by Gas; Cooking by Oil; the Arts of Roasting, Grilling, Boiling, Stewing, Braising, Frying.

Receipts for Dishes—Soups, Fish, Meat, Game, Poultry, Vegetables, Salads, Puddings, Pastry, Confectionery, Ices, etc., etc.; Foreign Dishes.

The Housewife's Room.—Testing Air, Water, and Foods; Cleaning and Renovating; Destroying Vermin.

Housekeeping, Marketing.

The Dining-Room.—Dietetics; Laying and Waiting at Table; Carving; Dinners, Breakfasts, Luncheons, Teas, Suppers, etc.

The Drawing-Room.—Etiquette; Dancing; Amateur Theatricals; Tricks and Illusions; Games (indoor).

The Bedroom and Dressing-Room; Sleep; the Toilet; Dress; Buying Clothes; Outfits; Fancy Dress.

The Nursery.—The Room; Clothing; Washing; Exercise; Sleep; Feeding, Teething; Illness; Home Training

The Sick-Room.—The Room; the Nurse; the Bed; Sick Room Accessories; Feeding Patients; Invalid Dishes and Drinks; Administering Physic; Domestic Remedies; Accidents and Emergencies; Bandaging; Burns; Carrying Injured Persons; Wounds; Drowning; Fits; Frost-bites; Poisons and Antidotes; Sunstroke; Common Complaints; Disinfection, etc.

The Bath-Room.—Bathing in General; Management of Hot-Water System.

The Laundry—Small Domestic Washing Machines, and methods of getting up linen Fitting up and Working a Steam Laundry.

The School-Room.—The Room and its Fittings; Teaching, etc

The Playground.—Air and Exercise; Training; Outdoor Games and Sports.

The Workroom.—Darning, Patching, and Mending Garments.

The Library.—Care of Books.

The Garden.—Calendar of Operations for Lawn, Flower Garden, and Kitchen Garden.

The Farmyard.—Management of the Horse, Cow, Pig, Poultry, Bees, etc., etc.

Small Motors.—A description of the various small Engines useful for domestic purposes, from 1 man to 1 horse power, worked by various methods, such as Electric Engines, Gas Engines, Petroleum Engines, Steam Engines, Condensing Engines, Water Power, Wind Power, and the various methods of working and managing them.

Household Law.—The Law relating to Landlords and Tenants, Lodgers, Servants, Parochial Authorities, Juries, Insurance, Nuisance, etc.

On Designing Belt Gearing. By E. J. COWLING
WELCH, Mem. Inst. Mech. Engineers, Author of 'Designing Valve Gearing.' Fcap. 8vo, sewed, 6d.

A Handbook of Formulæ, Tables, and Memoranda,
for Architectural Surveyors and others engaged in Building. By J. T. HURST, C.E. Fourteenth edition, royal 32mo, roan, 5s.

"It is no disparagement to the many excellent publications we refer to, to say that in our opinion this little pocket-book of Hurst's is the very best of them all, without any exception. It would be useless to attempt a recapitulation of the contents, for it appears to contain almost *everything* that anyone connected with building could require, and, best of all, made up in a compact form for carrying in the pocket, measuring only 5 in by 3 in., and about ½ in. thick, in a limp cover. We congratulate the author on the success of his laborious and practically compiled little book, which has received unqualified and deserved praise from every professional person to whom we have shown it."—*The Dublin Builder.*

Tabulated Weights of Angle, Tee, Bulb, Round,
Square, and Flat Iron and Steel, and other information for the use of Naval Architects and Shipbuilders. By C. H. JORDAN, M.I.N.A. Fourth edition, 32mo, cloth, 2s. 6d.

A Complete Set of Contract Documents for a Country
Lodge, comprising Drawings, Specifications, Dimensions (for quantities), Abstracts, Bill of Quantities, Form of Tender and Contract, with Notes by J. LEANING, printed in facsimile of the original documents, on single sheets fcap., in paper case, 10s.

A Practical Treatise on Heat, as applied to the
Useful Arts; for the Use of Engineers, Architects, &c. By THOMAS BOX. With 14 plates. Sixth edition, crown 8vo, cloth, 12s. 6d.

A Descriptive Treatise on Mathematical Drawing
Instruments: their construction, uses, qualities, selection, preservation, and suggestions for improvements, with hints upon Drawing and Colouring. By W. F. STANLEY, M.R.I. Sixth edition, *with numerous illustrations*, crown 8vo, cloth, 5s.

Quantity Surveying. By J. LEANING. With 42 illustrations. Second edition, revised, crown 8vo, cloth, 9*s.*

CONTENTS :

A complete Explanation of the London Practice.	Schedule of Prices
General Instructions.	Form of Schedule of Prices.
Order of Taking Off.	Analysis of Schedule of Prices.
Modes of Measurement of the various Trades	Adjustment of Accounts.
Use and Waste.	Form of a Bill of Variations.
Ventilation and Warming.	Remarks on Specifications.
Credits, with various Examples of Treatment.	Prices and Valuation of Work, with
Abbreviations.	Examples and Remarks upon each Trade.
Squaring the Dimensions.	The Law as it affects Quantity Surveyors,
Abstracting, with Examples in illustration of	with Law Reports.
each Trade.	Taking Off after the Old Method.
Billing.	Northern Practice.
Examples of Preambles to each Trade.	The General Statement of the Methods
Form for a Bill of Quantities. -	recommended by the Manchester Society
Do. Bill of Credits.	of Architects for taking Quantities.
Do. Bill for Alternative Estimate.	Examples of Collections.
Restorations and Repairs, and Form of Bill.	Examples of " Taking Off" in each Trade.
Variations before Acceptance of Tender.	Remarks on the Past and Present Methods
Errors in a Builder's Estimate.	of Estimating.

Spons' Architects' and Builders' Price Book, with useful Memoranda. Edited by W. YOUNG, Architect. Crown 8vo, cloth, red edges, 3*s.* 6*d.* *Published annually.* Seventeenth edition. *Now ready.*

Long-Span Railway Bridges, comprising Investigations of the Comparative Theoretical and Practical Advantages of the various adopted or proposed Type Systems of Construction, with numerous Formulæ and Tables giving the weight of Iron or Steel required in Bridges from 300 feet to the limiting Spans ; to which are added similar Investigations and Tables relating to Short-span Railway Bridges. Second and revised edition. By B. BAKER, Assoc. Inst. C.E. *Plates,* crown 8vo, cloth, 5*s.*

Elementary Theory and Calculation of Iron Bridges and Roofs. By AUGUST RITTER, Ph.D., Professor at the Polytechnic School at Aix-la-Chapelle. Translated from the third German edition, by H. R. SANKEY, Capt. R.E. With 500 *illustrations,* 8vo, cloth, 15*s.*

The Elementary Principles of Carpentry. By THOMAS TREDGOLD. Revised from the original edition, and partly re-written, by JOHN THOMAS HURST. Contained in 517 pages of letterpress, and *illustrated with* 48 *plates and* 150 *wood engravings.* Sixth edition, reprinted from the third, crown 8vo, cloth, 12*s.* 6*d.*

Section I. On the Equality and Distribution of Forces — Section II. Resistance of Timber — Section III. Construction of Floors — Section IV. Construction of Roofs — Section V. Construction of Domes and Cupolas — Section VI. Construction of Partitions — Section VII. Scaffolds, Staging, and Gantries — Section VIII Construction of Centres for Bridges — Section IX. Coffer-dams, Shoring, and Strutting — Section X. Wooden Bridges and Viaducts — Section XI. Joints, Straps, and other Fastenings — Section XII. Timber.

The Builder's Clerk : a Guide to the Management of a Builder's Business. By THOMAS BALES. Fcap. 8vo, cloth, 1*s.* 6*d.*

Practical Gold-Mining: a Comprehensive Treatise on the Origin and Occurrence of Gold-bearing Gravels, Rocks and Ores, and the methods by which the Gold is extracted. By C. G. WARNFORD LOCK, co-Author of 'Gold: its Occurrence and Extraction.' *With 8 plates and 275 engravings in the text,* royal 8vo, cloth, 2*l.* 2*s.*

Hot Water Supply: A Practical Treatise upon the Fitting of Circulating Apparatus in connection with Kitchen Range and other Boilers, to supply Hot Water for Domestic and General Purposes. With a Chapter upon Estimating. *Fully illustrated,* crown 8vo, cloth, 3*s.*

Hot Water Apparatus: An Elementary Guide for the Fitting and Fixing of Boilers and Apparatus for the Circulation of Hot Water for Heating and for Domestic Supply, and containing a Chapter upon Boilers and Fittings for Steam Cooking. 32 *illustrations,* fcap. 8vo, cloth, 1*s.* 6*d.*

The Use and Misuse, and the Proper and Improper Fixing of a Cooking Range. Illustrated, fcap. 8vo, sewed, 6*d.*

Iron Roofs: Examples of Design, Description. *Illustrated with* 64 *Working Drawings of Executed Roofs.* By ARTHUR T. WALMISLEY, Assoc. Mem. Inst. C.E. Second edition, revised, imp. 4to, half-morocco, 3*l.* 3*s.*

A History of Electric Telegraphy, to the Year 1837. Chiefly compiled from Original Sources, and hitherto Unpublished Documents, by J. J. FAHIE, Mem. Soc. of Tel. Engineers, and of the International Society of Electricians, Paris. Crown 8vo, cloth, 9*s.*

Spons' Information for Colonial Engineers. Edited by J. T. HURST. Demy 8vo, sewed.

No. 1, Ceylon. By ABRAHAM DEANE, C.E. 2*s.* 6*d.*

CONTENTS:

Introductory Remarks—Natural Productions—Architecture and Engineering—Topography, Trade, and Natural History—Principal Stations—Weights and Measures, etc., etc.

No. 2. Southern Africa, including the Cape Colony, Natal, and the Dutch Republics. By HENRY HALL, F.R.G.S., F.R.C.I. With Map. 3*s.* 6*d.* CONTENTS:

General Description of South Africa—Physical Geography with reference to Engineering Operations—Notes on Labour and Material in Cape Colony—Geological Notes on Rock Formation in South Africa—Engineering Instruments for Use in South Africa—Principal Public Works in Cape Colony: Railways, Mountain Roads and Passes, Harbour Works, Bridges, Gas Works, Irrigation and Water Supply, Lighthouses, Drainage and Sanitary Engineering, Public Buildings, Mines—Table of Woods in South Africa—Animals used for Draught Purposes—Statistical Notes—Table of Distances—Rates of Carriage, etc.

No. 3. India. By F. C. DANVERS, Assoc. Inst. C.E. With Map. 4*s.* 6*d.* CONTENTS:

Physical Geography of India—Building Materials—Roads—Railways—Bridges—Irrigation—River Works—Harbours—Lighthouse Buildings—Native Labour—The Principal Trees of India—Money—Weights and Measures—Glossary of Indian Terms, etc.

Our Factories, Workshops, and Warehouses: their
Sanitary and Fire-Resisting Arrangements. By B. H THWAITE, Assoc.
Mem. Inst. C.E.   *With* 183 *wood engravings*, crown 8vo, cloth, 9*s.*

A Practical Treatise on Coal Mining.   By GEORGE
G. ANDRÉ, F.G.S., Assoc. Inst. C.E., Member of the Society of Engineers.
With 82 *lithographic plates.*   2 vols., royal 4to, cloth, 3*l.* 12*s.*

A Practical Treatise on Casting and Founding,
including descriptions of the modern machinery employed in the art.   By
N. E. SPRETSON, Engineer.   Fifth edition, with 82 *plates* drawn to
scale, 412 pp., demy 8vo, cloth, 18*s.*

A Handbook of Electrical Testing.   By H. R. KEMPE,
M.S.T.E.   Fourth edition, revised and enlarged, crown 8vo, cloth, 16*s.*

The Clerk of Works: a Vade-Mecum for all engaged
in the Superintendence of Building Operations.   By G. G. HOSKINS,
F.R.I.B.A.   Third edition, fcap. 8vo, cloth, 1*s.* 6*d.*

American Foundry Practice: Treating of Loam,
Dry Sand, and Green Sand Moulding, and containing a Practical Treatise
upon the Management of Cupolas, and the Melting of Iron.   By T. D.
WEST, Practical Iron Moulder and Foundry Foreman.   Second edition,
with numerous illustrations, crown 8vo, cloth, 10*s.* 6*d.*

The Maintenance of Macadamised Roads.   By T.
CODRINGTON, M.I.C.E, F.G.S., General Superintendent of County Roads
for South Wales.   Second edition.   8vo.   [*Nearly ready.*

Hydraulic Steam and Hand Power Lifting and
Pressing Machinery.   By FREDERICK COLYER, M. Inst. C.E., M. Inst. M.E.
With 73 *plates*, 8vo, cloth, 18*s.*

Pumps and Pumping Machinery.   By F. COLYER,
M.I.C.E., M.I.M.E.   *With* 23 *folding plates*, 8vo, cloth, 12*s.* 6*d.*

Pumps and Pumping Machinery.   By F. COLYER.
Second Part.   *With* 11 *large plates*, 8vo, cloth, 12*s.* 6*d.*

A Treatise on the Origin, Progress, Prevention, and
Cure of Dry Rot in Timber; with Remarks on the Means of Preserving
Wood from Destruction by Sea-Worms, Beetles, Ants, etc.   By THOMAS
ALLEN BRITTON, late Surveyor to the Metropolitan Board of Works,
etc., etc.   *With* 10 *plates*, crown 8vo, cloth, 7*s.* 6*d.*

The Artillery of the Future and the New Powders.
By J. A. LONGRIDGE, Mem. Inst. C.E.   8vo, cloth, 5*s.*

Gas Works: their Arrangement, Construction, Plant, and Machinery. By F. COLYER, M. Inst. C.E. *With 31 folding plates*, 8vo, cloth, 12s. 6d.

The Municipal and Sanitary Engineer's Handbook. By H. PERCY BOULNOIS, Mem. Inst. C.E., Borough Engineer, Portsmouth. *With numerous illustrations.* Second edition, demy 8vo, cloth.

CONTENTS:

The Appointment and Duties of the Town Surveyor—Traffic—Macadamised Roadways—Steam Rolling—Road Metal and Breaking—Pitched Pavements—Asphalte—Wood Pavements—Footpaths—Kerbs and Gutters—Street Naming and Numbering—Street Lighting—Sewerage—Ventilation of Sewers—Disposal of Sewage—House Drainage—Disinfection—Gas and Water Companies, etc., Breaking up Streets—Improvement of Private Streets—Borrowing Powers—Artizans' and Labourers' Dwellings—Public Conveniences—Scavenging, including Street Cleansing—Watering and the Removing of Snow—Planting Street Trees—Deposit of Plans—Dangerous Buildings—Hoardings—Obstructions—Improving Street Lines—Cellar Openings—Public Pleasure Grounds—Cemeteries—Mortuaries—Cattle and Ordinary Markets—Public Slaughter-houses, etc.—Giving numerous Forms of Notices, Specifications,. and General Information upon these and other subjects of great importance to Municipal Engineers and others engaged in Sanitary Work.

Metrical Tables. By Sir G. L. MOLESWORTH, M.I.C.E. 32mo, cloth, 1s. 6d.

CONTENTS.

General—Linear Measures—Square Measures—Cubic Measures—Measures of Capacity—Weights—Combinations—Thermometers.

Elements of Construction for Electro-Magnets. By Count TH. DU MONCEL, Mem. de l'Institut de France. Translated from the French by C. J. WHARTON. Crown 8vo, cloth, 4s. 6d.

A Treatise on the Use of Belting for the Transmission of Power. By J. H. COOPER. Second edition, *illustrated*, 8vo, cloth. 15s.

A Pocket-Book of Useful Formulæ and Memoranda for Civil and Mechanical Engineers. By Sir GUILFORD L. MOLESWORTH, Mem. Inst. C.E. *With numerous illustrations*, 744 pp. Twenty-second edition, 32mo, roan, 6s.

SYNOPSIS OF CONTENTS:

Surveying, Levelling, etc.—Strength and Weight of Materials—Earthwor-., Brickwork Masonry, Arches, etc.—Struts, Columns, Beams, and Trusses—Flooring, Roofing, and Roof Trusses—Girders, Bridges, etc.—Railways and Roads—Hydraulic Formulæ—Canals, Sewers, Waterworks, Docks—Irrigation and Breakwaters—Gas, Ventilation, and Warming—Heat, Light, Colour, and Sound—Gravity: Centres, Forces, and Powers—Millwork, Teeth of Wheels, Shafting, etc.—Workshop Recipes—Sundry Machinery—Animal Power—Steam and the Steam Engine—Water-power, Water-wheels, Turbines, etc.—Wind and Windmills—Steam Navigation, Ship Building, Tonnage, etc.—Gunnery, Projectiles, etc—Weights, Measures, and Money—Trigonometry, Conic Sections, and Curves—Telegraphy—Mensuration—Tables of Areas and Circumference, and Arcs of Circles—Logarithms, Square and Cube Roots, Powers—Reciprocals, etc.—Useful Numbers—Differential and Integral Calculus—Algebraic Signs—Telegraphic Construction and Formulæ.

Hints on Architectural Draughtsmanship. By G. W.
TUXFORD HALLATT. Fcap. 8vo, cloth, 1s. 6d.

Spons' Tables and Memoranda for Engineers;
selected and arranged by J. T. HURST, C.E., Author of 'Architectural
Surveyors' Handbook,' 'Hurst's Tredgold's Carpentry,' etc. Eleventh
edition, 64mo, roan, gilt edges, 1s.; or in cloth case, 1s. 6d.

This work is printed in a pearl type, and is so small, measuring only 2½ in. by 1¾ in. by
¼ in. thick, that it may be easily carried in the waistcoat pocket.

"It is certainly an extremely rare thing for a reviewer to be called upon to notice a volume
measuring but 2½ in by 1¾ in, yet these dimensions faithfully represent the size of the handy
little book before us. The volume—which contains 118 printed pages, besides a few blank
pages for memoranda—is, in fact, a true pocket-book, adapted for being carried in the waistcoat pocket, and containing a far greater amount and variety of information than most people
would imagine could be compressed into so small a space. . . . . The little volume has been
compiled with considerable care and judgment, and we can cordially recommend it to our
readers as a useful little pocket companion."—*Engineering.*

A Practical Treatise on Natural and Artificial
Concrete, its Varieties and Constructive Adaptations. By HENRY REID,
Author of the 'Science and Art of the Manufacture of Portland Cement.'
New Edition, *with 59 woodcuts and 5 plates,* 8vo, cloth, 15s.

Notes on Concrete and Works in Concrete; especially
written to assist those engaged upon Public Works. By JOHN NEWMAN,
Assoc. Mem. Inst. C.E., crown 8vo, cloth, 4s. 6d.

Electricity as a Motive Power. By Count TH. DU
MONCEL, Membre de l'Institut de France, and FRANK GERALDY, Ingénieur des Ponts et Chaussées. Translated and Edited, with Additions, by
C. J. WHARTON, Assoc. Soc. Tel. Eng. and Elec. *With 113 engravings
and diagrams,* crown 8vo, cloth, 7s. 6d.

Treatise on Valve-Gears, with special consideration
of the Link-Motions of Locomotive Engines. By Dr. GUSTAV ZEUNER,
Professor of Applied Mechanics at the Confederated Polytechnikum of
Zurich. Translated from the Fourth German Edition, by Professor J. F.
KLEIN, Lehigh University, Bethlehem, Pa. *Illustrated,* 8vo, cloth, 12s. 6d.

The French-Polisher's Manual. By a French-
Polisher; containing Timber Staining, Washing, Matching, Improving,
Painting, Imitations, Directions for Staining, Sizing, Embodying,
Smoothing, Spirit Varnishing, French-Polishing, Directions for Re-
polishing. Third edition, royal 32mo, sewed, 6d.

Hops, their Cultivation, Commerce, and Uses in
various Countries. By P. L. SIMMONDS. Crown 8vo, cloth, 4s. 6d.

The Principles of Graphic Statics. By GEORGE
SYDENHAM CLARKE, Major Royal Engineers. *With 112 illustrations.*
Second edition, 4to, cloth, 12s. 6d.

Dynamo Tenders' Hand-Book. By F. B. BADT, late
1st Lieut. Royal Prussian Artillery. *With* 70 *illustrations.* Third edition,
18mo, cloth, 4*s.* 6*d.*

Practical Geometry, Perspective, and Engineering
Drawing; a Course of Descriptive Geometry adapted to the Requirements of the Engineering Draughtsman, including the determination of
cast shadows and Isometric Projection, each chapter being followed by
numerous examples; to which are added rules for Shading, Shade-lining,
etc., together with practical instructions as to the Lining, Colouring,
Printing, and general treatment of Engineering Drawings, with a chapter
on drawing Instruments. By GEORGE S. CLARKE, Capt. R.E. Second
edition, *with* 21 *plates.* 2 vols., cloth, 10*s.* 6*d.*

The Elements of Graphic Statics. By Professor
KARL VON OTT, translated from the German by G. S. CLARKE, Capt.
R.E., Instructor in Mechanical Drawing, Royal Indian Engineering
College. *With* 93 *illustrations,* crown 8vo, cloth, 5*s.*

A Practical Treatise on the Manufacture and Distri-
bution of Coal Gas. By WILLIAM RICHARDS. Demy 4to, with *numerous*
wood engravings and 29 *plates,* cloth, 28*s.*

SYNOPSIS OF CONTENTS:

Introduction — History of Gas Lighting — Chemistry of Gas Manufacture, by Lewis
Thompson, Esq., M R.C.S.—Coal, with Analyses, by J Paterson, Lewis Thompson, and
G. R. Hislop, Esqrs.—Retorts, Iron and Clay—Retort Setting—Hydraulic Main—Con-
densers — Exhausters — Washers and Scrubbers — Purifiers — Purification — History of Gas
Holder — Tanks, Brick and Stone, Composite, Concrete, Cast-iron, Compound Annular
Wrought-iron — Specifications — Gas Holders — Station Meter — Governor — Distribution—
Mains—Gas Mathematics, or Formulæ for the Distribution of Gas, by Lewis Thompson, Esq —
Services—Consumers' Meters—Regulators—Burners—Fittings—Photometer—Carburization
of Gas—Air Gas and Water Gas—Composition of Coal Gas, by Lewis Thompson, Esq.—
Analyses of Gas—Influence of Atmospheric Pressure and Temperature on Gas—Residual
Products—Appendix—Description of Retort Settings, Buildings, etc , etc.

The New Formula for Mean Velocity of Discharge
of Rivers and Canals. By W. R. KUTTER. Translated from articles in
the 'Cultur-Ingénieur,' by LOWIS D'A. JACKSON, Assoc. Inst. C.E.
8vo, cloth, 12*s.* 6*d.*

The Practical Millwright and Engineer's Ready
Reckoner; or Tables for finding the diameter and power of cog-wheels,
diameter, weight, and power of shafts, diameter and strength of bolts, etc.
By THOMAS DIXON. Fourth edition, 12mo, cloth, 3*s.*

Tin: Describing the Chief Methods of Mining,
Dressing and Smelting it abroad; with Notes upon Arsenic, Bismuth and
Wolfram. By ARTHUR G. CHARLETON, Mem. American Inst. of
Mining Engineers. *With plates,* 8vo, cloth, 12*s.* 6*d.*

Perspective, Explained and Illustrated. By G. S.
CLARKE, Capt. R.E. *With illustrations,* 8vo, cloth, 3s. 6d.

Practical Hydraulics; a Series of Rules and Tables
for the use of Engineers, etc., etc. By THOMAS BOX. Ninth edition,
numerous plates, post 8vo, cloth, 5s.

The Essential Elements of Practical Mechanics;
based on the Principle of Work, designed for Engineering Students. By
OLIVER BYRNE, formerly Professor of Mathematics, College for Civil
Engineers. Third edition, *with* 148 *wood engravings,* post 8vo, cloth,
7s. 6d.

CONTENTS:

Chap. 1. How Work is Measured by a Unit, both with and without reference to a Unit
of Time—Chap. 2. The Work of Living Agents, the Influence of Friction, and introduces
one of the most beautiful Laws of Motion—Chap. 3. The principles expounded in the first and
second chapters are applied to the Motion of Bodies—Chap 4. The Transmission of Work by
simple Machines—Chap. 5. Useful Propositions and Rules.

Breweries and Maltings: their Arrangement, Construction, Machinery, and Plant. By G. SCAMELL, F.R.I.B.A. Second
edition, revised, enlarged, and partly rewritten. By F. COLYER, M.I.C.E.,
M.I.M.E. *With* 20 *plates,* 8vo, cloth, 12s. 6d.

A Practical Treatise on the Construction of Hori-
zontal and Vertical Waterwheels, specially designed for the use of operative mechanics By WILLIAM CULLEN, Millwright and Engineer. *With*
11 *plates.* Second edition, revised and enlarged, small 4to, cloth, 12s. 6d.

A Practical Treatise on Mill-gearing, Wheels, Shafts,
Riggers, etc.; for the use of Engineers. By THOMAS BOX. Third
edition, *with* 11 *plates.* Crown 8vo, cloth, 7s. 6d.

Mining Machinery: a Descriptive Treatise on the
Machinery, Tools, and other Appliances used in Mining. By G. G.
ANDRÉ, F.G.S., Assoc. Inst. C.E., Mem. of the Society of Engineers.
Royal 4to, uniform with the Author's Treatise on Coal Mining, con-
taining 182 *plates,* accurately drawn to scale, with descriptive text, in
2 vols., cloth, 3l. 12s.

CONTENTS:

Machinery for Prospecting, Excavating, Hauling, and Hoisting—Ventilation—Pumping—
Treatment of Mineral Products, including Gold and Silver, Copper, Tin, and Lead, Iron,
Coal, Sulphur, China Clay, Brick Earth, etc.

Tables for Setting out Curves for Railways, Canals,
Roads, etc., varying from a radius of five chains to three miles. By A.
KENNEDY and R. W. HACKWOOD. *Illustrated* 32mo, cloth, 2s. 6d.

Practical Electrical Notes and Definitions for the use of Engineering Students and Practical Men. By W. PERREN MAYCOCK, Assoc. M. Inst. E.E., Instructor in Electrical Engineering at the Pitlake Institute, Croydon, together with the Rules and Regulations to be observed in Electrical Installation Work. Second edition. Royal 32mo, roan, gilt edges, 4s. 6d.

The Draughtsman's Handbook of Plan and Map Drawing; including instructions for the preparation of Engineering, Architectural, and Mechanical Drawings. *With numerous illustrations in the text, and 33 plates (15 printed in colours).* By G. G. ANDRÉ, F.G.S., Assoc. Inst. C.E. 4to, cloth, 9s.

CONTENTS:

The Drawing Office and its Furnishings—Geometrical Problems—Lines, Dots, and their Combinations—Colours, Shading, Lettering, Bordering, and North Points—Scales—Plotting —Civil Engineers' and Surveyors' Plans—Map Drawing—Mechanical and Architectural Drawing—Copying and Reducing Trigonometrical Formulæ, etc., etc.

The Boiler-maker's and Iron Ship-builder's Companion, comprising a series of original and carefully calculated tables, of the utmost utility to persons interested in the iron trades. By JAMES FODEN, author of 'Mechanical Tables,' etc. Second edition revised, *with illustrations*, crown 8vo, cloth, 5s.

Rock Blasting: a Practical Treatise on the means employed in Blasting Rocks for Industrial Purposes. By G. G. ANDRÉ, F.G.S., Assoc. Inst. C.E. *With 56 illustrations and 12 plates,* 8vo, cloth, 10s. 6d. .

Experimental Science: Elementary, Practical, and Experimental Physics. By GEO. M. HOPKINS. *Illustrated by* 672 *engravings.* In one large vol., 8vo, cloth, 18s.

A Treatise on Ropemaking as practised in public and private Rope-yards, with a Description of the Manufacture, Rules, Tables of Weights, etc., adapted to the Trade, Shipping, Mining, Railways, Builders, etc. By R. CHAPMAN, formerly foreman to Messrs. Huddart and Co., Limehouse, and late Master Ropemaker to H.M. Dockyard, Deptford. Second edition, 12mo, cloth, 3s.

Laxton's Builders' and Contractors' Tables; for the use of Engineers, Architects, Surveyors, Builders, Land Agents, and others. Bricklayer, containing 22 tables, with nearly 30,000 calculations. 4to, cloth, 5s.

Laxton's Builders' and Contractors' Tables. Excavator, Earth, Land, Water, and Gas, containing 53 tables, with nearly 24,000 calculations. 4to, cloth, 5s.

Egyptian Irrigation. By W. WILLCOCKS, M.I.C.E.,
Indian Public Works Department, Inspector of Irrigation, Egypt. With
Introduction by Lieut.-Col. J. C. ROSS, R.E., Inspector-General of
Irrigation. *With numerous lithographs and wood engravings*, royal 8vo,
cloth, 1*l.* 16*s.*

Screw Cutting Tables for Engineers and Machinists,
giving the values of the different trains of Wheels required to produce
Screws of any pitch, calculated by Lord Lindsay, M.P., F.R.S., F.R.A.S.,
etc. Cloth, oblong, 2*s.*

Screw Cutting Tables, for the use of Mechanical
Engineers, showing the proper arrangement of Wheels for cutting the
Threads of Screws of any required pitch, with a Table for making the
Universal Gas-pipe Threads and Taps. By W. A. MARTIN, Engineer.
Second edition, oblong, cloth, 1*s.*, or sewed, 6*d.*

A Treatise on a Practical Method of Designing Slide-
Valve Gears by Simple Geometrical Construction, based upon the principles
enunciated in Euclid's Elements, and comprising the various forms of
Plain Slide-Valve and Expansion Gearing ; together with Stephenson's,
Gooch's, and Allan's Link-Motions, as applied either to reversing or to
variable expansion combinations. By EDWARD J. COWLING WELCH,
Memb. Inst. Mechanical Engineers. Crown 8vo, cloth, 6*s.*

Cleaning and Scouring : a Manual for Dyers, Laun-
dresses, and for Domestic Use. By S. CHRISTOPHER. 18mo, sewed, 6*d.*

A Glossary of Terms used in Coal Mining. By
WILLIAM STUKELEY GRESLEY, Assoc. Mem. Inst. C.E., F.G.S., Member
of the North of England Institute of Mining Engineers. *Illustrated with*
numerous woodcuts and diagrams, crown 8vo, cloth, 5*s.*

A Pocket-Book for Boiler Makers and Steam Users,
comprising a variety of useful information for Employer and Workman,
Government Inspectors, Board of Trade Surveyors, Engineers in charge
of Works and Slips, Foremen of Manufactures, and the general Steam-
using Public. By MAURICE JOHN SEXTON. Second edition, royal
32mo, roan, gilt edges, 5*s.*

Electrolysis : a Practical Treatise on Nickeling,
Coppering, Gilding, Silvering, the Refining of Metals, and the treatment
of Ores by means of Electricity. By HIPPOLYTE FONTAINE, translated
from the French by J. A. BERLY, C.E., Assoc. S.T.E. *With engravings.*
8vo, cloth, 9*s.*

Barlow's Tables of Squares, Cubes, Square Roots,
Cube Roots, Reciprocals of all Integer Numbers up to 10,000. Post 8vo,
cloth, 6s.

A Practical Treatise on the Steam Engine, containing Plans and Arrangements of Details for Fixed Steam Engines,
with Essays on the Principles involved in Design and Construction. By
ARTHUR RIGG, Engineer, Member of the Society of Engineers and of
the Royal Institution of Great Britain. Demy 4to, *copiously illustrated
with woodcuts and 96 plates,* in one Volume, half-bound morocco, 2l. 2s.;
or cheaper edition, cloth, 25s.

This work is not, in any sense, an elementary treatise, or history of the steam engine, but
is intended to describe examples of Fixed Steam Engines without entering into the wide
domain of locomotive or marine practice   To this end illustrations will be given of the most
recent arrangements of Horizontal, Vertical, Beam, Pumping, Winding, Portable, Semi-
portable, Corliss, Allen, Compound, and other similar Engines, by the most eminent Firms in
Great Britain and America. The laws relating to the action and precautions to be observed
in the construction of the various details, such as Cylinders, Pistons, Piston-rods, Connecting-
rods, Cross-heads, Motion-blocks, Eccentrics, Simple, Expansion, Balanced, and Equilibrium
Slide-valves, and Valve-gearing will be minutely dealt with. In this connection will be found
articles upon the Velocity of Reciprocating Parts and the Mode of Applying the Indicator,
Heat and Expansion of Steam Governors, and the like. It is the writer's desire to draw
illustrations from every possible source, and give only those rules that present practice deems
correct

A Practical Treatise on the Science of Land and
Engineering Surveying, Levelling, Estimating Quantities, etc., with a
general description of the several Instruments required for Surveying,
Levelling, Plotting, etc. By H. S. MERRETT. Fourth edition, revised
by G. W. USILL, Assoc. Mem. Inst. C.E. 41 *plates, with illustrations
and tables,* royal 8vo, cloth, 12s. 6d.

PRINCIPAL CONTENTS :

Part 1 Introduction and the Principles of Geometry. Part 2. Land Surveying, com-
prising General Observations—The Chain—Offsets Surveying by the Chain only—Surveying
Hilly Ground—To Survey an Estate or Parish by the Chain only—Surveying with the
Theodolite—Mining and Town Surveying—Railroad Surveying—Mapping—Division and
Laying out of Land—Observations on Enclosures—Plane Trigonometry   Part 3 Levelling—
Simple and Compound Levelling—The Level Book—Parliamentary Plan and Section—
Levelling with a Theodolite—Gradients—Wooden Curves—To Lay out a Railway Curve—
Setting out Widths. Part 4 Calculating Quantities generally for Estimates—Cuttings and
Embankments—Tunnels—Brickwork—Ironwork—Timber Measuring.   Part 5. Description
and Use of Instruments in Surveying and Plotting—The Improved Dumpy Level—Troughton's
Level—The Prismatic Compass — Proportional Compass— Box Sextant—Vernier — Panta-
graph—Merrett's Improved Quadrant—Improved Computation Scale—The Diagonal Scale—
Straight Edge and Sector. Part 6. Logarithms of Numbers — Logarithmic Sines and
Co-Sines, Tangents and Co-Tangents—Natural Sines and Co-Sines—Tables for Earthwork,
for Setting out Curves, and for various Calculations, etc , etc , etc

Mechanical Graphics.   A Second Course of Mechanical Drawing. With Preface by Prof. PERRY, B.Sc., F.R.S.
Arranged for use in Technical and Science and Art Institutes, Schools
and Colleges, by GEORGE HALLIDAY, Whitworth Scholar. 8vo,
cloth, 6s.

The Assayer's Manual: an Abridged Treatise on
the Docimastic Examination of Ores and Furnace and other Artificial
Products. By BRUNO KERL. Translated by W. T. BRANNT. *With* 65
illustrations, 8vo, cloth, 12s. 6d.

Dynamo - Electric Machinery: a Text - Book for
Students of Electro-Technology. By SILVANUS P. THOMPSON, B.A.,
D.Sc., M.S.T.E.   [*New edition in the press.*

The Practice of Hand Turning in Wood, Ivory, Shell,
etc., with Instructions for Turning such Work in Metal as may be required
in the Practice of Turning in Wood, Ivory, etc. ; also an Appendix on
Ornamental Turning. (A book for beginners.) By FRANCIS CAMPIN.
Third edition, *with wood engravings*, crown 8vo, cloth, 6s.

CONTENTS :

On Lathes—Turning Tools—Turning Wood—Drilling—Screw Cutting—Miscellaneous
Apparatus and Processes—Turning Particular Forms—Staining—Polishing—Spinning Metals
—Materials—Ornamental Turning, etc.

Treatise on Watchwork, Past and Present. By the
Rev. H. L. NELTHROPP, M.A., F.S.A.   *With* 32 *illustrations*, crown
8vo, cloth, 6s. 6d.

CONTENTS :

Definitions of Words and Terms used in Watchwork—Tools—Time—Historical Sum-
mary—On Calculations of the Numbers for Wheels and Pinions; their Proportional Sizes,
Trains, etc.—Of Dial Wheels, or Motion Work—Length of Time of Going without Winding
up—The Verge—The Horizontal—The Duplex—The Lever—The Chronometer—Repeating
Watches—Keyless Watches—The Pendulum, or Spiral Spring—Compensation—Jewelling of
Pivot Holes—Clerkenwell—Fallacies of the Trade—Incapacity of Workmen—How to Choose
and Use a Watch, etc.

Algebra Self-Taught. By W. P. HIGGS, M.A.,
D.Sc, LL.D., Assoc. Inst. C.E., Author of ' A Handbook of the Differential Calculus,' etc.   Second edition, crown 8vo, cloth, 2s. 6d.

CONTENTS :

Symbols and the Signs of Operation—The Equation and the Unknown Quantity—
Positive and Negative Quantities—Multiplication—Involution—Exponents—Negative Expo-
nents—Roots, and the Use of Exponents as Logarithms—Logarithms—Tables of Logarithms
and Proportionate Parts—Transformation of System of Logarithms—Common Uses of
Common Logarithms—Compound Multiplication and the Binomial Theorem—Division,
Fractions, and Ratio—Continued Proportion—The Series and the Summation of the Series—
Limit of Series—Square and Cube Roots—Equations—List of Formulæ, etc.

Spons' Dictionary of Engineering, Civil, Mechanical,
Military, and Naval; with technical terms in French, German, Italian,
and Spanish, 3100 pp., and *nearly* 8000 *engravings*, in super-royal 8vo,
in 8 divisions, 5l. 8s.   Complete in 3 vols., cloth, 5l. 5s.   Bound in a
superior manner, half-morocco, top edge gilt, 3 vols., 6l. 12s.

Notes in Mechanical Engineering. Compiled principally for the use of the Students attending the Classes on this subject at the City of London College. By HENRY ADAMS, Mem. Inst. M.E., Mem. Inst. C.E., Mem. Soc. of Engineers. Crown 8vo, cloth, 2s. 6d.

Canoe and Boat Building: a complete Manual for Amateurs, containing plain and comprehensive directions for the construction of Canoes, Rowing and Sailing Boats, and Hunting Craft. By W. P. STEPHENS. *With numerous illustrations and* 24 *plates of Working Drawings.* Crown 8vo, cloth, 9s.

Proceedings of the National Conference of Electricians, *Philadelphia,* October 8th to 13th, 1884. 18mo, cloth, 3s.

Dynamo - Electricity, its Generation, Application, Transmission, Storage, and Measurement. By G. B. PRESCOTT. *With* 545 *illustrations.* 8vo, cloth, 1l. 1s.

Domestic Electricity for Amateurs. Translated from the French of E. HOSPITALIER, Editor of "L'Electricien," by C. J. WHARTON, Assoc. Soc. Tel. Eng. *Numerous illustrations.* Demy 8vo, cloth, 6s.

CONTENTS:

1. Production of the Electric Current—2. Electric Bells—3. Automatic Alarms—4. Domestic Telephones—5. Electric Clocks—6 Electric Lighters—7. Domestic Electric Lighting—8 Domestic Application of the Electric Light—9. Electric Motors—10. Electrical Locomotion—11. Electrotyping, Plating, and Gilding—12. Electric Recreations—13. Various applications—Workshop of the Electrician.

Wrinkles in Electric Lighting. By VINCENT STEPHEN. *With illustrations.* 18mo, cloth, 2s. 6d.

CONTENTS:

1. The Electric Current and its production by Chemical means—2. Production of Electric Currents by Mechanical means—3. Dynamo-Electric Machines—4. Electric Lamps—5. Lead—6. Ship Lighting.

Foundations and Foundation Walls for all classes of *Buildings,* Pile Driving, Building Stones and Bricks, Pier and Wall construction, Mortars, Limes, Cements, Concretes, Stuccos, &c. 64 *illustrations.* By G. T. POWELL and F. BAUMAN. 8vo, cloth, 10s. 6d.

Manual for Gas Engineering Students. By D. LEE. 18mo, cloth, 1s.

Telephones, their Construction and Management.
By F. C. ALLSOP. Crown 8vo, cloth, 5s.

Hydraulic Machinery, Past and Present. A Lecture
delivered to the London and Suburban Railway Officials' Association.
By H. ADAMS, Mem. Inst. C.E. *Folding plate.* 8vo, sewed, 1s.

Twenty Years with the Indicator. By THOMAS PRAY,
Jun., C.E., M.E., Member of the American Society of Civil Engineers.
2 vols., royal 8vo, cloth, 12s. 6d.

Annual Statistical Report of the Secretary to the
Members of the Iron and Steel Association on the Home and Foreign Iron
and Steel Industries in 1889. Issued June 1890. 8vo, sewed, 5s.

Bad Drains, and How to Test them ; with Notes on
the Ventilation of Sewers, Drains, and Sanitary Fittings, and the Origin
and Transmission of Zymotic Disease. By R. HARRIS REEVES. Crown
8vo, cloth, 3s. 6d.

Well Sinking. The modern practice of Sinking
and Boring Wells, with geological considerations and examples of Wells.
By ERNEST SPON, Assoc. Mem Inst. C.E., Mem. Soc Eng , and of the
Franklin Inst., etc. Second edition, revised and enlarged. Crown 8vo,
cloth, 10s. 6d.

The Voltaic Accumulator : an Elementary Treatise.
By ÉMILE REYNIER. Translated by J. A. BERLY, Assoc. Inst. E.E.
With 62 illustrations, 8vo, cloth, 9s.

Ten Years' Experience in Works of Intermittent
Downward Filtration. By J. BAILEY DENTON, Mem Inst. C.E.
Second edition, with additions. Royal 8vo, cloth, 5s.

Land Surveying on the Meridian and Perpendicular
System. By WILLIAM PENMAN, C.E. 8vo, cloth, 8s. 6d.

The Electromagnet and Electromagnetic Mechanism.
By SILVANUS P. THOMPSON, D.Sc., F.R S. 8vo, cloth, 15s.

Incandescent Wiring Hand-Book. By F. B. BADT, late 1st Lieut. Royal Prussian Artillery. *With* 41 *illustrations and* 5 *tables.* 18mo, cloth, 4s. 6d.

A Pocket-book for Pharmacists, Medical Prac-titioners, Students, etc., etc. (British, Colonial, and American). By THOMAS BAYLEY, Assoc. R. Coll. of Science, Consulting Chemist, Analyst, and Assayer, Author of a 'Pocket-book for Chemists,' 'The Assay and Analysis of Iron and Steel, Iron Ores, and Fuel,' etc., etc. Royal 32mo, boards, gilt edges, 6s.

The Fireman's Guide; a Handbook on the Care of Boilers. By TEKNOLOG, föreningen T. I. Stockholm. Translated from the third edition, and revised by KARL P. DAHLSTROM, M.E. Second edition. Fcap. 8vo, cloth, 2s.

A Treatise on Modern Steam Engines and Boilers, including Land Locomotive, and Marine Engines and Boilers, for the use of Students. By FREDERICK COLYER, M. Inst. C.E., Mem. Inst. M.E. *With* 36 *plates.* 4to, cloth, 12s. 6d.

CONTENTS:

1. Introduction—2. Original Engines—3. Boilers—4. High-Pressure Beam Engines—5. Cornish Beam Engines—6. Horizontal Engines—7. Oscillating Engines—8. Vertical High-Pressure Engines—9. Special Engines—10. Portable Engines—11. Locomotive Engines—12. Marine Engines.

Steam Engine Management; a Treatise on the Working and Management of Steam Boilers. By F. COLYER, M. Inst. C.E., Mem. Inst. M.E. 18mo, cloth, 2s.

A Text-Book of Tanning, embracing the Preparation of all kinds of Leather. By HARRY R. PROCTOR, F.C.S., of Low Lights Tanneries. *With illustrations.* Crown 8vo, cloth, 10s. 6d.

Aid Book to Engineering Enterprise. By EWING MATHESON, M. Inst. C.E. The Inception of Public Works, Parliamentary Procedure for Railways, Concessions for Foreign Works, and means of Providing Money, the Points which determine Success or Failure, Contract and Purchase, Commerce in Coal, Iron, and Steel, &c. Second edition, revised and enlarged, 8vo, cloth, 21s.

Pumps, Historically, Theoretically, and Practically Considered. By P. R. BJÖRLING. *With* 156 *illustrations.* Crown 8vo, cloth, 7*s.* 6*d.*

The Marine Transport of Petroleum. A Book for the use of Shipowners, Shipbuilders, Underwriters, Merchants, Captains and Officers of Petroleum-carrying Vessels. By G. H. LITTLE, Editor of the 'Liverpool Journal of Commerce.' Crown 8vo, cloth, 10*s.* 6*d.*

Liquid Fuel for Mechanical and Industrial Purposes. Compiled by E. A. BRAYLEY HODGETTS. *With wood engravings.* 8vo, cloth, 7*s.* 6*d.*

Tropical Agriculture: A Treatise on the Culture, Preparation, Commerce and Consumption of the principal Products of the Vegetable Kingdom. By P. L. SIMMONDS, F.L.S., F.R.C.I. New edition, revised and enlarged, 8vo, cloth, 21*s.*

Health and Comfort in House Building; or, Ventilation with Warm Air by Self-acting Suction Power. With Review of the Mode of Calculating the Draught in Hot-air Flues, and with some Actual Experiments by J. DRYSDALE, M.D., and J. W. HAYWARD, M.D. *With plates and woodcuts.* Third edition, with some New Sections, and the whole carefully Revised, 8vo, cloth, 7*s.* 6*d.*

Losses in Gold Amalgamation. With Notes on the Concentration of Gold and Silver Ores. *With six plates.* By W. McDERMOTT and P. W. DUFFIELD. 8vo, cloth, 5*s.*

A Guide for the Electric Testing of Telegraph Cables. By Col. V. HOSKIŒR, Royal Danish Engineers. Third edition, crown 8vo, cloth, 4*s.* 6*d.*

The Hydraulic Gold Miners' Manual. By T. S. G. KIRKPATRICK, M.A. Oxon. *With 6 plates.* Crown 8vo, cloth, 6*s.*

" We venture to think that this work will become a text-book on the important subject of which it treats. Until comparatively recently hydraulic mines were neglected This was scarcely to be surprised at, seeing that their working in California was brought to an abrupt termination by the action of the farmers on the *débris* question, whilst their working in other parts of the world had not been attended with the anticipated success."—*The Mining World and Engineering Record.*

The Arithmetic of Electricity. By T. O'CONOR SLOANE. Crown 8vo, cloth, 4*s.* 6*d.*

The Turkish Bath: Its Design and Construction for
Public and Commercial Purposes. By R. O. ALLSOP, Architect. *With
plans and sections.* 8vo, cloth, 6s.

Earthwork Slips and Subsidences upon Public Works:
Their Causes, Prevention and Reparation. Especially written to assist
those engaged in the Construction or Maintenance of Railways, Docks,
Canals, Waterworks, River Banks, Reclamation Embankments, Drainage
Works, &c., &c. By JOHN NEWMAN, Assoc. Mem. Inst. C.E., Author
of 'Notes on Concrete,' &c. Crown 8vo, cloth, 7s. 6d.

Gas and Petroleum Engines: A Practical Treatise
on the Internal Combustion Engine. By WM. ROBINSON, M.E., Senior
Demonstrator and Lecturer on Applied Mechanics, Physics, &c., City
and Guilds of London College, Finsbury, Assoc. Mem. Inst. C.E., &c.
Numerous illustrations. 8vo, cloth, 14s.

*Waterways and Water Transport in Different Coun-
tries.* With a description of the Panama, Suez, Manchester, Nicaraguan,
and other Canals. By J. STEPHEN JEANS, Author of 'England's
Supremacy,' 'Railway Problems,' &c. *Numerous illustrations.* 8vo,
cloth, 14s.

A Treatise on the Richards Steam-Engine Indicator
and the Development and Application of Force in the Steam-Engine.
By CHARLES T. PORTER. Fourth Edition, revised and enlarged, 8vo,
cloth, 9s.

CONTENTS.

The Nature and Use of the Indicator:
The several lines on the Diagram
Examination of Diagram No. 1.
Of Truth in the Diagram.
Description of the Richards Indicator.
Practical Directions for Applying and Taking
Care of the Indicator.
Introductory Remarks.
Units.
Expansion.
Directions for ascertaining from the Diagram
the Power exerted by the Engine.
To Measure from the Diagram the Quantity
of Steam Consumed.
To Measure from the Diagram the Quantity
of Heat Expended
Of the Real Diagram, and how to Construct it.
Of the Conversion of Heat into Work in the
Steam-engine.
Observations on the several Lines of the
Diagram.

Of the Loss attending the Employment of
Slow-piston Speed, and the Extent to
which this is Shown by the Indicator.
Of other Applications of the Indicator.
Of the use of the Tables of the Properties of
Steam in Calculating the Duty of Boilers.
Introductory.
Of the Pressure on the Crank when the Con-
necting-rod is conceived to be of Infinite
Length.
The Modification of the Acceleration and
Retardation that is occasioned by the
Angular Vibration of the Connecting-rod.
Method of representing the actual pressure
on the crank at every point of its revolu-
tion
The Rotative Effect of the Pressure exerted
on the Crank.
The Transmitting Parts of an Engine, con-
sidered as an Equaliser of Motion.
A Ride on a Buffer-beam (Appendix).

In demy 4to, handsomely bound in cloth, *illustrated with* **220** *full page plates*,
Price 15s.

ARCHITECTURAL EXAMPLES

IN BRICK, STONE, WOOD, AND IRON.

A COMPLETE WORK ON THE DETAILS AND ARRANGEMENT OF BUILDING CONSTRUCTION AND DESIGN.

By WILLIAM FULLERTON, Architect.

Containing 220 Plates, with numerous Drawings selected from the Architecture
of Former and Present Times.

The Details and Designs are Drawn to Scale, $\frac{1}{8}''$, $\frac{1}{4}''$, $\frac{1}{2}''$, *and Full size
being chiefly used.*

The Plates are arranged in Two Parts. The First Part contains Details of Work in the four principal Building materials, the following being a few of the subjects in this Part:—Various forms of Doors and Windows, Wood and Iron Roofs, Half Timber Work, Porches, Towers, Spires, Belfries, Flying Buttresses, Groining, Carving, Church Fittings, Constructive and Ornamental Iron Work, Classic and Gothic Molds and Ornament, Foliation Natural and Conventional, Stained Glass, Coloured Decoration, a Section to Scale of the Great Pyramid, Grecian and Roman Work, Continental and English Gothic, Pile Foundations, Chimney Shafts according to the regulations of the London County Council, Board Schools. The Second Part consists of Drawings of Plans and Elevations of Buildings, arranged under the following heads :—Workmen's Cottages and Dwellings, Cottage Residences and Dwelling Houses, Shops, Factories, Warehouses, Schools, Churches and Chapels, Public Buildings, Hotels and Taverns, and Buildings of a general character.

All the Plates are accompanied with particulars of the Work, with Explanatory Notes and Dimensions of the various parts.

Specimen Pages, reduced from the originals.

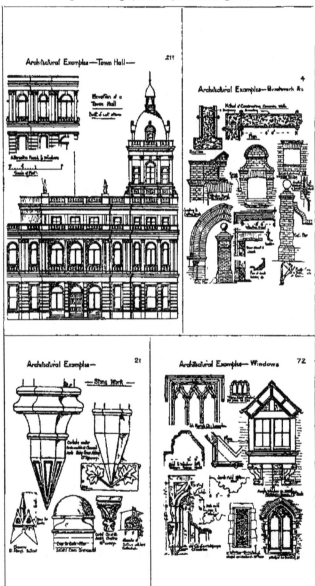

Crown 8vo, cloth, with illustrations, 5*s.*

WORKSHOP RECEIPTS,

FIRST SERIES.

By ERNEST SPON.

SYNOPSIS OF CONTENTS.

Bookbinding.
Bronzes and Bronzing.
Candles.
Cement.
Cleaning.
Colourwashing.
Concretes.
Dipping Acids.
Drawing Office Details.
Drying Oils.
Dynamite.
Electro - Metallurgy — (Cleaning, Dipping, Scratch-brushing, Batteries, Baths, and Deposits of every description).
Enamels.
Engraving on Wood, Copper, Gold, Silver, Steel, and Stone.
Etching and Aqua Tint.
Firework Making — (Rockets, Stars, Rains, Gerbes, Jets, Tourbillons, Candles, Fires, Lances, Lights, Wheels, Fire-balloons, and minor Fireworks).
Fluxes.
Foundry Mixtures.

Freezing.
Fulminates.
Furniture Creams, Oils, Polishes, Lacquers, and Pastes.
Gilding.
Glass Cutting, Cleaning, Frosting, Drilling, Darkening, Bending, Staining, and Painting.
Glass Making.
Glues.
Gold.
Graining.
Gums.
Gun Cotton.
Gunpowder.
Horn Working.
Indiarubber.
Japans, Japanning, and kindred processes.
Lacquers.
Lathing.
Lubricants.
Marble Working.
Matches.
Mortars.
Nitro-Glycerine.
Oils.

Paper.
Paper Hanging.
Painting in Oils, in Water Colours, as well as Fresco, House, Transparency, Sign, and Carriage Painting.
Photography.
Plastering.
Polishes.
Pottery—(Clays, Bodies, Glazes, Colours, Oils, Stains, Fluxes, Enamels, and Lustres).
Scouring.
Silvering.
Soap.
Solders.
Tanning.
Taxidermy.
Tempering Metals.
Treating Horn, Mother-o'-Pearl, and like substances.
Varnishes, Manufacture and Use of.
Veneering.
Washing.
Waterproofing.
Welding.

Crown 8vo, cloth, 485 pages, with illustrations, 5s.

WORKSHOP RECEIPTS,

SECOND SERIES.

By ROBERT HALDANE.

SYNOPSIS OF CONTENTS.

Acidimetry and Alkalimetry.	Disinfectants.	Iodoform.
Albumen.	Dyeing, Staining, and Colouring.	Isinglass.
Alcohol.	Essences.	Ivory substitutes.
Alkaloids.	Extracts.	Leather.
Baking-powders.	Fireproofing.	Luminous bodies.
Bitters.	Gelatine, Glue, and Size.	Magnesia.
Bleaching.	Glycerine.	Matches.
Boiler Incrustations.	Gut.	Paper.
Cements and Lutes.	Hydrogen peroxide.	Parchment.
Cleansing.	Ink.	Perchloric acid.
Confectionery.	Iodine.	Potassium oxalate.
Copying.		Preserving.

Pigments, Paint, and Painting : embracing the preparation of *Pigments*, including alumina lakes, blacks (animal, bone, Frankfort, ivory, lamp, sight, soot), blues (antimony, Antwerp, cobalt, cæruleum, Egyptian, manganate, Paris, Péligot, Prussian, smalt, ultramarine), browns (bistre, hinau, sepia, sienna, umber, Vandyke), greens (baryta, Brighton, Brunswick, chrome, cobalt, Douglas, emerald, manganese, mitis, mountain, Prussian, sap, Scheele's, Schweinfurth, titanium, verdigris, zinc), reds (Brazilwood lake, carminated lake, carmine, Cassius purple, cobalt pink, cochineal lake, colcothar, Indian red, madder lake, red chalk, red lead, vermilion), whites (alum, baryta, Chinese, lead sulphate, white lead—by American, Dutch, French, German, Kremnitz, and Pattinson processes, precautions in making, and composition of commercial samples—whiting, Wilkinson's white, zinc white), yellows (chrome, gamboge, Naples, orpiment, realgar, yellow lakes); *Paint* (vehicles, testing oils, driers, grinding, storing, applying, priming, drying, filling, coats, brushes, surface, water-colours, removing smell, discoloration ; miscellaneous paints—cement paint for carton-pierre, copper paint, gold paint, iron paint, lime paints, silicated paints, steatite paint, transparent paints, tungsten paints, window paint, zinc paints) ; *Painting* (general instructions, proportions of ingredients, measuring paint work ; carriage painting—priming paint, best putty, finishing colour, cause of cracking, mixing the paints, oils, driers, and colours, varnishing, importance of washing vehicles, re-varnishing, how to dry paint ; woodwork painting).

Crown 8vo, cloth, 480 pages, with 183 illustrations, 5s.

WORKSHOP RECEIPTS,

THIRD SERIES.

By C. G. WARNFORD LOCK.

Uniform with the First and Second Series.

SYNOPSIS OF CONTENTS.

Alloys.	Indium.	Rubidium.
Aluminium.	Iridium.	Ruthenium.
Antimony.	Iron and Steel.	Selenium.
Barium.	Lacquers and Lacquering.	Silver.
Beryllium.	Lanthanum.	Slag.
Bismuth.	Lead.	Sodium.
Cadmium.	Lithium.	Strontium.
Cæsium.	Lubricants.	Tantalum.
Calcium.	Magnesium.	Terbium.
Cerium.	Manganese.	Thallium.
Chromium.	Mercury.	Thorium.
Cobalt.	Mica.	Tin.
Copper.	Molybdenum.	Titanium.
Didymium.	Nickel.	Tungsten.
Electrics.	Niobium.	Uranium.
Enamels and Glazes.	Osmium.	Vanadium.
Erbium.	Palladium.	Yttrium.
Gallium.	Platinum.	Zinc.
Glass.	Potassium.	Zirconium.
Gold.	Rhodium.	

WORKSHOP RECEIPTS,

FOURTH SERIES,

DEVOTED MAINLY TO HANDICRAFTS & MECHANICAL SUBJECTS.

By C. G. WARNFORD LOCK.

250 Illustrations, with Complete Index, and a General Index to the Four Series, 5s.

Waterproofing — rubber goods, cuprammonium processes, miscellaneous preparations.

Packing and Storing articles of delicate odour or colour, of a deliquescent character, liable to ignition, apt to suffer from insects or damp, or easily broken.

Embalming and Preserving anatomical specimens.

Leather Polishes.

Cooling Air and Water, producing low temperatures, making ice, cooling syrups and solutions, and separating salts from liquors by refrigeration.

Pumps and Siphons, embracing every useful contrivance for raising and supplying water on a moderate scale, and moving corrosive, tenacious, and other liquids.

Desiccating—air- and water-ovens, and other appliances for drying natural and artificial products.

Distilling—water, tinctures, extracts, pharmaceutical preparations, essences, perfumes, and alcoholic liquids.

Emulsifying as required by pharmacists and photographers.

Evaporating—saline and other solutions, and liquids demanding special precautions.

Filtering—water, and solutions of various kinds.

Percolating and Macerating.

Electrotyping.

Stereotyping by both plaster and paper processes.

Bookbinding in all its details.

Straw Plaiting and the fabrication of baskets, matting, etc.

Musical Instruments—the preservation, tuning, and repair of pianos harmoniums, musical boxes, etc.

Clock and Watch Mending—adapted for intelligent amateurs.

Photography—recent development in rapid processes, handy apparatus, numerous recipes for sensitizing and developing solutions, and applications to modern illustrative purposes.

NOW COMPLETE.

With nearly 1500 *illustrations*, in super-royal 8vo, in 5 Divisions, cloth.
Divisions 1 to 4, 13s. 6d. each ; Division 5, 17s. 6d. ; or 2 vols., cloth, £3 10s.

SPONS' ENCYCLOPÆDIA

OF THE

INDUSTRIAL ARTS, MANUFACTURES, AND COMMERCIAL PRODUCTS.

EDITED BY C. G. WARNFORD LOCK, F.L.S.

Among the more important of the subjects treated of, are the following :—

Acids, 207 pp. 220 figs.
Alcohol, 23 pp. 16 figs.
Alcoholic Liquors, 13 pp.
Alkalies, 89 pp. 78 figs.
Alloys.   Alum.
Asphalt.   Assaying.
Beverages, 89 pp. 29 figs.
Blacks.
Bleaching Powder, 15 pp.
Bleaching, 51 pp. 48 figs.
Candles, 18 pp. 9 figs.
Carbon Bisulphide.
Celluloid, 9 pp.
Cements.   Clay.
Coal-tar Products, 44 pp.
    14 figs.
Cocoa, 8 pp.
Coffee, 32 pp. 13 figs.
Cork, 8 pp. 17 figs.
Cotton Manufactures, 62
    pp. 57 figs.
Drugs, 38 pp.
Dyeing and Calico
    Printing, 28 pp. 9 figs.
Dyestuffs, 16 pp.
Electro-Metallurgy, 13
    pp.
Explosives, 22 pp. 33 figs.
Feathers.
Fibrous Substances, 92
    pp. 79 figs.
Floor-cloth, 16 pp. 21
    figs.
Food Preservation, 8 pp.
Fruit, 8 pp.

Fur, 5 pp.
Gas, Coal, 8 pp.
Gems.
Glass, 45 pp. 77 figs.
Graphite, 7 pp.
Hair, 7 pp.
Hair Manufactures.
Hats, 26 pp. 26 figs.
Honey.   Hops.
Horn.
Ice, 10 pp. 14 figs.
Indiarubber Manufac-
    tures, 23 pp. 17 figs.
Ink, 17 pp.
Ivory.
Jute Manufactures, 11
    pp., 11 figs.
Knitted Fabrics —
    Hosiery, 15 pp. 13 figs.
Lace, 13 pp. 9 figs.
Leather, 28 pp. 31 figs.
Linen Manufactures, 16
    pp. 6 figs.
Manures, 21 pp. 30 figs.
Matches, 17 pp. 38 figs.
Mordants, 13 pp.
Narcotics, 47 pp.
Nuts, 10 pp.
Oils and Fatty Sub-
    stances, 125 pp.
Paint.
Paper, 26 pp. 23 figs.
Paraffin, 8 pp. 6 figs.
Pearl and Coral, 8 pp.
Perfumes, 10 pp.

Photography, 13 pp. 20
    figs.
Pigments, 9 pp. 6 figs.
Pottery, 46 pp. 57 figs.
Printing and Engraving,
    20 pp. 8 figs.
Rags.
Resinous and Gummy
    Substances, 75 pp. 16
    figs.
Rope, 16 pp. 17 figs.
Salt, 31 pp. 23 figs.
Silk, 8 pp.
Silk Manufactures, 9 pp.
    11 figs.
Skins, 5 pp.
Small Wares, 4 pp.
Soap and Glycerine, 39
    pp. 45 figs.
Spices, 16 pp.
Sponge, 5 pp.
Starch, 9 pp. 10 figs.
Sugar, 155 pp. 134
    figs.
Sulphur.
Tannin, 18 pp.
Tea, 12 pp.
Timber, 13 pp.
Varnish, 15 pp.
Vinegar, 5 pp.
Wax, 5 pp.
Wool, 2 pp.
Woollen Manufactures,
    58 pp. 39 figs.

In super-royal 8vo, 1168 pp., *with* 2400 *illustrations*, in 3 Divisions, cloth, price 13*s.* 6*d.*
each ; or 1 vol., cloth, 2*l.* ; or half-morocco, 2*l.* 8*s.*

A SUPPLEMENT

TO

SPONS' DICTIONARY OF ENGINEERING.

EDITED BY ERNEST SPON, MEMB. SOC. ENGINEERS.

Abacus, Counters, Speed Indicators, and Slide Rule.

Agricultural Implements and Machinery.

Air Compressors.

Animal Charcoal Machinery.

Antimony.

Axles and Axle-boxes.

Barn Machinery.

Belts and Belting.

Blasting. Boilers.

Brakes.

Brick Machinery.

Bridges.

Cages for Mines.

Calculus, Differential and Integral.

Canals.

Carpentry.

Cast Iron.

Cement, Concrete, Limes, and Mortar.

Chimney Shafts.

Coal Cleansing and Washing.

Coal Mining.

Coal Cutting Machines.

Coke Ovens. Copper.

Docks. Drainage.

Dredging Machinery.

Dynamo - Electric and Magneto-Electric Machines.

Dynamometers.

Electrical Engineering, Telegraphy, Electric Lighting and its practicaldetails,Telephones

Engines, Varieties of.

Explosives. Fans.

Founding, Moulding and the practical work of the Foundry.

Gas, Manufacture of.

Hammers, Steam and other Power.

Heat. Horse Power.

Hydraulics.

Hydro-geology.

Indicators. Iron.

Lifts, Hoists, and Elevators.

Lighthouses, Buoys, and Beacons.

Machine Tools.

Materials of Construction.

Meters.

Ores, Machinery and Processes employed to Dress.

Piers.

Pile Driving.

Pneumatic Transmission.

Pumps.

Pyrometers.

Road Locomotives.

Rock Drills.

Rolling Stock.

Sanitary Engineering.

Shafting.

Steel.

Steam Navvy.

Stone Machinery.

Tramways.

Well Sinking.

JUST PUBLISHED.

In demy 8vo, cloth, 600 pages, and 1420 Illustrations, 6s.

SPONS'
MECHANICS' OWN BOOK;
A MANUAL FOR HANDICRAFTSMEN AND AMATEURS.

CONTENTS.

Mechanical Drawing—Casting and Founding in Iron, Brass, Bronze, and other Alloys—Forging and Finishing Iron—Sheetmetal Working —Soldering, Brazing, and Burning—Carpentry and Joinery, embracing descriptions of some 400 Woods, over 200 Illustrations of Tools and their uses, Explanations (with Diagrams) of 116 joints and hinges, and Details of Construction of Workshop appliances, rough furniture, Garden and Yard Erections, and House Building—Cabinet-Making and Veneering — Carving and Fretcutting — Upholstery — Painting, Graining, and Marbling — Staining Furniture, Woods, Floors, and Fittings—Gilding, dead and bright, on various grounds—Polishing Marble, Metals, and Wood—Varnishing—Mechanical movements, illustrating contrivances for transmitting motion—Turning in Wood and Metals—Masonry, embracing Stonework, Brickwork, Terracotta, and Concrete—Roofing with Thatch, Tiles, Slates, Felt, Zinc, &c.— Glazing with and without putty, and lead glazing—Plastering and Whitewashing—Paper-hanging—Gas-fitting—Bell-hanging, ordinary and electric Systems — Lighting — Warming — Ventilating — Roads, Pavements, and Bridges — Hedges, Ditches, and Drains — Water Supply and Sanitation—Hints on House Construction suited to new countries.

E. & F. N. SPON, 125, Strand, London.
New York: 12, Cortlandt Street.